CLIMATE CHANGE AND THE CANADIAN CONSTITUTION:

Federalism and Charter Rights in a Warming World

CLIMATE CHANGE AND THE CANADIAN CONSTITUTION:

Federalism and Charter Rights in a Warming World

NATHALIE J. CHALIFOUR

UNIVERSITY OF TORONTO PRESS
Toronto Buffalo London

Climate Change and the Canadian Constitution:
Federalism and Charter Rights in a Warming World
© Nathalie Chalifour, 2025

Irwin Law
An imprint of University of Toronto Press
Toronto Buffalo London
utppublishing.com
Printed in Canada

ISBN 978-1-4875-7107-8 (paper) | ISBN 978-1-4875-7108-5 (PDF)
ISBN 978-1-4875-7109-2 (EPUB)

Library and Archives Canada Cataloguing in Publication

Title: Climate change and the Canadian constitution : federalism and Charter rights
    in a warming world / Nathalie J. Chalifour.
Names: Chalifour, Nathalie J., author
Description: Includes bibliographical references and index.
Identifiers: Canadiana (print) 20250228718 | Canadiana (ebook) 20250228742 |
    ISBN 9781487571078 (paper) | ISBN 9781487571085 (PDF) | ISBN 9781487571092 (EPUB)
Subjects: LCSH: Climatic changes—Law and legislation—Canada. |
    LCSH: Constitutional law—Canada. | LCSH: Environmental law—Canada.
Classification: LCC KE3619 .C43 2025 | LCC KF3819 .C43 2025 kfmod |
    DDC 344.7104/6342—dc23

Cover design: Kristjian Buckingham

We wish to acknowledge the land on which the University of Toronto Press operates. This land is the traditional territory of the Wendat, the Anishnaabeg, the Haudenosaunee, the Métis, and the Mississaugas of the Credit First Nation.

University of Toronto Press acknowledges the financial support of the Government of Canada, the Canada Council for the Arts, and the Ontario Arts Council, an agency of the Government of Ontario, for its publishing activities.

SUMMARY TABLE OF CONTENTS

DETAILED TABLE OF CONTENTS

DEDICATION

To Mom and Dad for your constant love, support, and belief in me, which have never wavered over the years. It has meant the world to me.

FOREWORD

"This challenge of pollution of our rivers, and lakes, of our farmlands and forests, and of the very air we breathe, cannot be met effectively in our federal state without some constitutional reforms or clarification."

Pierre Elliott Trudeau, 1969

More than fifty years ago, Prime Minister Trudeau identified the lack of environmental provisions in the 1867 British North America Act (the then heart of Canada's Constitution) as a major barrier to addressing pollution. Despite numerous proposals during the 1970s and early 1980s, the repatriation of the Constitution in 1982 ultimately failed to address Trudeau's critique. The division of powers between federal and provincial governments drafted in the 1860s was only modestly modified by giving the provinces additional power over natural resources. The 1982 *Charter of Rights and Freedoms* was a much-ballyhooed step forward at the time, but has massive gaps when it comes to social, economic and environmental rights. The *Charter*, to the surprise of many Canadians, contains no right to health, no right to water and no right to a healthy environment. While understandably focused on air, water and soil pollution in the late 1960s, Trudeau's insight seems to apply with full force to the immense challenge of addressing today's climate crisis.

The ongoing consequences of the environmental hole in Canada's fundamental law are enormous and deeply troubling. Canada has weak environmental laws, made worse by poor implementation and weak enforcement.[1] When the federal government attempts to strengthen the legal framework, it is routinely challenged in court by provincial governments and businesses who argue that new or improved climate and environmental laws and regulations are beyond federal jurisdiction. Provisions in the *Ocean Dumping Control Act*, the *Fisheries Act*, the *Canadian Environmental Assessment Act*, the *Canadian Environmental Protection Act*, the *Species at Risk Act*, and the *Greenhouse Gas Pollution Pricing Act* have all been subjected to lawsuits, in some cases successful, attacking their constitutional bona fides.[2] For example, in the past three years courts have struck down parts of the federal *Impact Assessment Act* and elements of federal efforts to address plastic pollution.[3]

Unfortunately, the prospects for constitutional reform in Canada are bleak. A proposed amendment to address the environmental gap in the Constitution would require the approval of the House of Commons, the Senate, and at least seven provinces, representing at least 50% of Canada's population. To make matters even more difficult, the federal government passed a law requiring constitutional amendments to be supported by Ontario, Quebec, BC, at least two of the prairie provinces, and at least two of the Atlantic provinces. Alberta and British Columbia each enacted legislation requiring public referendums before those governments can approve constitutional amendments. As well, the scars from

---

1   David R. Boyd, *Unnatural Law: Rethinking Canadian Environmental Law and Policy* (UBC Press, 2003). David R. Boyd, *Cleaner, Greener, Healthier: A Prescription for Stronger Canadian Environmental Laws and Policies* (UBC Press, 2015).

2   See for example: *R v Crown Zellerbach Canada Ltd*, 1988 CanLII 63 (SCC); *Friends of the Oldman River v Canada (Minister of Transport)*, [1992] 1 SCR 3; *R v Hydro-Québec*, 1997 CanLII 318 (SCC); *Groupe Maison Candiac Inc v Procureur general du Canada*, 2020 FCA 88, leave to appeal to SCC refused *Groupe Maison Candiac Inc v Procureur general du Canada*, 2020 CanLII 97859 (SCC); and *Reference re Greenhouse Gas Pollution Pricing Act*, 2021 SCC 11.

3   *Reference re Impact Assessment Act*, 2023 SCC 23.

the constitutional train wrecks of Meech Lake and the Charlottetown Accord remain viscerally fresh for older politicians. The threat of Quebec separatism persists. In light of these factors, it is difficult to envision an environmental amendment to the Constitution in the short-term, despite the reality that it is urgently needed.[4]

Therefore, the questions confronted by Professor Nathalie Chalifour in this timely and insightful book are vitally important: Is Canada's Constitution fit for purpose in the 21st century? Is it capable of meeting the daunting challenges posed by the accelerating climate emergency? Can Canadian courts navigate the inevitable climate lawsuits in a fair, principled and effective manner despite the constraints of an outdated constitution? Can Canadian courts reconcile the antiquated division of powers doctrine with the concurrent imperatives of addressing the climate emergency and maintaining the delicate balance of our federal nation? Will Canadian courts find a way to recognize Indigenous law as one of Canada's three founding legal systems, weaving it together with common law and civil law to achieve forward-looking legal pluralism? Will Canadian courts finally recognize that Canadians have a constitutionally protected right to live in a healthy environment, a human right that is widely recognized to include a safe, livable climate?

The Canadian Constitution has long been conceived of as a living tree.[5] With Canada's vast expanses of forests facing dire threats from climate-exacerbated wildfires, the living tree metaphor takes on renewed relevance.[6] Recent decisions have cleared out some of the dead wood that was impeding progress, such as obstacles related to justiciability and very restrictive readings of the right to life, liberty and security of the person (s. 7 of the *Charter*). Remaining barriers include narrow interpretations of equality rights under the *Charter* (s. 15) and the reluctance of courts to craft effective and equitable remedies in climate cases based on their

---

4   David R. Boyd, *The Right to a Healthy Environment: Revitalizing Canada's Constitution* (UBC Press, 2012).

5   *Reference re Same-Sex Marriage*, 2004 SCC 79 at paras 22-30.

6   John Vaillant, *Fire Weather: The Making of a Beast*, (Penguin Random House Canada, 2023). Larissa Parker, "Let Our Living Tree Grow – Beyond Non-Justiciability for the Adjudication of Wicked Problems" (2022) 81(1) U of T Fac of L Rev. 54.

concerns about transgressing the separation of powers between the legislative, executive and judicial branches of government.

Climate cases have reached Canadian courtrooms for almost twenty years now. The earliest cases were disposed of dismissively, with judges ruling that the applicants raised political rather than legal issues and therefore the lawsuits were not justiciable, i.e. not appropriate for judicial resolution.[7] While justiciability continues to be an issue in Canadian climate litigation, it is no longer an insurmountable obstacle. Canadian judges have joined their colleagues around the world in recognizing that there is nothing inherently non-justiciable about lawsuits involving the climate crisis. This is particularly true when parties raise arguments grounded in human rights, as it is widely acknowledged that one of the judiciary's fundamental duties is to hold governments accountable when they violate or fail to fulfil their human rights obligations. As detailed in this book, ongoing Charter challenges led by youth and Indigenous Peoples, both of whom are disproportionately impacted by the climate crisis, will answer some of the outstanding questions identified by Professor Chalifour.

In repeated cases dating back to 1995, the Supreme Court of Canada has made statements endorsing the human right to a healthy environment. For example, in 2023, Chief Justice Richard Wagner wrote that "The Canadian judiciary, in tandem with the other branches of government, has an important role to play in protecting the 'right to a safe environment'."[8] Yet this right is not found, explicitly, in the text of Canada's Constitution. Youth climate activists and Indigenous Peoples are arguing that the right to a healthy environment is implicit in the right to life (s. 7), a legal conclusion already reached by courts in at least two dozen other nations.[9]

---

7   *Friends of the Earth v Canada (Governor in Council)*, 2008 FC 1183; *Turp v Canada (Minister of Justice)*, 2012 FC 893.

8   *Reference re Impact Assessment Act*, 2023 SCC 23 at para 1, citing *Ontario v Canadian Pacific Ltd*, [1995] 2SCR 1031 at para 55, which in turn quoted the Law Reform Commission of Canada, *Crimes Against the Environment* (1985), Working Paper 44, at p. 8.

9   David R. Boyd, "The Implicit Constitutional Right to a Healthy Environment," (2011) Review of European Community and International Environmental Law 20, 2: 171-79.

As described in this book, in leading climate cases from all over the world courts have taken essential steps to hold governments accountable, pushing them to take more urgent and ambitious action in order to fulfill their obligations pursuant to legislation, constitutions, and international law. Relying heavily on climate science, courts have ordered governments to take stronger action (e.g., setting higher targets for emissions reduction), but generally have refrained from prescribing the specific steps. Strong precedents have been set by courts in countries with diverse legal systems including the Netherlands, Brazil, Germany, Belgium, Colombia, France, India, Mexico, and the United States.[10]

The scientific evidence about the acceleration of the climate emergency is truly frightening. Humanity has departed from the Holocene epoch, 10,000 years of climatic stability that enabled the emergence of agriculture and the evolution of human civilization. Dangerous tipping points lay at unknown distances on the horizon. As scientists have exhaustively documented, the magnitude of chaotic climate disruption and associated human misery depends very much upon short-term actions taken by wealthy, high emitting nations like Canada. In the words of the Intergovernmental Panel on Climate Change, "There is a rapidly closing window of opportunity to secure a liveable and sustainable future for all."[11]

---

10   See for example: *Netherlands v Urgenda Foundation*, ECLI:NL:HR:2019:2007, 19/00135 (Supreme Court Netherlands); *Future Generations v Ministry of the Environment and Others*, [2018] STC4360-2018 (Supreme Court, Columbia); *Neubauer and others v. Germany* (Federal Constitutional Court, 24 March 2021); *Klimaatzaak v Kingdom of Belgium and Others*, 2021/AR/1589 2022/AR/737 2022/AR/891 (Brussels Court of Appeal, Belgium, 30 November 2023); *MK Ranjitsinh & Ors v Union of India & Others*, Writ Petition (Civil) No. 838 of 2019 with Civil Appeal No. 3570 of 2022 (Supreme Court of India, 21 March 2024); Partido Socialista Brasileiro, *Partido Socialismo e Liberdade, Partido dos Trabalhadores e Rede Sustentabilidade v. Brazil* (on Climate Fund), ADPF 708, (Federal Supreme Court of Brazil, 1 July 2022); and *In the matter of Hawai'i Electric Light Company*, 2023, SCOT—22—0000418, Supreme Court of Hawai'i, 13 March 2023.

11   Intergovernmental Panel on Climate Change, "2023: Summary for Policymakers", in *Climate Change 2023: Synthesis Report. Contribution of Working Groups I, II and III to the Sixth Assessment Report of the Intergovernmental Panel on Climate Change.* Geneva, Switzerland, pp. 1-34, at 24.

For Canada, the climate emergency, with its interconnected economic, social and environmental complexities, may seem like an unsolvable puzzle. Yet with this eloquent, passionate and impeccably researched book, Professor Nathalie Chalifour has provided a clear and compelling pathway out of the maze. Whether Canadian courts and policymakers will adopt this wise guidance is a question that will be answered in the years ahead.

David R. Boyd

ACKNOWLEDGMENTS

This book would not have not been possible without the support, guidance, and inspiration of many remarkable individuals. I am deeply grateful to each of you for your generosity, encouragement, and wisdom.

I wrote this manuscript in Ottawa, on the unceded, unsurrendered territory of the Anishinaabe Algonquin Nation. With deep respect, I honour them and all First Nations, Inuit, and Métis Peoples whose enduring stewardship, knowledge, and cultural traditions have long sustained these lands and waters. It is a privilege to work and write in a place shaped by such profound histories.

To those who reviewed draft chapters, your thoughtful and generous feedback strengthened this work in more ways than I can express. I extend special thanks to Professor Heather McLeod-Kilmurray, Professor David R Boyd, Professor Lynda Collins, Professor Martha Jackman, Professor Aimée Craft, Professor Anne Levesque, Professor Martin Olszynski, Professor Chris Tollefson, Professor Ryan Katz-Rosene, Larissa Parker, and Joshua Ginsberg. You each took time from your full lives to engage with my writing, and your insights were invaluable. I remain humbled by your generosity and grateful for your example of collegiality and care.

The ideas in this book were also shaped through collaborations and conversations with many generous colleagues. I wish to acknowledge those with whom I have had the privilege to work directly on related publications and legal memoranda: Professor Peter Oliver, Professor Lynda Collins, Professor Anne Levesque, Erin Dobbelsteyn, Taylor Wormington, Laura McIntyre, and Jessica Earle. Thank you for the thoughtfulness, rigour, and generosity you brought to our work together.

I also wish to recognize colleagues whose intellectual leadership have helped shape this book. Professor Jamie Benidickson has long been a pillar in Canadian environmental law, and I have been fortunate to count on his wisdom, humour, and support since I began teaching at uOttawa. The late Professor Meinhard Doelle's brilliance, humility, and compassion continue to inspire me — his legacy is a lasting light in this field. Professor David Boyd's scholarship and his global leadership as United Nations Special Rapporteur on Human Rights and the Environment (2018–24) set a standard of excellence I aspired to throughout the writing of this book. I am especially honoured and grateful that he wrote the Foreword, — an act of great generosity for someone with innumerable demands on his time.

To my students, who continually inspire me with their excellence and hope, keep reaching for your dreams and embrace an open mind and heart, especially in difficult times.

I am also indebted to the outstanding research assistance of Gérick Girard, Alexandria Peacock, and Ronald Cheung, whose diligence and attention to detail helped bring this book to completion. Thanks as well for their dedicated work, both in conducting research and formatting references. My thanks as well to the Social Sciences and Humanities Research Council (SSHRC) for supporting much of the research underpinning this project and to the wonderful editorial team at University of Toronto Press-especially Josephine Mo, Tina Eng, Kai Toh, and Aditi Parikh for your care and guidance throughout the shepherding publication process.

This work is inspired by the courage and commitment of youth, Indigenous Peoples, and public interest lawyers who tirelessly advocate for a vision of the Constitution grounded in justice, sustainability, and a liveable future for all.

I am fortunate to be surrounded by some of the kindest, most generous colleagues at uOttawa. Heather McLeod-Kilmurray, Lynda Collins,

Anne Levesque, Natasha Bakht, Penelope Simons, Martha Jackman, Constance Backhouse — thank you for enriching my life in so many ways.

Finally, to my family. Lorne — thank you for your love, patience, and the many ways you held things together so I could write — Lucie and Keelan—you are my heart. I thought about you—and your futures—continually as I wrote. Thank you for your understanding during the many hours I worked late, even though I would have rather spent it with you. If this book plays even a small role in helping secure a safe and vibrant planet for you to grow old on, then it will have been worth it.

There are so many more people to recognize than I can mention here. To all who have been part of this journey, my heartfelt thanks.

INTRODUCTION*

Climate change is an urgent, existential planetary crisis unlike anything humanity has faced before. Its impacts, many of which are irreversible on a human timescale, threaten the very systems that sustain life.[1] In 2021, the Supreme Court of Canada recognized the climate crisis for the first time, describing it as a grave threat to humanity.[2] This acknowledgement underscores the staggering implications: Earth is our only home — a planet that provides fresh water, fertile soil, clean air, and a climate capable of sustaining life. These systems, which many take for granted, are now in peril.

The climate crisis is not just an environmental challenge; it is an equity problem of unprecedented magnitude at a global, national and local scale. Its harms are disproportionately borne by those least responsible, including Indigenous Peoples, millions of people in Small Island

---

*   The author wishes to thank reviewers of this chapter, including Professor Heather McLeod-Kilmurray, Professor Lynda Collins, Professor David Boyd, and students who provided research assistance, including Gérick Girard (JD/LLL Candidate, uOttawa) and Ronald Cheung (JD Candidate, uOttawa).

1   See Intergovernmental Panel on Climate Change, *Climate Change 2023: Synthesis Report*, Contribution of Working Groups I, II and III to the Sixth Assessment Report of the IPCC, Hoesung Lee et al, eds (Geneva: IPCC, 2023).

2   *References re Greenhouse Gas Pollution Pricing Act*, 2021 SCC 11 at para 12 [*GGPPA References*].

Developing States, seniors, people living in poverty, children, and future generations who will inherit a destabilized climate.[3] The injustices ripple beyond human communities, devasting the more-than-human world with irreversible consequences. How did we reach this point?

THE ANTHROPOCENE AND THE GREAT ACCELERATION

After World War II, humanity entered a period of exponential growth – socially, economically, and technologically – that has left an ecological footprint far exceeding Earth's carrying capacity. This "Great Acceleration" has reshaped the planet's systems so profoundly that scientists now propose a new geological epoch: the Anthropocene.[4] We are leaving behind the stable Holocene epoch of the last 12,000 years in which agriculture developed and human societies thrived and entering an uncertain future defined by humanity's overwhelming impact on the planet.[5]

Against this backdrop, scientists have developed a framework identifying nine planetary boundaries — thresholds that must not be crossed if we are to maintain a safe operating space for humanity.[6] This framework was further refined to identify not only safe, but also just, boundaries for

---

3   Marina Romanello et al, "The 2024 Report of the *Lancet* Countdown on Health and Climate Change: Facing Record-breaking Threats from Delayed Action" (2024) 404:10465 Lancet 1847.

4   While geologists recently voted against officially declaring the start of the Anthropocene, critics emphasize that regardless of the definition, there is consensus that humans have an undeniably major and widespread impact on the planet at a scale that is altering Earth systems. See Editorial, "Are We in the Anthropocene yet? Measurement Matters But Should Not Detract from the Reality That Humans Are Altering Earth Systems" (2024) 627:8004 Nature 466.

5   See Colin N Waters et al, "Candidate Sites and Other Reference Sections for the Global Boundary Stratotype Section and Point of the Anthropocene Series" (2023) 10:1 Anthropocene Rev 3; Anthony Barnosky & Mary Ellen Hannibal, "Despite Official Vote, the Evidence of the Anthropocene is Clear" (2 April 2024), *YaleEnvironment360*, online: <e360.yale.edu> [perma.cc/5ZFZ-S7ZL]. For more on the rapid, widespread increase in human activity and its impacts on the planet since the industrial age, see also Globaia, "The Great Acceleration" (last visited 17 January 2025), online: <globaia. org> [perma.cc/X55T-DYDF].

6   See Johan Rockström et al, "Planetary Boundaries: Exploring the Safe Operating Space for Humanity" (2009) 14:2 Ecology & Society.

Earth systems that aim to avoid significant harm to people. These just boundaries consider interspecies justice, intergenerational justice, and intragenerational justice between countries.[7] Today, six of the nine safe and just boundaries – including climate - have been exceeded.[8]

The impact of crossing the climate threshold is already evident. Almost every day I spent writing this book, there was news of a major flood, wildfire, drought, or severe storm wreaking havoc on the lives of people somewhere in the world.[9] Even as we bear witness to the unfolding of the climate crisis, humans are skilled at ignoring, deflecting, denying, and questioning what is happening. It is overwhelming and easier to turn to the next distraction. But we must be accountable. Indeed, we are the generation that knows enough about the magnitude of the threat we face while there is still time to avoid the most devastating consequences. Time is of the essence. As the Intergovernmental Panel on Climate Change (IPCC) has noted, "[t]here is a rapidly closing window of opportunity to secure a liveable and sustainable future for all."[10] Our decisions in the next decade will determine the fate of the planet and humanity's future on it.

NAVIGATING THE CONSTITUTION IN THE CLIMATE CRISIS

This book aims to discharge a small part of our responsibility by considering how the Canadian Constitution[11] can and should be interpreted in the era of the climate emergency. Climate change will test and

---

7   Johan Rockström et al, "Safe and Just Earth System Boundaries" (2023) 619:7968 Nature 102.

8   The other boundary is aerosols. Climate is one of only two boundaries where just boundaries are stricter than the safe ones. In other words, people are harmed before the climate system has breached the destabilization threshold (see *ibid*).

9   See, e.g., "Summer 2024 Shatters Records for Severe Weather Damage: Over $7 Billion in Insured Losses from Floods, Fires and Hailstorms," *Insurance Board of Canada* (24 September 2024), online: <ibc.ca> [perma.cc/JM2W-5UV4].

10   Intergovernmental Panel on Climate Change, "2023: Summary for Policymakers" in *Climate Change 2023: Synthesis Report*. Contribution of Working Groups I, II and III to the Sixth Assessment Report of the Intergovernmental Panel on Climate Change. Geneva, Switzerland, pp 1-34 at 24.

11   *Constitution Act, 1867* (UK), 30 & 31 Vict, c 3, reprinted in RSC 1985, Appendix II, No 5; *Constitution Act, 1982*, being Schedule B to the *Canada Act 1982* (UK), 1982, c 11.

challenge every part of society, including our legal systems and the rule of law. This book takes on some of the questions and issues that climate change will pose for lawyers, judges, law students, policymakers, and the public as we confront one of the greatest challenges of our time. It explores how the Constitution must and can withstand the increasing social, economic and political instability that will result from the ecological impacts of the climate crisis. It recognizes that we must address the Constitution's colonial roots and repair past and ongoing harms of colonialism. Indigenous Peoples are at the forefront of the climate crisis. Though they bear little responsibility for creating the problem, they are bearing and will continue to bear a disproportionate burden of its harms.[12] Indigenous Peoples must be leaders in the response to climate change. The Constitution must reflect and restore a Nation-to-Nation relationship between the Crown and Indigenous Peoples who have been stewards of the land and waters for millennia with their own legal traditions and constitutions.[13]

The judiciary needs to be ready to interpret and apply legal principles in a way that ensures that the very foundations upon which our legal system is based will endure. The judiciary will face a growing number of cases linked to climate change in the coming years. From challenges to jurisdictional authority to enact climate legislation and cases engaging with Indigenous rights and reconciliation to claims of rights infringements under the *Canadian Charter of Rights and Freedoms*,[14] judges will be increasingly called upon to address many novel legal issues with great significance. This book aims to unpack some of the questions that will arise at the intersection of climate change and the Constitution.

---

12 See *GGPPA References*, above note 2 at paras 11, 12, 187, and 206; United Nations Department of Economic and Social Affairs "Climate Change" (last visited 17 January 2025) online: <un.org> [perma.cc/DU2W-MEDK].

13 See, e.g., John Borrows, *Canada's Indigenous Constitution* (Toronto: University of Toronto Press, 2010).

14 *Canadian Charter of Rights and Freedoms*, Part I of the *Constitution Act, 1982*, being Schedule B to the *Canada Act 1982* (UK), 1982, c 11.

HOW THE BOOK IS STRUCTURED

Chapter 1 outlines essential climate science facts and Canada's legal obligations related to climate change. The chapter highlights the most important and relevant points to inform the rest of the book, with ample footnotes and additional resources to guide readers who wish to access a deeper level of information.

Chapter 2 engages with the topic of Indigenous rights, legal orders, self-governance, and reconciliation in the context of climate change. The deep wrongness of climate change's disproportionate burden on Indigenous Peoples casts a powerful shadow over any analysis of the Constitution that does not actively find ways to remedy, repair, reconcile, and prevent further harms. I acknowledge my position as a settler descended from francophone immigrants who arrived on this land in the sixteenth century as carpenters and approach the work in this book with humility. The analysis in subsequent chapters of this book does not actively engage with the formidable question of how to create a Constitution that repairs, rebuilds, and reconciles the relationship between the Crown and Indigenous Peoples – a Constitution that reflects the legal pluralism of three legal systems (common law, civil law and Indigenous laws) in our country. However, I believe that an interpretation of the Constitution that promotes accountability and responsibility on all governments to address climate change and safeguards *Charter* rights is aligned with a vision of reconciliation and climate justice. I encourage readers to seek to learn about Indigenous perspectives about climate change and the Constitution beyond this book with opens minds and open hearts.[15]

Chapter 3 tackles what it means to create a regulatory regime to address climate change in a federation. The division of legislative powers that produces shared, overlapping, and sometimes unclear jurisdiction

---

15   See, e.g., Borrows, above note 13; Mario Blaser et al, eds, *Indigenous Peoples and Autonomy: Insights for a Global Age* (Vancouver: UBC Press, 2010); Eriel Tchekwie Deranger et al, "Decolonizing Climate Research and Policy: Making Space to Tell Our Own Stories, in Our Own Ways" (2022) 57:1 Community Development J 52; Aimee Craft & Paulette Regan, eds, *Pathways of Reconciliation: Indigenous and Settler Approaches to Implementing the TRC's Calls to Action* (Winnipeg: University of Manitoba Press, 2020).

adds an additional challenge to the already difficult task of developing this framework. The chapter will explore what the emerging jurisprudence, including the Supreme Court of Canada's decisions in *GGPPA References*, *Reference re Impact Assessment Act*[16] and other relevant case law, tell us with respect to the scope of key powers such as provincial authority over property and civil rights and natural resources and federal jurisdiction under the criminal law power and the National Concern branch of POGG. In a federation marked by increasingly contentious and polarized climate politics, the chapter examines the role of doctrines such as pith and substance and double aspect and approaches such as cooperative federalism in the ongoing jurisdictional challenges to federal climate policies.

Chapter 4 examines the emerging *Charter* cases brought by youth and Indigenous Peoples claiming infringements of their rights to life, security of the person, and equality. Can governments be held accountable through rights-based litigation for not doing enough to reduce greenhouse gas (GHG) emissions and protect the public from the impacts of climate change? Can the government be held responsible for failing to act or setting weak targets? These questions and more are now being posed in (and answered by) a variety of Canadian courts. So far, the decisions lay bare the reality that *Charter* rights were drafted in an era that did not contemplate the kind of widespread, systemic harms provoked by the global threat of climate change. Courts in climate litigation cases are grappling with the role of the judiciary in the face of government inaction, in the context of rapidly evolving science, and an unprecedented level of threat that could undermine the ability of humanity to thrive and exercise human rights. Chapter 4 unpacks issues ranging from justiciability and causation to the implications of framing cases as negative versus positive claims. It also looks at how the temporal quality of climate harms impacts the interpretation of age-related discrimination.

Finally, the book ends with a short conclusion that pulls together the earlier chapters and tries to answer the central question of this book: "Is the Canadian Constitution up to the challenge of climate change?" Will

---

16 *Reference re Impact Assessment Act*, 2023 SCC 23.

it evolve to support human and ecological survival or remain stagnant and risk condoning harm? Will the judiciary favour cautious, incremental interpretations that will render it a bystander to the unfolding climate crisis? Or will it adapt and evolve constitutional jurisprudence in a way that enables each order of government to use its jurisdictional authority to address the climate threat and safeguard the rights of all Canadians, especially Indigenous Peoples, children, and future generations?

HOW TO USE THIS BOOK

This book can be read from start to finish, but it can also be used as a reference for particular subjects, important cases, or the relevance of principles and approaches. Each chapter can be read alone and in any order. I have provided a detailed Table of Contents, comprehensive index, and ample footnotes and supplementary resources. My hope is that the book will serve as a practical tool for scholars, students, practitioners, law clerks, judges, policymakers, the media, and the general public, while also offering thought-provoking points of reflection. I also hope it will be a launch pad for more detailed research as this legal space develops.

WHAT THE BOOK DOES NOT DO

Given the enormity of climate instability and how it intersects with virtually all parts of life, it would be impossible to cover all aspects of constitutional law and climate change. The number of climate-related laws, regulations and court decisions are growing exponentially in Canada and abroad, and each of these has the potential to intersect with the Constitution. The book covers the subjects identified above, which reflect some of the main ways in which the Constitution is already engaging with climate change. However, there are many important topics that are not covered or not fully explored. The book is focused mainly on GHG mitigation efforts, rather than adaptation or other climate policies. While I occasionally identify topics that are outside the scope of the book, I am reassured by the mainstreaming of climate issues into the lives of lawyers from all disciplines. No lawyer, judge, law professor, or law student will be able to avoid climate change in their study or practice as we move further into the heart of the unfolding climate emergency.

MY PERSPECTIVE

The analysis in this book is intended to be objective and is based on a rigorous foundation of reserach. However, I am of the view that everyone — no matter how well-intentioned in terms of objectivity — brings their own perspectives, ideologies, biases, and experiences to their work. This includes academics, scientists, bureaucrats, judges and members of the public. While my aim is to offer sound legal analysis and a balanced perspective, I am comfortable wearing a particular bias openly: I am in favour of supporting the natural conditions that have allowed human civilization to prosper and the Earth's magnificent, albeit besieged, biodiversity to evolve. As such, I have a bias toward a Constitution that supports living conditions on Earth that can sustain flourishing, fair and just human societies, and vibrant, healthy ecosystems. When the Constitution is open to interpretation and one interpretation points us in the direction of human survival and prosperity and one forks toward death, destruction, suffering and inequality, I will be unapologetic in favouring the interpretation that is aligned with survival and well-being. Some may debate those interpretations, and that is fair play. Indeed, this book aims to contribute to legal and policy debates about how our Constitution will shape our futures. But let us at least be honest about the vision we uphold when we hold fast to narrow, formalistic interpretations of the Constitution that lock us in the past. Let us pose difficult questions, explore the consequences of various approaches, and be truthful about what interests we are advancing by supporting particular positions. Let us forge a path forward that allows us to continue to debate these issues on a healthy, habitable planet for an indefinite future.

chapter 1

CLIMATE SCIENCE AND THE LAW AND POLICY FRAMEWORK FOR CLIMATE CHANGE[*]

"Despite the initial hope inspired by the 2015 Paris Agreement, the world is now dangerously close to breaching its target of limiting global multi-year mean heating to 1.5°C. Annual mean surface temperature reached a record high of 1.45°C above the pre-industrial baseline in 2023, and new temperature highs were recorded throughout 2024. The resulting climatic extremes are increasingly claiming lives and livelihoods worldwide."[1]

"The effects of climate change have been and will be particularly severe and devastating in Canada. Temperatures in this country have risen by 1.7°C since 1948, roughly double the global average rate of increase, and are expected to continue to rise faster than that rate."[2]

---

[*] The author wishes to thank the following reviewers of this chapter, including Professor Ryan Katz-Rosene, Professor David Boyd, and Professor Heather McLeod-Kilmurray, and research assistants Ronald Cheung (JD Candidate, uOttawa), Gérick Girard (JD/LLL Candidate, uOttawa), and Alexandria Peacock (JD Candidate, uOttawa).

[1] Marina Romanello et al, "The 2024 Report of the *Lancet* Countdown on Health and Climate Change: Facing Record-breaking Threats from Delayed Action" (2024) 404 Lancet 1847 at 1847.

[2] *References re Greenhouse Gas Pollution Pricing Act*, 2021 SCC 11 at para 10 [*GGPPA References*].

The Supreme Court of Canada has often recognized that "the Constitution must be interpreted in a manner that is fully responsive to emerging realities."[3] Climate change is one of those realities. In fact, the Earth's climate and the Canadian Constitution have a few important things in common.

First, they are both foundational to society's well-being and ability to prosper into the future. A stable climate is a precondition for life on Earth. Living beings today evolved to exist upon a planet supported by an atmosphere that keeps temperatures within a given range and with ecological functions that provide clean air, water, and nourishing soil. Similarly, the Constitution is the bedrock of a nation.[4] It reflects the values of society, establishing and safeguarding foundational principles of democracy, equality, and fundamental freedoms. It is the source of stability and predictability that enables society to flourish and prosper. Its principles are a pre-condition to societal stability and identity.

Second, both climate change and the Canadian Constitution are currently in a fragile state that poses a threat to the future of modern human society as we know it. Atmospheric concentrations of greenhouse gas (GHG) emissions are at their highest levels in some three to five million years, resulting in climatic instability and planetary-scale disruptions that endanger the lives and well-being of all beings who inhabit the planet.[5] As the Supreme Court of Canada recently recognized, the threat of climate change to humanity is grave and existential.[6] The Canadian Constitution is similarly in a fragile place, with Indigenous reconciliation forcing a much needed and overdue reckoning with the harms caused by colonialism and confederation, calling the legitimacy of the formative document into question. Increasing use of the

---

3   *R v Hydro-Quebec*, 1997 CanLII 318 at para 86 (SCC).

4   The Canadian Constitution exists within a multi-juridical context that includes Indigenous constitutions, legal traditions, self-determination and self-governance. See, e.g., John Borrows, *Canada's Indigenous Constitution* (Toronto: University of Toronto Press, 2010). See also Chapter 2 of this book, as well as sources cited therein.

5   NASA, "Graphic: The Relentless Rise of Carbon Dioxide" (29 August 2013), online: science.nasa.gov [https://perma.cc/28W6-UU3K].

6   *GGPPA References*, above note 2 at para 2.

notwithstanding clause (e.g., in Ontario),[7] unilateral amendments to the Constitution (Quebec),[8] and unilateral enactments of powers aimed at immunizing a province from federal laws it views as a threat to its interests (Alberta) all undermine the stability of the Constitution.[9]

All of our legal institutions, including the Constitution and the rule of law, will be tested as we grapple to address and adapt to an unstable climate.[10] A context-driven interpretation of the Constitution that takes into account climate disruption will help support its stability and longevity by ensuring fundamental Constitutional values are maintained as society faces the destabilizing erosion of living conditions on our collective planetary home. Our Constitution can and must evolve toward reconciliation with Indigenous Peoples and provide stability, resilience, and promote equality in the face of the climate crisis.

This chapter aims to provide essential background about climate change science, including eight unique features of climate change that are relevant to constitutional analysis and the legal framework to address

---

7   *Protecting Elections and Defending Democracy Act, 2021*, SO 2021, c 31; *Keeping Students in Class Act, 2022*, SO 2022, c 19, as repealed by *Keeping Students in Class Repeal Act, 2022*, SO 2022, c 20.

8   Bill 96, *An Act respecting French, the official and common language of Quebec*, 2nd Sess, 42nd Leg, Quebec, 2021 (assented to 1 June 2022), SQ 2022, c 14. The Bill unilaterally amends s 90 of the *Constitution Act, 1867* (UK), 30 & 31 Vict, c 3, reprinted in RSC 1985, Appendix II, No 5, by declaring that French is the common and *only* language of the province.

9   See Tanzim Rashid, "The Slow-Moving, Silent, and Creeping Constitutional Crises facing Canada" (1 November 2022), online: yorku.ca [perma.cc/98C8-ZAF4] ("[t]here is no precedent in the history of Canada where a Provincial Government has actively, knowingly, and enthusiastically breached its obligations under the constitutionally determined division of powers"). A further example is Saskatchewan's decision not to remit the carbon levy owing to the federal government under the *Greenhouse Gas Pollution Pricing Act* (see Alexander Quon, "Sask. Faces Uphill Battle in Carbon Tax Lawsuit with CRA, Legal Experts Say" (9 July 2024), online: cbc.ca [perma.cc /MK7B-HFKX]).

10   In *Reference re Secession of Quebec*, 1998 CanLII 793 at para 32 (SCC), the Supreme Court stated that Constitutionalism and the Rule of Law are fundamental principles of the Constitution. The Rule of Law means that "the law is supreme over the acts of both government and private persons" and that the "exercise of all public power must find its ultimate source in a legal rule" (*ibid* at para 71) The principle of Constitutionalism requires that "all government action must comply with the law, including the Constitution" (see *ibid* at para 72).

it. I have attempted to provide ample resources in footnotes to guide readers interested in diving deeper into the many aspects of climate disruption that are relevant to the constitutional questions addressed in this book.

1.1 WHAT IS CLIMATE CHANGE?

Climate science shows unequivocally that the Earth is warming primarily because of anthropogenic (human-caused) GHG emissions and that this warming is harmful.[11] Anyone who has been inside of a greenhouse can quickly grasp the basic mechanism of climate or global warming. The Supreme Court of Canada explains it simply: GHGs "trap solar energy from the sun's incoming radiation in the atmosphere instead of allowing it to escape, thereby warming the planet."[12] As the atmospheric layer blanketing the planet thickens, more of the sun's energy is absorbed within the Earth system, causing a warming effect.[13]

The gradual warming of land, water, and air temperatures causes widespread and rapid changes to the atmosphere, ocean, cryosphere, and biosphere. The impacts of these changes are extensive, including many adverse impacts and related losses and damages to nature and people.[14] These impacts are often divided into two (sometimes three) broad categories. The more immediate impacts include more frequent and severe weather events, such as floods, wildfires, heat domes, droughts, and severe storms. Slow-onset changes include rising sea levels and ocean acidification, which lead to loss of coastal zones, fish stocks, and coral reefs, as well

---

11 Intergovernmental Panel on Climate Change, *Climate Change 2023: Synthesis Report*, Contribution of Working Groups I, II and III to the Sixth Assessment Report of the Intergovernmental Panel on Climate Change, Core Writing Team et al, eds (Geneva: IPCC, 2023) [IPCC, "AR6"].

12 *GGPPA References*, above note 2 at para 7.

13 Approximately half of human-caused CO_2 emissions is absorbed by the oceans through diffusion (26 percent) and by ecosystems on land (29 percent). The rest accumulates in the atmosphere contributing to warming (see Climate Central, "The Global Carbon Budget" (1 February 2023), online: climatecentral.org [perma.cc /A8GH-W2KS]).

14 IPCC, "AR6," above note 11 at 3.

as changing disease vectors. A third category includes systemic changes caused by tipping points, discussed in Section 1.5.2.3 below.

1.2 MEASURING CLIMATE CHANGE

As per the adage "you cannot manage what you do not measure," climate scientists have identified a variety of thresholds and mechanisms for measuring the extent of climate change. There are two main ways in which climate change can be measured that are relatively simple to understand and provide useful baselines against which to measure mitigation efforts. The first is by global average surface temperature and the second is by concentrations of GHGs in the atmosphere.

1.2.1 Global Average Surface Temperatures

Measurements of global average surface temperatures provide an objective way to calculate changes in the global climate system. The IPCC estimates that global average surface temperatures reached 1.19 degrees Celsius above pre-industrial levels (1850–1900) between 2014 and 2023.[15] By 2024, the annual average crossed 1.5 degrees Celsius.[16] A global average temperature rise of more than 1.5 to two degrees Celsius above pre-industrial levels represents a dangerous and irreversible level of climate change, with potentially catastrophic consequences.[17] As knowledge has evolved,

---

15   Piers M Forster et al, "Indicators of Global Climate Change 2023: Annual Update of Key Indicators of the State of the Climate System and Human Influence" (2024) 16:6 Earth System Science Data 2625 at 2637, online: essd.copernicus.org [perma.cc/A284-8D5T]. See also IPCC, "AR6," above note 11 at 4.

16   World Meteorological Organization, *State of the Global Climate 2023*, (Geneva: World Meteorological Organization, 2024) at ii, online: wmo.int [perma.cc/F68Y-BWJM], documenting 1.45 degrees of average warming in 2023; World Meteorological Organization, "2024 is on Track to Be Hottest Year on Record as Warming Temporarily Hits 1.5° C" (11 November 2024), online: wmo.int [perma.cc/SN7J-EG5W].

17   David G Victor et al, "Introductory Chapter" in Ottmar Edenhofer et al, eds, *Climate Change 2014: Mitigation of Climate Change*, Contribution of Working Group III to the Fifth Assessment Report of the Intergovernmental Panel on Climate Change (New York: Cambridge University Press, 2014) 111 at 113. See also Myles R Allen et al, "Summary for Policymakers" in Valerie Masson-Delmotte et al, eds, *Global Warming*

the scientific community has ratcheted down the temperature threshold to 1.5 degrees of warming to avert catastrophic levels of climate disruption. The difference between 1.5 and two degrees of warming is significant enough to determine the fate of coral reefs.[18] Warming produces harms at a rise of even one degree Celsius,[19] a threshold the planet has already crossed.[20] Indeed, we are starting to experience some of the more frequent and extreme weather events associated with a warming world.[21] Unfortunately, this is the tip of the proverbial iceberg.

The United Nations Framework Convention on Climate Change (UNFCCC) regime (described below) and many courts around the world have used these temperature thresholds as a way of setting policy goals and evaluating whether policy actions are sufficient to keep emissions from causing warming to surpass these thresholds.[22] The Supreme Court of Canada referenced temperature thresholds to describe the impacts of climate change and the importance of reducing GHG emissions.[23]

---

of 1.5 C: An IPCC Special Report on the Impacts of Global Warming of 1.5 C above Pre-Industrial Levels and Related Global Greenhouse Gas Emission Pathways, in the Context of Strengthening the Global Response to the Threat of Climate Change, Sustainable Development, and Efforts to Eradicate Poverty (New York: Cambridge University Press, 2018) at 3; David Armstrong McKay et al, "Exceeding 1.5°C Global Warming Could Trigger Multiple Climate Tipping Points" (2022) 377:6611 Science 1 at 7. See also Climate Action Tracker, *Warming Projections Global Update* (December 2023), online: climateactiontracker.org [perma.cc/7BFD-W53G].

18   See below note 54.

19   See James Hansen et al, "Young People's Burden: Requirement of Negative CO_2 Emissions" (2017) 8:3 Earth System Dynamics 577 at 578.

20   Myles R Allen et al, "Chapter 1: Framing and Context" in Valerie Masson-Delmotte et al, eds, *Global Warming of 1.5 C: An IPCC Special Report on the Impacts of Global Warming of 1.5 C above Pre-Industrial Levels and Related Global Greenhouse Gas Emission Pathways, in the Context of Strengthening the Global Response to the Threat of Climate Change, Sustainable Development, and Efforts to Eradicate Poverty* (New York: Cambridge University Press, 2018) 49.

21   See Romanello et al, above note 1 at 1852.

22   See, e.g., The Hague District Court, Den Haag, 24 June 2015, *Urgenda Foundation v the State of the Netherlands* (2015), C/09/456689 (The Netherlands) [*Urgenda* Trial Court]; The Hague Court of Appeal, Den Haag, 9 October 2018, *Urgenda Foundation v the State of the Netherlands* (2018), 200.178.245/01 (The Netherlands); Supreme Court of the Netherlands, Den Haag, 13 January 2020, *Urgenda Foundation v the State of the Netherlands* (2020), 19/00135 (The Netherlands) [*Urgenda* Supreme Court].

23   *GGPPA References*, above note 2 at paras 8–11.

1.2.2   Atmospheric Concentrations of CO2 (Parts per Million or PPM)

A second method of measuring warming and identifying dangerous thresholds is by assessing atmospheric concentrations of carbon dioxide (the predominant GHG in the atmosphere). Ice core data dating back some four hundred thousand years shows that carbon dioxide concentrations have generally varied between 180 parts per million (ppm) (during ice ages) and 280 ppm (during warmer interglacial periods).[24] In 2013, carbon dioxide levels crossed the threshold of 400 ppm for the first time in human history. At the time of writing, carbon dioxide levels were at 430 ppm.[25] Given the current rate of growth in carbon dioxide, we are likely to reach 500 ppm in approximately fifty years.[26] The last time the Earth's atmosphere contained 400 ppm of carbon dioxide, the global average surface temperature was two to three degrees Celsius warmer than it is today, with sea levels ten to twenty metres higher than they are now.[27] Scientists have identified 350 ppm (which corresponds to approximately one degree Celsius of warming above the pre-Industrial baseline) as a safe threshold for atmospheric concentrations of carbon dioxide.[28]

While fewer cases use concentrations of CO_2 as the central metric for measuring climate change, a set of cases including the landmark *Juliana v United States* youth climate lawsuit focused on this measurement.[29] In a favourable ruling on behalf of youth plaintiffs, a Montana court described

---

24   Co2.earth, Numbers for Living on Earth, online: co2.earth/daily-co2.

25   *Ibid.*

26   Nicola Jones, "How the World Passed a Carbon Threshold and Why It Matters" (26 January 2017), online: e360.yale.edu [perma.cc/K57T-NWF2].

27   World Meteorological Organization, "WMO Statement on the State of the Global Climate in 2017" (2018) World Meteorological Organization, Working Paper No 1212 at 8, online: library.wmo.int [perma.cc/BFD9-2YQ2].

28   James Hansen et al, "Target Atmospheric CO_2: Where Should Humanity Aim?" (2008) 2 Open Atmospheric Science J 217.

29   See Our Children's Trust, "*Juliana v United States*" (last visited 14 January 2024), online: ourchildrenstrust.org [perma.cc/C9JR-H7WH].

the accumulation of CO_2 in the atmosphere, referencing the increase in PPMs.[30]

1.2.3 Geographic Variations

While climate science is often presented in terms of global averages, there are important geographic variations to consider. Canada, for example, is warming at roughly twice the global average.[31] The most pronounced warming is in the North, where warming is approximately three times the global average.[32] Dramatic reductions in Arctic sea ice cover have already been documented,[33] with over 75 percent of the North Pole's summer ice disappearing over the last thirty years.[34] The Greenland ice sheet has been losing an average of 286 billion tons of ice per year since 2002.[35] In August 2020, Canada's last fully intact ice sheet — four thousand years old and as large as the island of Manhattan — collapsed into the sea.[36] The Supreme Court of Canada took note of this accelerated warming in Canada in

---

30 *Held v State*, CDV-2020-307 (Mont Dist Ct 2023) [*Held*]. Judge Seely also referenced the concept of energy imbalance (the energy arriving to Earth from the sun minus the energy radiated back to space) as a critical metric for climate change in the decision (see *ibid*).

31 Elizabeth Bush et al, *Canada's Changing Climate Report – Executive Summary* (Gatineau: Environment and Climate Change Canada, 2019) at 5, online: changingclimate.ca [perma.cc/5CLN-JGQW]. See also *GGPPA References*, above note 2.

32 Bush et al, above note 31. The Mackenzie River delta is projected to warm up to six degrees (see Environment and Climate Change Canada, *Temperature Change in Canada: Canadian Environmental Sustainability Indicators* (Gatineau: Environment and Climate Change Canada, 2024) at 8, online: canada.ca [perma.cc/TF6T-CDHT]). See also *GGPPA References*, above note 2 at para 11.

33 Copernicus Climate Change Service, "Sea Ice Cover for September 2023" (September 2023), online: climate.copernicus.eu [perma.cc/S63J-ANAT].

34 Chris Mooney, "The Arctic Ocean Has Lost 95 Percent of Its Oldest Ice: A Startling Sign of What's to Come," *Washington Post* (11 December 2018), online: washingtonpost.com [perma.cc/TLA5-5KTK]; National Oceanic and Atmospheric Administration Arctic, "Arctic Report Card" (2022), online: arctic.noaa.gov [perma.cc/M9C7-PRPD].

35 Alexandre Witze, "Climate Change: Losing Greenland" (2008) 452 Nature 798. See also Michalea D King et al, "Dynamic Ice Loss from the Greenland Ice Sheet Driven by Sustained Glacier Retreat" (2020) 1 Nature Communications Earth & Environment 1.

36 Jordan Davidson, "Canada's Last Intact Ice Shelf the Size of Manhattan Collapses Due to Global Warming" (10 August 2020), online: ecowatch.com [perma.cc/8Q5X-4Q7Z].

GGPPA References, characterizing the effects of climate change for Canada as "particularly severe and devastating."[37]

1.3   THE EFFECTS AND IMPACTS OF CLIMATE CHANGE

The image of a pot of boiling water is often used to explain two aspects of climate change. The depiction of a frog sitting in the pot impervious to the gradual rise in temperature is used to illustrate the point that as passengers on the journey, we may not notice the impacts of climate change as they unfold. The pot of boiling water also helps illustrate why a warming climate creates volatility in weather patterns.[38] Imagine watching a pot of water start warming over a stove. As the temperature of the water rises, small bubbles start to appear in the water. As the water heats, the bubbles become bigger and appear more frequently, until the water is so volatile that one may need to take a step back to avoid getting splashed by droplets of agitated water. Similarly, as global average temperatures rise, the climate system is more volatile, with more frequent and severe weather events and extremes, including cold snaps in unexpected places.[39]

1.3.1   More Frequent and Extreme Weather Events

Twenty years ago, predictions of more frequent and severe meteorological events seemed somewhat theoretical — something to imagine, perhaps inspired by films about extreme weather and post-apocalyptic scenarios. Over the period in which I have written this book, every part of the world has experienced a heat wave. In 2021, temperatures in British Columbia reached over 49 degrees Celsius in an unprecedented heat wave that caused some 740 deaths,

---

37   *GGPPA References*, above note 2 at paras 10–11.
38   Credit for this analogy goes to Professor John Stone, Carleton University, who shared it with students in a guest lecture many years ago.
39   See United Nations Environment Programme, "Debunking Eight Common Myths About Climate Change" (4 June 2024), online: unep.org [perma.cc/LV3C-D79D].

one of the deadliest weather events in Canadian history.[40] In 2023, marine temperatures in Florida reached thirty-eight degrees Celsius.[41] Temperatures in California's Death Valley reached fifty-four degrees Celsius on July 7, 2024.[42] European heatwaves in 2022 caused an estimated 62,862 deaths[43] and in 2023 between 55,000 and 72,000.[44] East Antarctica experienced a record-breaking heat wave in 2022 with temperatures reaching thirty to forty degrees Celsius above average.[45] Over 900 wildfires were burning across Canada at one point in 2023.[46] The town of Fort McMurray was devastated by a raging wildfire in 2016, and in 2024, half of the historic town of Jasper in Alberta was destroyed by wildfire.[47] Lungs across Canada and parts of the United States were irritated from pervasive wildfire smoke inhaled when major Canadian cities such as Toronto, Montreal, Ottawa and Vancouver experienced some of the world's worst air quality in 2023.[48] The entire city of Yellowknife and community of Hay River had to be evacuated for weeks while firefighters fought back wildfires. A 9-year-old boy died in BC from asthma triggered by wildfire smoke.[49]

---

40 Michael J Lee et al, "Chronic Diseases Associated with Mortality in British Columbia, Canada During the 2021 Western North America Extreme Heat Event" (2023) 7:3 GeoHealth.

41 Eric Zerkel, "Ocean Heat Around Florida is 'Unprecedented,' and Scientists Are Warning of Major Impacts," *CNN* (16 July 2023), online: cnn.com [perma.cc/HJ3A-2X72].

42 Mary Cunningham, "Death Valley National Park Just Had its Hottest Summer on Record," *CBS News* (6 September 2024), online: cbsnews.com [perma.cc/AG35-A329].

43 Joan Ballester et al, "Heat-Related Mortality in Europe During the Summer of 2022" (2023) 29 Nature Medicine 1857.

44 United Nations, Climate and Environment, "Heatwave Deaths Increased Across Almost All Europe in 2023, says UN Weather Agency" (22 April 2024), online: news. un.org [perma.cc/R7Q8-ZGM3].

45 British Antarctic Survey, "Extreme Heat Wave in East Antarctica Driven by Record-breaking 'Atmospheric River,' Analysis Finds" (10 January 2024), online: phys.org [perma.cc/B4W2-U82Q].

46 Canadian Interagency Forest Fire Centre, "Fire Information Interactive Map" (2023), online: www.ciffc.ca/ [perma.cc/RMW7-S9H6].

47 See John Vaillant, *Fire Weather: The Making of a Beast* (Toronto: Penguin Random House Canada, 2023); Parks Canada, *Jasper Wildfire 2024*, online: parks.canada.ca [perma.cc/CDG4-J86A].

48 Government of Canada, "Canada's Record-breaking Wildfires in 2023: A Fiery Wake-up Call" (last modified 19 August 2024), online: natural-resources.canada.ca [perma.cc/Q9XB-FMR6].

49 Helen Regan et al, "July 18, 2023 - Millions Face Extreme Heat in the US, Europe and China," *CNN World* (18 July 2023), online: cnn.com [perma.cc/59BY-TPY7].

In 2023, extreme flooding events caused damage around the world, from Africa, Greece and Nepal to China, Central Europe and Canada.[50] Extensive flooding continued in 2024, with Spain experiencing several extreme downpours that killed over 200 people and displaced many more.[51] In Ontario, flooding in August 2024 caused over $100 million in insurance damage.[52] Extreme weather events are becoming an every day phenomenon across the globe.

1.3.2   Slow Onset Changes

Climate change also produces slow-onset changes, many of which are interconnected. Slow-onset changes include melting permafrost and ice sheets, retreating glaciers, rising sea levels, coastal erosion, and ocean acidification.[53] One of the most startling statistics is that coral reefs are projected to experience a 70 percent decline at 1.5 degrees of warming and to virtually disappear at two degrees of warming.[54] We are on a steady march toward eradicating an incredibly diverse, rich, and beautiful part of our tropical biosphere.

Another aspect of slow-onset changes that is difficult to fathom is the warming of oceans. Over 91 percent of the warming created by humans

---

50   World Meteorological Organization, "Exceptional Heat and Rain, Wildfires and Floods Mark Summer of Extremes" (3 August 2023), online: wmo.int [perma.cc /HW7P-386Z].

51   World Weather Attribution, "Extreme Downpours Increasing in Southeastern Spain as Fossil Fuel Emissions Heat the Climate" (4 November 2024), online: worldweather-attribution.org [perma.cc/8LVS-UD5S].

52   Insurance Bureau of Canada, "Summer 2024 Shatters Records for Severe Weather Damage: Over $7 Billion in Insured Losses from Floods, Fires and Hailstorms" (24 September 2024), online: ibc.ca [perma.cc/N3HA-3E69].

53   The slow, insidious harm of environmental impacts has been described as slow vio-lence (see Rob Nixon, *Slow Violence and the Environmentalism of The Poor* (Cambridge: Harvard University Press 2011) at 2–3).

54   Ove Hoegh-Guldberg et al, "Chapter 3: Impacts of 1.5°C Global Warming on Natural and Human Systems" in Valerie Masson-Delmotte et al, eds, *Global Warming of 1.5 C: An IPCC Special Report on the Impacts of Global Warming of 1.5 C above Pre-Industrial Levels and Related Global Greenhouse Gas Emission Pathways, in the Context of Strengthening the Global Response to the Threat of Climate Change, Sustainable Development, and Efforts to Eradicate Poverty* (New York: Cambridge University Press, 2018) 175.

has been absorbed by oceans.[55] This excess heat leads to thermal expansion, rising sea levels, coastal erosion, and marine heat waves. The temperature of oceans warm and cool very slowly, which means that the oceans will continue warming for centuries even once GHG emissions reach net-zero, which is why sea level rise is also locked in for hundreds of years.[56] Scientists have raised concern about changes in ocean currents that play an important role in moving heat around the Earth, including the potential collapse of an important current known as the Atlantic Meridional Ocean Circulation.[57] These extraordinary changes are transforming our planet in profound ways.

1.3.3 Weather Attribution Science and Causation

Not every extreme weather event is directly attributable to climate change. However, a body of rapidly evolving science is helpful in understanding the extent to which warming temperatures are contributing to these more frequent and extreme episodes.[58] Recent studies help to explain the role climate change is playing in causing these weather extremes, from heat domes, cold spells, extreme rainfalls, powerful storms, droughts, and wildfires.[59] Canada has developed a *Rapid Weather Event Attribution* system that has started analyzing the links between heat waves and human-caused

---

55 Piers Forster et al, "The Earth's Energy Budget, Climate Feedback and Climate Sensitivity" in Valerie Masson-Delmotte et al, eds, *Climate Change 2021: The Physical Science Basis*, Working Group I Contribution to the Sixth Assessment Report of the Intergovernmental Panel on Climate Change (New York: Cambridge University Press, 2018) 923 at 930.

56 Piers Forster et al, above note 55 at 925, 930.

57 See, e.g., René M Van Westen et al, "Physics-Based Early Warning Signal Shows that AMOC is on Tipping Course" (2024) 10:6 Science Advances 1. Tipping points are discussed in Section 1.5.2.3 below.

58 See, e.g., World Weather Attribution, "Methods" (last accessed 14 January 2024), online: worldweatherattribution.org [perma.cc/WB8A-P22N].

59 Ben Clarke et al, "Extreme Weather Impacts of Climate Change: An Attribution Perspective" (2022) 1:1 Environmental Research: Climate 1; Luke Kemp et al, "Climate Endgame: Exploring Catastrophic Climate Change Scenarios" (2022) 119:34 Proceedings National Academy Sciences United States of America 1; Matthew W Jones et al, "Global and Regional Trends and Drivers of Fire Under Climate Change" (2022) 60:3 Rev Geophysics 1.

climate change, adding to the developing repository of research.[60] The body of science on attribution is helpful in climate litigation in the context of causation.[61]

1.3.4 Disproportionate Impacts

Groups, communities, individuals and regions who face discrimination or disadvantage are disproportionately impacted by climate change, even though they have generally contributed the least historically to the GHG emissions that cause the problem.[62] The Supreme Court of Canada recognized this in *GGPPA References*, noting that the irreversible harm from climate change would be "borne disproportionately by vulnerable communities and regions, with profound effects on Indigenous Peoples, on the Canadian Arctic and on Canada's coastal regions."[63] The pioneers of the planetary boundaries framework[64] recently integrated justice considerations into their framework. They noted that climate change is one of the few planetary boundaries where vulnerable communities will begin experiencing harm before the planet reaches critical thresholds.[65] As a high emitting country, Canada bears responsibility for the harms of climate change not only in Canada but also in small island developing states, low-lying coastal regions, and drought-, storm-, and flood-stricken African nations. It is essential that fair and equitable climate policies are rapidly deployed to minimize the grave injustices of climate change in Canada and globally.

---

60   Government of Canada, "Extreme Weather Event Attribution" (last modified 25 October 2024), online: canada.ca [perma.cc/U9T9-YFQM].

61   See Climate Attribution, online: climateattribution.org [perma.cc/8F2A-VF8F]. See also Rupert Stuart-Smith et al, "Filling the Evidentiary Gap in Climate Litigation" (2021) 11 Nature Climate Change 651.

62   IPCC, "AR6," above note 11 at 5.

63   *GGPPA References*, above note 2 at paras 11 and 206.

64   See Introductory Chapter of this book, notes 6–8.

65   Johan Rockstrom et al, "Safe and Just Earth System Boundaries" (2023) 619 Nature 102.

Equality-seeking groups[66] are not only at greater risk from the effects of climate change, but they may also be disproportionately impacted by the policies enacted to address climate change. In other words, the impacts of climate change are multiplied not only on identity grounds but also in the way in which harms ensue. Unless mitigation, adaptation, and loss and damage policies are crafted deliberately to ensure they do not further entrench or exacerbate inequalities, they can impose an additional burden on individuals and groups already facing systemic discrimination.[67] For instance, carbon pricing is an effective mitigation strategy that can be equality promoting *if designed properly* (i.e., with exceptions, credits or rebates for those in lower income brackets and funding from the revenues directed toward equality-promoting programs).[68] Adaptation policies can be designed in a way that is equality promoting, such as by ensuring core services like cooling centres and air filtration are accessible and affordable to all and designing evacuation plans in a way that actively assists those with disabilities.[69] Policies that respond to the inevitable losses and damages of climate change that are not avoided through mitigation or adaptation must similarly be designed fairly to reach equality-seeking groups first.[70] Policy-makers must be aware of, and actively work to counter, this potential for a **double burden of injustice**.

There is also the possibility of a **triple burden of injustice** due to structural shifts in geopolitical stability that could result from the ongoing and compounding emergencies resulting from climate change. From

---

66   Equality here refers to substantive equality, as the term is explained by the Supreme Court of Canada in *Fraser v Canada (Attorney General)*, 2020 SCC 28 at paras 40–42.

67   See generally Farhana Sultana, *Confronting Climate Coloniality: Decolonizing Pathways for Climate Justice* (Taylor & Francis Group, 2024).

68   See, e.g., Nathalie Chalifour, "A Feminist Perspective on Carbon Taxes" (2010) 22:1 CJWL 169 [Chalifour, "Feminist Perspective"]; Karen Bubna-Litic & Nathalie Chalifour, "Are Climate Change Policies Fair to Vulnerable Communities? The Impact of British Columbia's Carbon Tax and Australia's Carbon Pricing Policy on Indigenous Communities" (2012) 35:1 Dal LJ 127.

69   See, e.g., Sébastien Jodoin et al, "A Disability Rights Approach to Climate Governance" (2020) 47:1 Ecology L Quarterly 1.

70   See, e.g., Nathalie Chalifour, "Equity Considerations in Loss and Damage" in Meinhard Doelle & Sara L Seck, eds, *Research Handbook on Climate Change Law and Loss & Damage* (Cheltenham: Edward Elgar Publishing, 2021) 18.

droughts and wildfires to widespread flooding and destructive storms, climate change may lead to the dismantling of social institutions (in the face of budgets diverted to dealing with the crises, for instance), increased migration from affected communities, and conflicts that result from infrastructure damage and water and food insecurity.[71] While it is difficult to predict the precise contours of this potential for geopolitical instability and disruption to social institutions, it is clear that those with socio-economic disadvantages, health inequities, and other structural and systemic disadvantages will be disproportionately impacted by such disruption.[72]

1.4 RESPONDING TO CLIMATE CHANGE: MITIGATION, ADAPTATION, AND ADDRESSING LOSS AND DAMAGE

The core responses to climate change generally fall into three broad categories. First, we must reduce GHG emissions and increase sinks (mitigation). Second, we must take steps to adapt to the changes that are here and headed our way (resilience and adaptation). Third, we must address the loss and damage that has occurred and will ensue despite mitigation and adaptation.

1.4.1 Mitigation

The level of mitigation required to avoid dangerous levels of warming has become increasingly clear as science has evolved. While warming has already locked in a number of unavoidable and irreversible changes, deep, rapid, and sustained GHG reductions can limit the extent of those changes. The best available science today suggests that warming could

---

71   See, e.g., United Nations, Meetings Coverage and Press Releases, "With Climate Crisis Generating Global Threats to Global Peace, Security Council Must Ramp Up Efforts, Lessen Risk of Conflicts, Speakers Stress in Open Debate" (13 June 2023), 9345th Meeting, SC/15318, online: press.un.org [perma.cc/98G9-RMYP].

72   For a discussion of catastrophic climate change scenarios, see Luke Kemp et al, "Climate Endgame: Exploring Catastrophic Climate Change Scenarios" (2022) 119:34 Proceedings National Academy Sciences United States of America 1, online: doi .org/10.1073/pnas.2108146119 [perma.cc/U9L2-VDAF].

be halted if the world reaches net-zero emissions.[73] Since the likelihood of abrupt and/or irreversible changes increases with every increment of warming, the ultimate goal must be to achieve net-zero CO_2 emissions as quickly as possible.[74]

Science increasingly shows that climate change causes harms in all regions at one degree of warming (or 350 ppm of CO_2 concentrations), a threshold we have already well surpassed.[75] The changes required to ensure a liveable and sustainable future for all, and to limit the disproportionate harms we are already imposing on vulnerable communities, are rapid and far-reaching transitions across all sectors and systems.[76]

1.4.2 Resilience and Adaptation

Adaptation receives relatively less attention in climate research and policy circles compared to mitigation. Adaptation needs to be a priority alongside mitigation since options for adaptation that are feasible, effective now, and relatively affordable will become less so as warming continues.[77] Governments are often reacting to specific damaging weather events, such as wildfires, floods, or severe storms, rather than addressing adaptation proactively. As disasters unfold around the world at an increasing pace, building resilience and adapting to the inevitable impacts of climate change has become more important. With "mitigation action … woefully underachieving on the scale and ambition needed," the necessity of adaption is even greater and more urgent.[78] From building and designing infrastructure

---

73 If GHG emissions reached net-zero, the concentration of atmospheric CO_2 would start to fall as CO_2 is absorbed by land and oceans and eventually stabilize at a lower level (see Zeke Hausfather, "Explainer: Will Global Warming 'Stop' as Soon as Net-Zero Emissions Are Reached?" (29 April 2021), online: carbonbrief.org [perma.cc /RDP6-D4FH].

74 *Ibid* at 20.

75 Rockstrom et al, above note 65.

76 IPCC, "AR6," above note 11 at 28. See also International Energy Agency, "World Energy Outlook 2022" (November 2022), online: iea.org [perma.cc/TUV5-SZKV].

77 IPCC, "AR6," above note 11 at 19.

78 United Nations Environment Programme, *Adaptation Gap Report 2024* (Nairobi: United Nations Environment Programme, 2024), online: unep.org [perma.cc /DMF2-UKNG].

to withstand the forthcoming weather extremes to adjusting agricultural and forest management practices to align with changing precipitation and temperatures, there is a lot that adaptation policies can do to reduce the amount of loss and damage caused by climate change.

1.4.3 Loss and Damage

Loss and damage refer to the harms resulting from climate change that is not mitigated or that ensue when adaptation efforts are insufficient or non-existent. It includes permanent or irrecoverable harms, such as the loss of coastlines due to sea-level rise, and reparable harms such as the damage inflicted by storms.[79] It can be quantified in terms of economic and non-economic losses and also categorized in terms of slow-onset or extreme weather events.[80] The extensive losses caused by climate change, often at an unimaginable scale, are difficult to fathom and are among some of the most challenging moral questions our society has faced given their existential quality. It was only in 2022 at UNFCCC COP27 that countries agreed to establish a loss and damage fund, now officially called the Fund for Responding to Loss and Damage, to be hosted by the World Bank and housed in the Philippines. A number of developed countries concerned about the prospect of compensation for loss and damage triggering their legal liability led to a provision in the funding mechanism's enabling document that states that the funding is based on cooperation and does not entail liability or a legal responsibility for compensation.

A final category of harm not captured within loss and damage, but which is important to bear in mind, is the potential for harm caused by

---

79 Meinhard Doelle & Sara L Seck, "Introducing Loss and Damage" in Meinhard Doelle & Sara L Seck, eds, *Research Handbook on Climate Change Law and Loss & Damage* (Cheltenham: Edward Elgar Publishing, 2021) 1 at 1 [Doelle & Seck, "Introducing Loss and Damage"].

80 Meinhard Doelle & Sara L Seck, *Research Handbook on Climate Change Law and Loss & Damage* (UK: Edward Elgar Publishing, 2021).

mitigation and adaptation responses themselves, such as geoengineering mishaps.[81]

1.5 CLIMATE CHANGE SCIENCE IN THE COURTROOM

Because of the political and socio-economic ramifications of climate change and the scale of undertakings required to address it, the policy response to climate change is unsurprisingly highly politicized. Debates were initially focused on establishing the link between anthropogenic GHG emissions and climate change. Today, debates often centre around the type of policy responses required, the urgency and speed with which they should be deployed, and geo-political disagreements about responsibility and accountability for harms. Throughout, the debate has been fraught with ideological differences and vitriol.

Fortunately, there are robust, credible sources of climate science that the public and courts can safely depend upon. Even though climate science is complex, multi-faceted and dynamic, the *Intergovernmental Panel on Climate Change* plays a critical role in providing the scientific basis to support the work of policy-makers and judges.

1.5.1 The *Intergovernmental Panel on Climate Change* – Why the Judiciary Can and Does Rely Upon IPCC Science

Climate science is a diversified collection of information that is rapidly evolving. To best understand the culmination of knowledge emerging from the evolving literature, the United Nations Environmental Program (UNEP) and the World Meteorological Organization (WMO) created the Intergovernmental Panel on Climate Change (IPCC) in 1988. The IPCC's objective is to provide a clear scientific view on the state of climate science and knowledge about the impacts of climate change.

---

81 Doelle & Seck, "Introducing Loss and Damage," above note 79 at 2. See also Bubna-Litic & Chalifour, above note 68 at 127–78; Chalifour, "Feminist Perspective," above note 68 at 169 (examining the implications of carbon pricing for gender and Indigenous perspectives and the importance of careful design to avoid creating a double burden of harm for such groups).

The IPCC is comprised of thousands of scientists from around the world who voluntarily review and evaluate the most recent scientific, technical, and socio-economic information about climate change. The IPCC then produces assessment reports that document, assess, evaluate, and synthesize the latest science, GHG mitigation efforts, climate change impacts, adaptation approaches, and vulnerability to the effects of climate change. The IPCC produces a Summary for Policy Makers for its most important reports that is subject to detailed, line-by-line discussion and agreement by governments. The most recent assessment (sixth) was published in stages between 2021 and 2023.[82]

Given the widespread basis of knowledge combined with the rigorous assessment process including governmental sign-off, which signals government acceptance of the science, courts can confidently rely upon the IPCC's findings. Indeed, IPCC science has been the uncontested foundation of judicial decisions from around the world, including the Supreme Court of Canada.[83]

1.5.2 Unique Features of Climate Science That Are Relevant to Constitutional Interpretation and Analysis

Climate change is unlike any other issue modern society has faced. While it is quite a simple problem in principle that can be solved by reducing a known set of GHG emissions, it is in practice a very difficult problem to solve since fossil fuels are integral to contemporary energy systems. It has been conceptualized as a "super wicked" problem for this and other reasons.[84] For instance, those who cause the climate problem are

---

82 IPCC, "AR6," above note 11.

83 See, e.g., *GGPPA References*, above note 2 at paras 7–12. See also *Urgenda* Supreme Court, above note 22 at paras 48–51.

84 Richard J Lazarus, "Super Wicked Problems and Climate Change: Restraining the Present to Liberate the Future" (2009) 94 Cornell L Rev 1153 (asserting that this issue "defies resolution because of the enormous interdependencies, uncertainties, circularities, and conflicting stakeholders implicated by any effort to develop a solution" at 1159–60); Kelly Levin et al, "Overcoming the Tragedy of Super Wicked Problems: Constraining Our Future Selves to Ameliorate Global Climate Change" (2012) 45 Policy Sciences 123 at 124.

also those seeking to provide a solution; the central authority needed to address the climate crisis is weak or non-existent; irrational discounting pushes responses too far into the future; and time to address the problem is rapidly running out.

These unique characteristics are relevant to legal analyses. Canadian courts have been consistent in determining that purpose and context are relevant to legal questions. The purposive approach is a cornerstone of *Charter* analysis, requiring a generous and liberal interpretation aimed at achieving the purpose of the right in question.[85] The contextual approach requires that the *Charter* be interpreted against the backdrop of current social, political, and legal realities.[86] A contextual analysis may involve consideration of the principles underlying an area of law and suggests a flexible approach when striking a balance between individual and societal interests.[87] Contextual factors, which include the "matrix of legislative and social facts," may require adapting or updating a legal test and may even justify the Supreme Court of Canada departing from its own precedents.[88] Context may help determine the appropriate remedy.[89] In other words, facts are essential to a proper consideration of *Charter* issues.[90]

The living tree doctrine explicitly recognizes the need for constitutional doctrine to evolve according to changing modern realities. It was introduced in the *Persons* case almost a century ago, when the Court described the Constitution as having "planted in Canada a living tree capable of growth and expansion within its natural limits."[91] The doctrine in that case allowed the Court to interpret the word "persons" as including women along with men. The doctrine has continued to help courts interpret constitutional rights and principles against changing social norms.

---

85   See *Hunter v Southam Inc*, 1984 CanLII 33 (SCC); *R v Big M Drug Mart Ltd*, 1985 CanLII 69 at para 117 (SCC); *Irwin Toy Ltd v Quebec (AG)*, 1989 CanLII 87 (SCC).

86   See *Thomson Newspapers Ltd v Canada (Director of Investigation and Research, Restrictive Trade Practices Commission)*, 1990 CanLII 135 (SCC); *R v Wholesale Travel Group Inc*, 1991 CanLII 39 (SCC).

87   *RJR Macdonald Inc v Canada (Attorney General)*, 1995 CanLII 64 (SCC).

88   See, e.g., *Carter v Canada (AG)*, 2015 SCC 5.

89   *Schachter v Canada*, 1992 CanLII 74 (SCC).

90   *Mackay v Manitoba*, 1989 CanLII 26 at 361 (SCC).

91   *Edwards v Canada (Attorney General)*, 1929 CanLII 438 at 106–7 (UK JCPC).

In the *Reference re Same-Sex Marriage*, the Court referred to the living tree metaphor to explain how the Constitution is progressively interpreted to accommodate contemporary values, such as the recognition that marriage can include same-sex unions.[92]

The following eight features of climate change are an essential part of the context that should inform any climate-related constitutional analyses undertaken by Canadian courts.

1.5.2.1 Climate Change Poses an Existential Threat

The Earth's increasingly disrupted climate system is destabilizing human, ecological, socio-economic, and cultural systems across the planet. Climate change is altering the living conditions on Earth in such a profound way that it presents an existential threat to modern civilization and a "grave threat to humanity's future,"[93] as the Court unequivocally stated for the first time in *GGPPA References* in 2021.[94] We face a future of unprecedented wildfires, flooding, severe storms, droughts, and heat domes that kill and injure people, destroy the infrastructure that we take for granted, and threaten the ecological systems that sustain us. These extreme events will become more frequent and more severe for decades to come.

As will be discussed in chapters 3 and 4, the existential nature of climate change is relevant to whether climate change is a national concern or emergency, to whether it infringes on the right to life or security of the person, and to considerations of intergenerational equity in equality rights.

1.5.2.2 Climate Change is Irreversible on a Human Timescale Relevant to Modern Society

One of the most daunting features of climate change is its permanence, or irreversibility, at least on a timescale relevant to modern society. While

---

92 *Reference Re Same-Sex Marriage*, 2004 SCC 79.

93 *Ibid* at para 2.

94 Courts around the world have noted climate change's existential nature (see, e.g., *Urgenda* Supreme Court, above note 22 at paras 4.1–4.8; German Constitutional Court, 24 March 2021, *Neubauer v Germany* (2021), 1 BvR 2656/18 (Germany) at para 2 [*Neubauer*].

methane emissions dissipate within a decade or so, NO2 emissions take a century, and carbon dioxide can persist in the atmosphere for hundreds and even thousands of years.[95] Thus, at a fixed atmospheric concentration of CO_2 (say, today's rate of 430 ppm), the climate continues to warm for several hundred years given the thermal inertia of the ocean.[96] If we maintained today's atmospheric concentrations of GHGs until the system reached a point of equilibrium (thousands of years), warming could be up to eight to ten degrees Celsius.[97] The good news is that rapidly and drastically reducing emissions can avoid this level of warming. When emissions stop, atmospheric CO_2 starts to decline due to absorption of CO_2 by the land and oceans.[98]

Even if GHG emissions are sharply reduced in the near future, sea levels will continue to rise for centuries due to the built-in inertia of the global climate system.[99] Sea level rise is already twice as fast today as it was in pre-industrial times, rising by 3.6 millimetres per year and accelerating.[100] The US National Oceanic and Atmospheric Administration predicted that, under a high emissions pathway, sea levels could rise up to 2.5 metres (8.2 feet) by 2100.[101] Sea level rise will reach thirty to sixty centimetres by 2100 even if warming is kept to well below two degrees

---

95 Environmental Protection Agency, "Understanding Global Warming Potentials" (last modified 8 January 2025), online: epa.gov [perma.cc/38D8-FVVC].

96 See Andrew Dessler, "Stop Emissions, Stop Warming: A Climate Reality Check," *The Climate Brink* (2 December 2024), online: theclimatebrink.com [perma.cc /7M3D-LVQC].

97 James E Hansen et al, "Global Warming in the Pipeline" (2023) 3:1 Oxford Open Climate Change 1, online: doi.org/10.1093/oxfclm/kgad008 [perma.cc/XB6G-EZ4S].

98 Dessler, above note 96.

99 Intergovernmental Panel on Climate Change, "Choices Made Now Are Critical for the Future of Our Ocean and Cryosphere" (25 September 2019), online: ipcc.ch [perma.cc /T7MV-3KHW] [IPCC, "Ocean News Release"].

100 *Ibid.*

101 Rebecca Lindsey, "Climate Change: Global Sea Level" (22 August 2023), online: climate.gov [perma.cc/NT8Z-NX72]. This projection was based on a high emissions scenario described in IPCC RCP 8.5 report. A new methodology, the Shared Socio-economic Pathways, or SSP, used in the IPCC's sixth assessment report offers a more integrated and nuanced modelling which accounts for factors like changes in use of coal-fired electricity generation. As such, because of the turn away from coal-fired electricity, the concerning scenario is now less likely to materialize.

Celsius.[102] Even a modest sea level rise will have major implications for the hundreds of millions of people who live on islands and in coastal areas.[103] Small island states are being forced to consider futures in which their territories are submerged by rising seas.

The Supreme Court of Canada recognized the irreversibility of climate change in 2021.[104] Combined with its existential impacts, the irreversibility of climate change underscores the importance of prevention. A precautionary approach is vital. Canada has recognized the importance of taking a precautionary approach in the face of irreversible consequences related to the environment in past decisions. It also said a limited impact on provincial jurisdiction by federal laws was justified on the basis of the irreversible harms of climate change in *GGPPA References*.[105]

1.5.2.3 Thresholds and Tipping Points

As concentrations of GHGs in the atmosphere increase, we approach thresholds that cannot be uncrossed. A particular kind of threshold called a tipping point in Earth's natural systems is one that will rapidly accelerate global warming and cause devastating, large-scale impacts.[106] The precise point at which these thresholds may be crossed is difficult to predict but the consequences of crossing them are serious. Once an ice sheet melts, it cannot be restored in a timeframe relevant to our society. Once coral reefs are eradicated — which the IPCC has predicted is likely at two degrees of warming — they could be gone for good, unless research to develop heat-resistant corals is successful and deployed in time. Recent science also suggests that abrupt and irreversible changes in the climate system at lower global average temperatures are more probable than initially thought.[107] There is growing evidence that some

---

102 *Ibid.*

103 IPCC, "Ocean News Release," above note 99.

104 *GGPPA References*, above note 2 at para 206. See, e.g., *Mathur v Ontario*, 2020 ONSC 6918 at para 149 [*Mathur* ONSC 2020]; *Held*, above note 30 at paras 6–7.

105 See, e.g., *Castonguay Blasting Ltd v Ontario*, 2013 SCC 52 at para 20; *114957 Canada Ltée (Spraytech, Société d'arrosage) v Hudson (Town)*, 2001 SCC 40.

106 *Mathur* ONSC 2020, above note 104 at para 24.

107 Timothy M Lenton et al, "Climate Tipping Points: Too Risky to Bet Against" (2019) 575 Nature 592.

tipping points are already underway, which will set in motion a range of other consequences.[108] For instance, melting ice sheets and permafrost (which release methane, a powerful GHG) could cause a cascade of largely irreversible effects.[109]

The gravity and irreversibility of these thresholds require the courts to take a careful and cautious approach in constitutional interpretation and mandate application of the precautionary principle. The Ontario Court of Appeal in *Mathur v Ontario*, discussed in chapter 4, recognized the significance of tipping points, noting that "every incremental increase in global temperature increases the likelihood of large-scale, devastating climate tipping points being crossed."[110]

1.5.2.4 Urgency – The Window of Time to Respond Is Rapidly Closing

UNEP captured the urgency well in its recent Emissions Gap Report: "The future of our planet is at stake. We are in the midst of a climate emergency, and the window to act is closing fast."[111] In 2023, the IPCC reported that emissions must be reduced by 43 to 45 percent below 2010 levels by 2030 and reach net-zero by 2050 to limit warming to 1.5 degrees.[112] Continuation of current policies, without ramping up the ambition or delivery of such policies, will lead to a temperature increase of 2.6 to 3.1 degrees this century.[113] Global GHG emissions in 2023 increased by 1.3 percent compared to 2022 and are up 60 percent since 1990.[114] To achieve the IPCC target of 43 to 45 percent

---

108   *Ibid.*

109   *Ibid.*

110   *Mathur v Ontario*, 2024 ONCA 762 at para 12 [*Mathur* ONCA].

111   United Nations Environment Program, *The Emissions Gap Report 2024* (Nairobi: United Nations Environment Programme, 2024), online: unep.org [perma.cc/BHB9-2LB5] [UNEP, *Emissions Gap 2024*].

112   IPCC, "AR6," above note 11.

113   United Nations Environment Program, "Nations Must Close Huge Emissions Gap in New Climate Pledges and Deliver Immediate Action, or 1.5 Degrees C is Lost" (24 October 2024), online: unep.org [perma.cc/6PBL-ZX3X]; See also UNEP, *Emissions Gap 2024*, above note 111.

114   *Ibid* at IV. See also Statista, *Annual Carbon Dioxide (CO2) emissions worldwide 1940-2024*, online: statista.com [perma.cc/9HMU-68H8].

below 2010 levels by 2030, annual emissions would need to decline by approximately 7 percent annually starting in 2025, a rate of emissions decline never intentionally achieved by any nation over an extended period, let alone the whole world.

The legal implications of the time crunch for addressing climate change are extensive. Delays in reducing GHGs creates a greater burden of responsibility for youth and future generations, as they will have to make more radical and difficult reductions, something the German Constitutional Court recognized as contrary to human rights in the *Neubauer* case.[115] The urgency may also justify expediting proceedings since it can take years for legal matters to work their way through the courts. In *Friends of the Irish Environment*, the case was appealed directly from a lower court to the Irish Supreme Court in an expedited manner in light of the time-sensitivity inherent in climate cases.[116] Similarly, the European Court of Human Rights (ECHR) decided in 2020 to expedite three climate cases: the youth case of *Duarte Agostinho and Others v Portugal and 32 Others, KlimaSeniorinnen v Switzerland,* and *Carême v France,* allowing them to proceed to the Grand Chamber of the ECHR before exhausting domestic remedies.[117]

1.5.2.5 Time Lag Between Emissions and Locked-In Harm

The urgency of climate change is compounded by the time lag between the emission of GHGs and when their enduring climate destabilizing effects will be felt. The time lag between a unit of GHG and its effects may not be experienced for decades or more, depending on the gas.[118] For instance, once CO_2 is emitted, it has a latent effect that lasts millennia, rendering the changes largely irreversible for modern

---

115   *Neubauer*, above note 94.

116   *Friends of the Irish Environment v Ireland*, [2020] IESC 49.

117   See European Court of Human Rights, "Grand Chamber rulings in the climate change cases", (April 9, 2024) online: <echr.coe.int [perma.cc/6H6J-39NK].

118   This is a complex area, as explained here: Jessica McKenzie, "'Mass Delusion and Wishful Thinking': Why Everything You Think You Know About Methane Is Probably Wrong," *Bulletin of the Atomic Scientists* (18 December 2023), online: thebulletin.org [perma.cc/8XPZ-FLLJ].

society.[119] So even if or when we reach net-zero emissions, the effects of past emissions will continue.[120] The impact of today's emissions on the atmosphere (and the consequential impacts on the Earth) is in this sense locked in — but there is a time lag between cause and effect, meaning that the changes experienced today are linked to past emissions. Because of this climate inertia, a certain level of future warming is unavoidable, given existing emissions and their locked-in Earth System changes.[121]

This time lag between present-day actions and known future impacts is relevant in constitutional analyses, especially when the rights of youth and/or future generations are at play. Because of the time lag, we need to slow emissions well before we sense their full impact. Not only is this necessary to avert catastrophic climate change, but delaying action is much less cost-effective in the long run.[122] As such, this time lag requires courts to take a precautionary approach in their analyses of constitutional rights, particularly in considerations of intergenerational equity. The time lag is also relevant to the analysis of causation, to justify remedies for future harms that are locked in.

---

119   *Ibid.*

120   See Intergovernmental Panel on Climate Change, *Climate Change 2022: Mitigation to Climate Change*, Working Group III Contribution to the Sixth Assessment Report of the Intergovernmental Panel on Climate Change, Priyadarshi R Shukla et al, eds (Cambridge & New York: Cambridge University Press, 2022). See also Sam Fankhauser et al, "The Meaning of Net Zero and How to Get it Right" (2022) 12 Nature Climate Change 15 at 15.

121   There are heated debates about some of the nuances of locked-in warming versus locked-in changes, though there is agreement about the length of time CO_2 persists in the atmosphere and its warming potential. See, e.g., Hansen, above note 97; Dessler, above note 96. See also Intergovernmental Panel on Climate Change, *Climate Change 2001: Synthesis Report*, A Contribution of Working Groups I, II and III to the Third Assessment Report of the Intergovernmental Panel on Climate Change, Robert T Watson et al, eds (Cambridge: Cambridge University Press, 2001) at 89.

122   Dave Sawyer et al, *Damage Control: Reducing the Costs of Climate Impacts in Canada* (Canadian Climate Institute, 2022), online: climateinstitute.ca [perma.cc/4S7Z-WU7P]; Dave Sawyer et al, *Tip of the Iceberg: Navigating the Known and Unknown Costs of Climate Change for Canada* (Canadian Institute for Climate Choices, 2020), online: climateinstitute.ca [perma.cc/ZZS3-G2TK].

1.5.2.6 A Global, Collective Action Problem
(Multiple Contributors, Free Riders)

Every unit of GHG emitted contributes to worsening the climate change problem.[123] Concomitantly, every unit of GHG emissions averted contributes to solving or lessening the problem. When a given jurisdiction chooses to emit more than is scientifically justifiable in a global context, this directly contributes to causing the harms of climate change. Collective action across jurisdictions is needed to effectively reduce global GHG emissions. No single jurisdiction can stop the climate problem alone.

Courts around the world have recognized this contextual feature of climate change and used it to justify rejecting arguments that one jurisdiction's actions are irrelevant when other jurisdictions are not reducing their emissions (especially if they are major emitters). From the Dutch *Urgenda* decision to the United States *Massachusetts v Environmental Protection Agency* case, this *de minimis* defence has been rejected.[124] The Supreme Court of Canada recognized the collective action nature of climate change at both an international and national level in *GGPPA References*.[125] The Ontario Superior Court in *Mathur v Ontario* joined in, holding that every unit of GHG emitted contributes to the problem, rejecting claims that Ontario's relatively small overall contributions globally excuse it from responsibility.[126]

1.5.2.7 Regional Variability

While climate change causes an average global rise in temperature, some jurisdictions are warming at a faster rate. The Supreme Court of Canada has recognized these geographic variations, noting that the effects of climate change have been and will continue to be especially severe in Canada, where temperatures have risen at approximately double the global

---

123   While this is true, it is important to differentiate between anthropogenic and non-anthropogenic GHG emissions and biogenic emissions that may be emitted by human activities that are difficult to avoid (e.g., emissions from decomposition and waste, carbon cycling of trees and organic matter, and even mammalian respiration).

124   *Urgenda* Supreme Court, above note 22 at para 4.79; *Massachusetts v Environmental Protection Agency*, 549 US 497 (2007) at 526.

125   *GGPPA References*, above note 2 at paras 12 and 24.

126   *Mathur v Ontario*, 2023 ONSC 2316 at para 148. This finding was confirmed on appeal, *Mathur v Ontario*, 2024 ONCA 762. The case is discussed in greater detail in Chapter 4.

average rate (by 1.7 degree Celsius) and are expected to continue to rise faster.[127] The Court noted that warming in the Canadian Arctic is even more rapid, with average temperature increasing at a rate nearly three times the global average.[128]

This contextual factor may be legally relevant in multiple contexts, including assessments of present and future risks of harm in the assessment of rights and in evaluating whether climate change is a national concern or emergency in a division of powers analysis.

1.5.2.8 Disproportionate Intergenerational, Intragenerational, and Inter-Species Impacts at Multiple Scales

Climate change impacts communities and regions differently, with disproportionate impacts on communities that already face discrimination and inequality. For instance, Indigenous and racialized communities, women, children, seniors, people with disabilities, and those living in poverty are often the most impacted by severe weather events.[129] These communities may also be negatively impacted by the responses to climate change if those policies are not drafted carefully to avoid this, which can create a double burden of injustice.[130] As elaborated in Chapter 2, climate change has a particularly serious effect on Indigenous Peoples in Canada.[131] From the erosion of hunting, gathering and spiritual practices on lands and waters affected by climate change to the loss of roads built on ice or permafrost that reduce access to communities, climate change has grave implications for constitutionally recognized and protected Indigenous rights.

Climate injustices are perpetuated not only intragenerationally (within a current generation), but also intergenerationally (between different generations), with future generations facing existential consequences

---

127 Rates of warming are generally higher than average for land masses since the rate of warming is slower for oceans.

128 *GGPPA References*, above note 2 at paras 10–11.

129 World Bank Group, "Social Dimensions of Climate Change" (last accessed 14 January 2024), online: worldbank.org [perma.cc/7U9F-BB8C]; Romanello et al, above note 1 at 1869, 1875.

130 See discussion in Section 1.3.4 above.

131 *GGPPA References*, above note 2 at paras 10 –11, 187, and 206.

at multiple levels.[132] The impacts of climate change also have grave consequences for the more-than-human world, with implications for biodiversity and ecological systems. The disproportionate impacts exist at multiple scales, from global to regional, national to local.

The legal relevance of these disproportionate impacts is broad, including interpretations of equality rights, balancing impacts on jurisdiction in a division of powers analysis, and legal frameworks to safeguard biodiversity and ecological systems. These impacts make Unwritten Constitutional Principles of Ecological Sustainability or protection of minority rights highly relevant.[133]

1.6   THE POLICY RESPONSE AND LEGAL FRAMEWORK FOR CLIMATE CHANGE

Climate change is not new nor a modern issue. The greenhouse effect was first demonstrated by scientist and women's right advocate Eunice Foote in 1856.[134] In 1896, Swedish scientist and professor Svend Arrhenius warned that the accumulation of carbon in the atmosphere from burning of coal would cause warming that would bake the earth close to the boiling point.[135] News articles dating from the early 1900s cautioned about the increase in temperature and associated harms that would ensue from the CO_2 emissions associated with burning a lot of coal.[136]

---

132   See discussion in Chapter 4 related to age-based discrimination for a discussion of climate change's intergenerational implications.

133   See Lynda M Collins, "The Unwritten Constitutional Principle of Ecological Sustainability: A Solution to the Pipelines Puzzle?" (2019) 70 UNBLJ 30; Mari Galloway, "The Unwritten Constitutional Principles and Environmental Justice: A New Way Forward?" (2021) 52:2 Ottawa L Rev. The proposed UCP of Ecological Sustainability is discussed further in Chapter 4.

134   Sylvia G Dee, "Scientists Understood Physics of Climate Change in the 1800s – Thanks to a Woman named Eunice Foote," *The Conversation* (22 July 2021), online: theconversation.com [perma.cc/LV2M-4QV6].

135   This article is referenced by Justice Seely in *Held*, above note 30 at para 74.

136   Linden Ashcroft, "For 110 Years, Climate Change Has Been in the News. Are We Finally Ready to Listen?" *The Conversation* (15 August 2022), online: theconversation. com [perma.cc/43LU-47TQ].

BOX 1   FACTUAL AND CONTEXTUAL FINDINGS RELATED TO CLIMATE
CHANGE SCIENCE IN CANADIAN JURISPRUDENCE TO DATE

Finding	Reference
Climate change is real.	*GGPPA References*, para 2.
Climate change is caused by GHG emissions from human activities.	*GGPPA References*, para 2.
The impacts of climate change are existential and pose a grave threat to humanity.	*GGPPA References*, para 2.
The impacts of climate change are irreversible on a human time scale.	*GGPPA References*, para 13.
Every increment of warming brings us closer to tipping points that can unleash large-scale devastating impacts.	*Mathur v Ontario*, ONCA, para 12.
The risks and adverse impacts of climate change escalate with every increment of warming.	*Mathur v Ontario*, ONCA, para 12.
The impacts of climate change on Canada are particularly severe, with temperature rise of roughly double the global average.	*GGPPA References*, para 10.
The issue is urgent; the window of time to address it is rapidly diminishing.	*GGPPA References*, paras 13 and 401.
The only way to address anthropogenic climate change is to reduce GHGs.	*GGPPA References*, para 2. *Mathur v Ontario*, ONCA, para 9.
Climate change is a global collective action problem that requires collective national and international action.	*GGPPA References*, para 12; 190.
Climate change causes disproportionate harms to particular regions and groups.	*GGPPA References*, paras 11, 12, 187, and 206.
The failure to act threatens Canada's ability to meet its international obligations.	*GGPPA References*, para 189.
Each jurisdiction's emissions are measurable and contribute to climate change.	*GGPPA References*, para 188. *Mathur v Ontario*, para 9.

It is difficult to pinpoint precisely when discussions about climate change began at the diplomatic level, but an official international dialogue commenced in the 1980s. In 1988, Canada hosted the World Conference on the Changing Atmosphere in Toronto, which convened delegates (including scientists and policymakers) from several countries to discuss an international response to climate change.[137]

1.6.1 The International Response

1.6.1.1 The United Nations Framework Convention on Climate Change (UNFCCC) (1992)

The first major international policy milestone was the 1992 *United Nations Framework Convention on Climate Change (UNFCCC)*.[138] The *UNFCCC* established the goal of "stabilizing greenhouse gas concentrations in the atmosphere at a level that would prevent dangerous anthropogenic interference with the climate system."[139] What constitutes dangerous levels of GHGs was later quantified as two degrees Celsius of average warming, a number that has shifted to 1.5 degrees Celsius as scientific knowledge has evolved.

Parties to the *UNFCCC* recognized the importance of addressing climate change for future generations, agreeing that they "should protect the climate system for the benefit of present and future generations of humankind, on the basis of equity."[140] It incorporated the principle of common but differentiated responsibilities.[141]

Despite the commitments made, more GHGs have been emitted since the *UNFCCC* was signed than in all the years before the Agreement.[142]

---

137 This conference is said to have put climate change on the international agenda. See "The Changing Atmosphere: Implications for Global Security - Conference Statement" (1988), Canadian Meteorological and Oceanographic Society Archives, online (pdf): <cmosarchives.ca [perma.cc/C7UB-CGCN].

138 *United Nations Framework Convention on Climate Change*, 9 May 1992, 1771 UNTS 107 (entered into force 21 March 1994), online (pdf): <unfccc.int [perma.cc/R33T-5TZX] [*UNFCCC*]. Canada signed the treaty on 12 June 1992 and ratified the treaty on 4 December 1992 (see United Nations Framework Convention on Climate Change, "Status of Ratification of the Convention" (last accessed 15 January 2025), online: unfccc.int [perma.cc/6AM5-Y968].

139 *UNFCCC*, above note 138, art 2.

140 *Ibid*, art 3(1).

141 *Ibid*. The concept of "common but differentiated responsibilities" is a principle of international environmental law that refers to a shared duty that is met by actions adapted to a given state's particular circumstances.

142 IPCC, "AR6," above note 11 at 4.

1.6.1.2 Kyoto Protocol (Signed 1997, Ratified 2005)

Since the *UNFCCC* was essentially a framework agreement that set out general principles and objectives, the next step was to negotiate more specific targets. Parties negotiated the *Kyoto Protocol* in 1997, which established GHG emissions reduction targets for developed (Annex I) countries to be met between 2008 and 2012.[143] The *Kyoto Protocol* was meant to be the instrument through which a greater number of *UNFCCC* nations (including some developing countries) would negotiate increasingly ambitious targets. This never materialized, however, for a variety of geopolitical reasons, not the least of which was the fact that the United States never adopted the *Kyoto Protocol* and several large emitters, including Canada, eventually withdrew from it.[144] Despite these retreats, the original Kyoto targets were essentially met largely due to reductions within the EU.

While diplomacy faltered under the pressure of the growing imperative to reduce emissions and finger pointing among emitting countries, a few milestones still managed to emerge during the challenging Kyoto period. One of these developments, which proved important in climate litigation, is that the previously undefined goal of "avoiding dangerous levels of climate change" in the *UNFCCC* was transformed and quantified into a collective policy goal of limiting global warming to two degrees Celsius above pre-industrial levels.[145] This more measurable benchmark was set out in the 2010 *Cancun Agreement*.[146] In an effort to translate this benchmark

---

143 *Kyoto Protocol to the United Nations Framework Convention on Climate Change*, 11 December 1997, 2023 UNTS 162 (entered into force 16 February 2005) [*Kyoto Protocol*].

144 United Nations Framework Convention on Climate Change, "The Kyoto Protocol - Status of Ratification" (last accessed 14 January 2025), online: unfccc.int [perma.cc /N4XW-2KKE]; "Canada Pulls out of Kyoto Protocol," *CBC News* (12 December 2011), online: cbc.ca [perma.cc/M7QU-K25H].

145 Avoiding Dangerous Climate Change: A Scientific Symposium on Stabilization of Greenhouse Gases (International Conference, Exeter, United Kingdom, 1–3 February 2004), online: stabilisation.metoffice.com [perma.cc/XVA4-MPLL].

146 The *Cancun Agreement* of 2010 states that nations should work together in "reducing global greenhouse gas emissions so as to hold the increase in global average temperatures below 2 degrees Celsius above pre-industrial levels, and that Parties should take urgent action to meet this long-term goal" (*Report of the Conference of the Parties on its sixteenth session, held in Cancun from 29 November to 10 December 2010*, UNFCCC, 2011 FCCC/CP/2010/7/Add.1 Dec 1/CP.16 at 3).

into actionable targets, the international community further determined that keeping warming below two degrees Celsius required industrialized (Annex I) states to reduce their emissions by 25 to 40 percent below 1990 levels by 2020 and by 80 to 95 percent by 2050.[147] This is the threshold against which the Dutch court measured the adequacy of the government's climate goals in *Urgenda v State of the Netherlands (Ministry of Infrastructure and the Environment)*.[148] This objective has been superseded by the goal established most recently as part of the Glasgow Pact, discussed below.

1.6.1.3   The Paris Agreement (2015)

To avoid the political deadlock created by the top-down approach taken in the *Kyoto Protocol*, which lasted many years, *UNFCCC* parties pivoted to a bottom-up approach in the lead up to the twenty-first Conference of the Parties (COP21) in Paris 2015. Under this approach, states put forth their own nationally determined contributions (NDCs) for GHG reductions rather than negotiating targets.[149] This more flexible approach is often credited for the fact that parties were able to reach agreement and sign the 2015 *Paris Agreement*.[150]

In addition to agreeing to meet their individual NDC targets, the *Paris Agreement* committed signatories to ratchet up their NDCs on a periodic

---

147   *Report of the Conference of the Parties Serving as the Meeting of the Parties to the Kyoto Protocol on Its Sixth Session, held in Cancun from 29 November to 10 December 2010*, UNFCCC, 6th Sess, Doc FCCC/KP/CMP/2010/12/Add.1, Dec 1/CMP.6, art 4, pre-amble, online (pdf): <unfccc.int [perma.cc/T3KE-G8WE].

148   *Urgenda* Trial Court, above note 22.

149   United Nations Framework Convention on Climate Change, "Nationally Determined Contributions (NDCs)" (last accessed 14 January 2024), online: unfccc.int [perma.cc /FUP9-2AVC].

150   *Paris Agreement*, 22 April 2016, Can TS 2016 No 9, art 2, para 1 (entered into force 4 November 2016) [*Paris Agreement*]. The *Paris Agreement* was opened for signature on 22 April 2016, at which time 174 states and the European Union signed the agreement and fifteen states also ratified the agreement. The agreement came into force on 4 November 2016, thirty days after which at least fifty-five parties accounting for at least 55 percent of total GHG emissions ratified it. Canada has signed and ratified the Paris Agreement (see United Nations Framework Convention on Climate Change, "Paris Agreement: Status of Ratification," online: unfccc.int [perma.cc/PF5D-X5Y7]).

basis and implement domestic measures to meet their targets.[151] Additionally, states agreed to aim to peak their emissions as soon as possible and achieve net-zero emissions (a balance between emissions and sinks) in the second half of the century.[152] Since then, many countries have established net-zero goals.[153] Parties agreed to take stock of progress on mitigation, adaptation, and emerging science beginning in 2023 and every five years thereafter.[154]

Because of intense pressure from small island developing states, the *Paris Agreement* also ratcheted up the ambition of the *UNFCCC*'s objective, aiming to hold "the increase in the global average temperature to well below two degrees Celsius above pre-industrial levels and [pursue] efforts to limit the temperature increase to 1.5 degrees Celsius above pre-industrial levels, recognizing that this would significantly reduce the risks and impacts of climate change."[155] The post-2015 conversation has increasingly focused on the 1.5 degree Celsius target, given the ongoing evolution of scientific understanding about the severity of implications at even one degree Celsius of warming.

1.6.1.4 Glasgow Climate Pact (2021)

At COP26 in Glasgow, the Parties agreed to a Pact that reaffirmed and refined commitments made in the Paris Agreement, with some added elements. In addition to renewed commitments to finance the transition to net-zero, double financing for adaptation and support renewables, [156] over 103 countries committed to the Global Methane Pledge to reduce methane emissions by 30 percent below 2020 levels by 2030. The Glasgow Pact also included a pledge to phase down coal, with a subset of countries, including Canada, agreeing to a full phase-out of coal.[157]

---

151 *Paris Agreement*, above note 150, art 4, paras 2–3, and 9. For an overview of the *Paris Agreement*, see Meinhard Doelle, "The Paris Agreement: Historic Breakthrough or High Stakes Experiment?" (2016) 6:1–2 Climate L 1 [Doelle, "The Paris Agreement"].

152 *Paris Agreement*, above note 150, art 4, para 1.

153 Doelle, "Paris Agreement," above note 151.

154 *Paris Agreement*, above note 150, art 14, paras 1–2.

155 *Ibid*, art 2.1(a).

156 Glasgow Financial Alliance for Net Zero, "Accelerating the Transition to a Net-Zero Global Economy" (last accessed 14 January 2025), online: gfanzero.com [perma. cc/6GRE-SPBF].

157 United Nations Framework Convention on Climate Change, *Report of the Conference of the Parties Serving as the Meeting of the Parties to the Paris Agreement on its*

Subsets of countries also promised to halt and reverse deforestation and land degradation and end the sale of internal combustion engines by 2035 in leading markets and globally by 2040.[158]

The Glasgow Climate Pact also articulated the level of CO_2 emissions reductions needed to limit warming to 1.5 degrees Celsius, drawn from the latest IPCC assessment at the time: 45 percent below 2010 levels by 2030, along with deep reductions in other GHGs and a net-zero target at mid-century.[159] The Superior Court in *Mathur v Ontario* accepted this metric as the scientific benchmark for determining what is expected of jurisdictions in terms of climate targets.[160]

1.6.1.5 Progress to Date

Unfortunately, even if the signatories to the *Paris Agreement* all fully meet their unconditional pledges — a feat that is in itself doubtful given the past failure of countries to meet their GHG reduction commitments — temperatures are slated to rise 2.6 to 3.1 degrees Celsius by 2100.[161] To remain with the 1.5 to two degree threshold, nations need to commit to reducing annual GHG emissions by 57 percent by 2035 in the next round of NDC commitments and meet those commitments.[162]

Despite its current and historical role as a significant contributor to GHG emissions, Canada has a perfect track record of failing to meet its previous climate commitments. The next section considers Canada's emerging policy and legal framework for climate change.

    Third Session, held in Glasgow from 31 October to 13 November 2021, FCCC/PA/CMA/2021/10/Add.1 Dec 1/CMA.3 (UNFCCC, 2022), online: unfccc.int [perma.cc/VN3Y-9EQT] [Glasgow Climate Pact].

158   Deirdre Cogan et al, "Where Do We Stand on COP26 Climate Promises? A Progress Report" (13 October 2022), online: wri.org [perma.cc/6VGM-CGG6]; Government of UK, "COP26 Declaration on Accelerating the Transition to 100% Zero Emission Cars and Vans" (last modified 17 November 2022), online: gov.uk [perma.cc/D6BS-RY4R].

159   Glasgow Climate Pact, above note 157 at para 22.

160   *Mathur* ONCA, above note 110 at para 23.

161   United Nations Environment Programme, "Executive Summary" in *The Closing Window: Climate Crisis Calls for Rapid Transformation of Societies* (Nairobi: United Nations Environment Programme, 2022) at XVI.

162   UNEP, *Emissions Gap 2024*, above note 111.

1.6.2 Canadian Policy and Legal Framework for Climate Change

Canada has always been an important GHG emitting country. It has one of the most carbon-intensive economies in the world.[163] Canada remains one of the top ten emitters globally in absolute terms, having emitted approximately 2.5 percent of global cumulative emissions.[164] In 2022, it was the eleventh largest emitter of CO_2 from fuel combustion.[165] In terms of emissions per capita, Canada ranked thirteenth globally in 2022.[166] Including emissions from land use and forestry, Canada ranked second in terms of global per capita (or carbon intensity) emissions among the top ten emitters in 2021.[167] Accounting for historical cumulative emissions between 1850 and 2021, Canada ranked tenth globally with 2.6 percent of historical emissions and first or second in per capita historical emissions, depending on the methodology.[168]

The Canadian legal framework for climate change is rapidly evolving. Most efforts to date have focused on mitigation though there is increasing effort on adaptation, especially by municipalities in recent years. Much of the action at the federal level has taken place since 2015. The following section describes Canada's climate targets, a brief survey of the legal framework for addressing those targets, and projections for where we are headed in terms of reductions.

1.6.2.1 Canadian Targets

Creating the legal and policy infrastructure needed to address climate change within the timeframe required presents a major public policy challenge for any country. The deep, rapid interventions needed at multiple levels and in many domains call for considerable intergovernmental coordination and cooperation in largely unchartered territory. While Canadian governments have abundant experience with intergovernmental

---

163 See Kathryn Harrison, "Climate Governance and Federalism in Canada" in Alan Fenna et al eds, *Climate Governance and Federalism* (Cambridge University Press, 2023) at 64.

164 Simon Evans, "Analysis: Which Countries Are Historically Responsible for Climate Change?" (5 October 2021), online: carbonbrief.org [perma.cc/5VDG-J7GF].

165 International Energy Agency, "How Much CO_2 Does Canada Emit?" (last accessed 14 January 2025), online: iea.org [perma.cc/8BBP-BBET].

166 *Ibid.*

167 Environment and Climate Change Canada, "Global greenhouse gas emissions" (last accessed 16 January 2025), online: canada.ca [perma.cc/BP3K-RJYY].

168 Evans, above note 164.

coordination in our federation, multi-dimensional issues such as climate change have proven to be challenging. The particular importance of the oil and gas sector in some regions of the country, combined with increasingly divisive ideologies in politics, makes it ever more contentious.

Canada has been internationally committed to taking steps to mitigate GHG emissions since 1992, when it signed the UNFCCC. In the early years following signature of the UNFCCC, Canada's response signalled that it would take climate action seriously and establish effective policies under the *UNFCCC*.[169] Canada enthusiastically signed the Kyoto Protocol,[170] under which it committed to reduce GHG emissions by 6 percent below 1990 levels by the year 2012. However, ratification of the Protocol took over five years, as many provinces were not on board with the commitment; in the absence of political consensus on climate action, domestic efforts to address climate change in Canada developed slowly and in an uncoordinated fashion in the period between 1992 and 2015, resulting in a patchwork of climate-related legislation and policy at multiple levels.

Canadian governments developed various plans and strategies in the late 1990s and early 2000s, but there was little effort at intergovernmental coordination and the plans were generally weak.[171] Between 1990 and 2008 — the beginning of the first Kyoto commitment period — GHG emissions in Canada had increased by 24 percent (rather than decreasing by 6 percent, as per the Kyoto commitment).[172] With no plan in place to meaningfully change this trajectory, Canada, under the Stephen Harper administration, became the first signatory country to withdraw from the Kyoto Protocol in 2011.

---

169 Environment and Climate Change Canada, "United Nations Framework Convention on Climate Change: Plain Language Summary" (last modified 2 September 2022), online: canada.ca [perma.cc/ACY6-UFQK].

170 Under the Chretien government, Canada was one of the first signatories.

171 With emissions at 592 metric tons of carbon dioxide equivalents in 1990, this target translated to 557 metric tons of carbon dioxide equivalents (see Environment Canada, *A Climate Change Plan for the Purposes of the Kyoto Protocol Implementation Act* (Gatineau: Environment Canada, 2012), online: <publications.gc.ca [perma.cc /FQ7M-ML44]).

172 Environment Canada, *Canada's 2008 Greenhouse Gas Inventory: A Summary of Trends: 1990–2008* (Gatineau: Environment Canada, 2010), online: <publications.gc.ca [perma. cc/63GH-TXV9].

Despite its withdrawal from Kyoto, Canada continued to participate in the UNFCCC proceedings. In 2009, Canada decreased the ambition of its target, promising to reduce emissions by 17 percent from 2005 levels by 2020 in the UNFCCC Copenhagen Accord.[173] Unfortunately, Canada continued its perfect track record of failing to meet its climate targets, surpassing its 2020 target by 123 Mt of carbon dioxide equivalent.[174]

In 2015, after the election of Prime Minister Justin Trudeau, who had campagned on a climate action platform, Canada renewed its commitments to take serious action on climate change. Canada signed the Paris Agreement, committing in its first NDC to reduce emissions 30 percent below 2005 levels by 2030 (bringing Canadian emissions down to 511 metric tons).[175] This represents a 6 percent increase from 1990 levels, so is a far less ambitious goal than the initial Kyoto target[176] and far from the 25–40 percent reduction by Annex I countries that was identified in Cancun in 2010, the target which served as the scientific benchmark against which the Dutch government was held accountable by the court in the *Urgenda case.*[177] In 2021, as depicted in Figure 1, Canada ratcheted up its

---

173   With emissions in 2005 at 736 metric tons of carbon dioxide equivalents, Canada's Copenhagen target translates to 611 metric tons of carbon dioxide equivalents in 2020 (see Letter from Michael Martin to Yvo de Boer (29 January 2010), Submission of Canada to UNFCCC Copenhagen Accord, Government of Canada, online: unfccc.int [perma.cc/7K9L-HJ73]).

174   Kimberley Leach et al, "Canada's Commitments and Actions on Climate Change" in *Report 5 - Lessons Learned From Canada's Record on Climate Change*, Reports of the Commissioner of the Environment and Sustainable Development to the Parliament of Canada (Office of the Auditor General of Canada, 2021) online: oag-bvg.gc.ca [perma.cc/28F7-DL42].

175   Government of Canada, "Canada's INDC Submission to the UNFCCC" (May 2015), online: <unfccc.int [perma.cc/QXQ3-B9B4]. This target was created by the then Conservative federal government. A majority Liberal government was elected to power in the fall of 2015. The target represents 523 metric tons of carbon dioxide equivalents. See Andrew Hayes et al, "Report 1: Progress on Reducing Greenhouse Gases — Environment and Climate Change Canada" (12 June 2017) at exhibit 1.4, online: oag-bvg.gc.ca [perma.cc/NE3U-KDP8].

176   Sophie Yeo, "Mismatched Graph Creates Confusion in Canada's UN Climate Pledge" (20 May 2015), online: carbonbrief.org [perma.cc/4E6S-YX38].

177   See Nathalie J Chalifour & Jessica Earle, "Feeling the Heat: Climate Litigation under the Charter's Right to Life, Liberty and Security of the Person" (2018) 42:2 Vermont LR 690 at 705 for a table comparing levels of ambitions of the various targets.

national target to 40 to 45 percent below 2005 levels (which translates to a target of 406–443 MT.[178] Additionally, in the 2021 *Canadian Net-Zero Emissions Accountability Act*, Canada set a target to reduce national net GHG emissions to zero by 2050.[179] In 2024, Canada released a new target of reducing emissions by 45 to 50 percent below 2005 levels by 2035, which is not shown on Figure 1.[180] The Net-Zero Advisory Body to the government had recommended a 50 to 55 percent target and, failing that, urged the government to strive for the upper end of its new target (i.e., 50 percent).[181]

Even though we have consistently failed to meet our climate goals, these targets are far too low considering Canada's historical and current responsibility for GHG emissions. While there are a variety of ways to evaluate and fairly allocate carbon budgets among states, every method yields the same result: Canadian targets fall short of what is required for it to do its fair share in averting catastrophic climate change. The long delays in action make it much harder to do what is required now — had we acted responsibly in the early days of the *UNFCCC*, we would not have such a daunting task ahead.

Canada's target is less ambitious than that of every other G7 nation. The European Union's current pledge, for example, is to reduce its emissions by 40 percent below 1990 levels by 2030.[182] Under the Biden administration, the US target was to reduce emissions by 50 to 52 percent below 2005 levels by 2030. However, the United States announced in 2024 its withdrawal from the Paris Agreement following the election of Donald Trump,

GHG emissions in Canada have increased since 1990. From 1990 to 2021, they increased by 14 percent.[183] Relative to a 2005 baseline, emissions

---

178 Graham Harris, "Has the Future of Carbon Pricing in Canada Just Become More Certain?" (31 March 2022), online: fireflyghg.eco [perma.cc/U6SC-DD4M].

179 *Canadian Net-Zero Emissions Accountability Act*, SC 2021, c 22.

180 Government of Canada, "Canada's Next Net-Zero Milestone: The 2035 Emissions Reduction Target" (last modified 6 January 2025), online: www.canada.ca [perma.cc /UQ3Z-V2CS].

181 Net-Zero Advisory Body, "Net-Zero Advisory Body Responds to Today's Announcement of Canada's Greenhouse Gas Emissions Target for 2035" (12 December 2024), online: nzab2050.ca [perma.cc/JL9A-FBDW].

182 Margo McDiarmid, "Canada Sets Carbon Emission Reduction Target of 30% by 2030," *CBC News* (15 May 2015), online: cbc.ca [perma.cc/C6YT-V3SM].

183 Office of the Auditor General of Canada, *Report 7 – Canadian Net-Zero Emissions Accountability Act – 2024 Report*, Reports of the Commissioner of the Environment

in 2022 decreased by 7 percent.[184] The goal is now to reduce emissions by 45 to 50 percent below 2005 by 2035.

FIGURE 1   CANADA'S PERFORMANCE IN REDUCING GHG EMISSIONS 1990–2021[185]

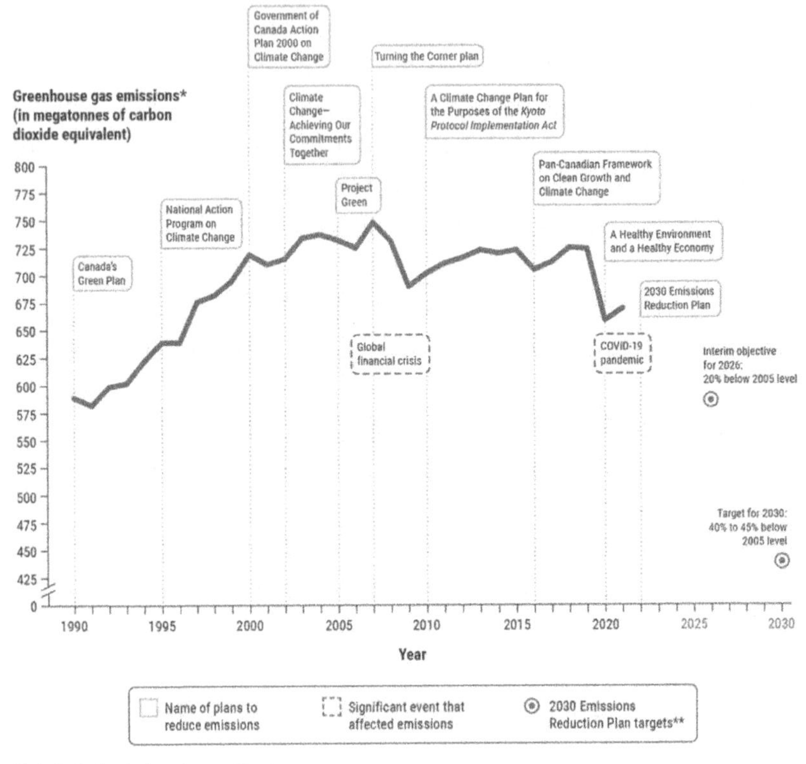

and Sustainable Development to the Parliament of Canada (Office of the Auditor General of Canada, 2024), online: oag-bvg.gc.ca [perma.cc/SF3D-28UR] [Office of the Auditor General of Canada, *Report 7*].

184 *Ibid.*

185 Figure is a direct reproduction of Figure 6.2 from the Office of the Auditor General of Canada, *Report 6 - Canadian Net-Zero Emissions Accountability Act – 2030 Emissions Reduction Plan*, Reports of the Commissioner of the Environment and Sustainable Development to the Parliament of Canada (Office of the Auditor General of Canada, 2023), online: oag-bvg.gc.ca [perma.cc/4N27-ZBLZ].

1.6.2.2 The Legal Framework for Climate Change

As a dualist country, signing a treaty in the Canadian federation does not translate into domestic obligations until the treaty is enacted into domestic law. Following the *UNFCCC* and *Kyoto Protocol*, a patchwork of climate policies emerged at multiple levels across the country over the years, including the phase out of coal in Ontario, several distinct carbon pricing systems (e.g., British Columbia's carbon tax,[186] Quebec's cap and trade system),[187] initiatives aimed at promoting the development of clean energy (e.g., Ontario's former Feed-In Tariff Program),[188] and some support for adaptation and resilience (e.g., infrastructure funding).[189]

After years of what the Auditor-General's office found in 2018[190] to be ineffective and insufficient climate policies, Canada seemed poised to move forward in a more coordinated way under the 2015 Paris Agreement. Intergovernmental meetings immediately following the Paris Agreement led to federal, provincial and territorial consensus on core tenets of a national climate plan, including carbon pricing, that was reflected in the 2016 *Vancouver Declaration on Clean Growth and Climate Change*.[191] The Vancouver Declaration was followed by a detailed plan of action agreed to by all but the province of Saskatchewan in the 2016

---

186 *Carbon Tax Act*, SBC 2008, c 40 [*Carbon Tax Act*].

187 *Regulation respecting a cap-and-trade system for greenhouse gas emissions allowances*, CQLR, c Q-2, r 46.1 [Cap and Trade System].

188 Government of Ontario, "4.0 Feed-In Tariff Program" in *Renewable Energy Development in Ontario: A Guide for Municipalities* (last modified 10 June 2022), online: ontario. ca [perma.cc/BA53-35AW].

189 See, e.g., Infrastructure Canada, "Investing in Canada Infrastructure Program" (last modified 19 September 2023), online: infrastructure.gc.ca [perma.cc/6LGB-QNSS], which includes initiatives such as the Green Infrastructure Fund that support efforts to reduce GHG emissions.

190 Office of the Auditor General of Canada, "Perspectives on Climate Change Action in Canada - A Collaborative Report from Auditors General - March 2018" (March 2018), online: oag-bvg.gc.ca [https://perma.cc/GY8L-9K88].

191 Canadian Intergovernmental Conference Secretariat, "Vancouver Declaration on Clean Growth and Climate Change" (3 March 2016), online: scics.ca [perma.cc /PAZ2-YAX4].

Pan-Canadian Framework on Clean Growth and Climate Change (PCF).[192] It is worth noting that when the PCF was signed, over 80 percent of Canada's GHG emissions were subject to a provincial carbon pricing system since the four biggest emitting provinces each had a system (BC and Alberta had a carbon tax and levy, respectively, while both Ontario and Quebec had an emissions trading system).[193] However, the fragile post-Paris consensus rapidly eroded with changes in provincial leadership and has been replaced by highly politicized rhetoric and divisive discourse.

1.6.2.3 National Carbon Pricing Framework

At the centre of the political debate was the federal government's centre-piece climate legislation, the *Greenhouse Gas Pollution Pricing Act (GGP-PA)*.[194] The *GGPPA* came into force in June 2018 and established a national minimum standard for carbon pricing across the country. Modelled after Alberta's carbon pricing system at the time, the *GGPPA* establishes a federal backstop carbon pricing system that entails two parts. Part 1 establishes a fuel charge applied to the production, distribution, and importation of a broad range of GHG emitting fuels, at prices specified in a schedule of rates. Part 2 establishes an Output-Based Pricing System (OBPS) for registered industrial facilities in covered sectors, which are exempt from paying the charge on fuel purchases but are required to pay for the portion of their emissions that exceed their annual facility emissions limit. The OBPS is aimed at recognizing that certain large industrial emitters are trade exposed. Together, Parts 1 and 2 form a national carbon price,

---

192 Environment and Climate Change Canada, *Pan-Canadian Framework on Clean Growth and Climate Change: Canada's Plan to Address Climate Change and Grow the Economy* (Gatineau: Environment and Climate Change Canada, 2016).

193 Note that Ontario abolished its emissions trading system in 2018. See *The Cap and Trade Program*, O Reg 144/16 [*Cap and Trade Program*]. This program was cancelled on July 3, 2018 by the Ford government (see *Prohibition Against the Purchase, Sale, and Other Dealings with Emission Allowances and Credits*, O Reg 386/18, s 2 [*Prohibition Against Emission Allowances*]; *Cap and Trade Cancellation Act*, SO 2018, c 13 [*Cap and Trade Cancellation Act*]).

194 *Greenhouse Gas Pollution Pricing Act*, SC 2018, c 12, s 186.

implementing the polluter-pays principle, which has been endorsed by the Supreme Court of Canada multiple times, including recently.[195]

One of the most important design features of the *GGPPA* is that it acts as a backstop. The *GGPPA* permits provinces and territories to continue or enact their own pricing systems, as long as they meet the national minimum standards of stringency. Jurisdictions may also request that the federal price apply to them, instead of enacting their own policies. Revenues generated by the carbon price must be returned to the province in which the revenues were collected or directly to the taxpayers.[196] The *GGPPA* achieves this "backstop" approach by authorizing the Governor in Council to list, in Schedule 1, the provinces and territories that have either opted to have the federal government implement the price in whole or in part within their jurisdiction or that were determined not to have a carbon pricing system that meets the minimum standards of stringency.

In 2018, the province of Saskatchewan was the first of three provinces to launch a constitutional challenge to the *GGPPA*. Ontario and Alberta joined Saskatchewan in each challenging Parliament's jurisdictional authority to pass the *GGPPA* in respective reference questions posed to their appeal courts. Ultimately, in 2021, the Supreme Court of Canada affirmed the constitutionality of the legislation. The litigation is discussed in detail in chapter 3.

Building on the PCF, Canada released an updated national climate plan in 2020, *A Healthy Environment and a Healthy Economy*. This plan included sixty-four federal measures building upon the foundation of the

---

195　The Supreme Court first confirmed the importance of the polluter-pays principle in 2003 in the Imperial Oil case, where an analysis of federal and provincial environmental laws led the court to conclude that this principle "has become firmly entrenched in environmental law in Canada" (see *Imperial Oil Ltd v Quebec (Minister of the Environment)*, 2003 SCC 58 at para 23). More recently, the majority of the Supreme Court in the *Orphan Well* decision highlighted that Alberta's established "cradle-to-grave licensing regime" (see *Orphan Well Association v Grant Thornton Ltd*, 2019 SCC 5 at para 1).

196　According to recent analysis, most Canadian households are financially better off after the carbon levy.

PCF.[197] In 2022, Canada published the *2030 Emissions Reduction Plan*,[198] the first policy roadmap issued under the *Canadian Net-Zero Emissions Accountability Act*, discussed next.

In early 2025, when Mark Carney replaced Justin Trudeau as Prime Minister, he suspended Part 1 of the *GGPPA* (the Fuel Charge).[199]

1.6.2.4   *Canadian Net-Zero Emissions Accountability Act*

In 2021, Parliament passed the *Canadian Net-Zero Emissions Accountability Act (CNZEAA)* which enshrined the initial 2030 target under the Paris Agreement (40 to 45 percent below 2005 levels by 2030), now updated to 45 to 50 percent below 2005 by 2035, establishes a series of interim targets at five-year intervals starting in 2030, and establishes a goal of acheiving net-zero GHG emissions by 2050.[200] The law creates a set of accountability measures that include producing a transparent process through which to plan, reporting on, assess, and adjust the government's progress on achieving its targets. The legislation requires regular progress reports (2023, 2025 and 2027) and additional targets and plans for 2035 to 2050.

Under the *CNZEAA*, the government published its first *Emissions Reduction Plan* (ERP) in 2022. The plan sets out the mitigation and funding measures that are intended to meet its goals. For the purposes of this book, I will highlight only a few of the main climate laws emerging under the ERP. It is worth noting, however, that the legal framework for climate change is becoming increasingly complex and integrated into many areas of law. From insurance and corporate disclosure to agricultural policies and taxation, there are few areas of law now untouched by climate change.

The federal government's most recent emissions reduction plan includes a variety of proposed policies, including economy-wide initiatives such as establishing a national carbon price, clean fuel standards, methane regulations

---

197   Environment and Climate Change Canada, *A Healthy Environment and a Healthy Economy* (Gatineau: Environment and Climate Change Canada, 2020), online: <canada.ca [perma.cc/Z595-VBQ9].

198   Environment and Climate Change Canada, *2030 Emissions Reduction Plan: Canada's Next Steps for Clean Air and a Strong Economy* (Gatineau: Environment and Climate Change Canada, 2022), online: canada.ca [perma.cc/YP7D-FTK7].

199   Canada Gazette Part II, SOR/2025-107 (March 15, 2025).

200   *Canadian Net-Zero Emissions Accountability Act*, SC 2021, c 22 [*CNZEAA*].

and a cap on oil and gas emissions. The national carbon price (implemented through the *GGPPA*) started at $20 per tonne in 2019 and will rise to $170 per tonne in 2030. The impact of suspending Part 1 of *GGPPA* is estimated to be modest, since a significant proportion of the emissions reductions from the carbon price are from the large emitters system in Part 2 of the legislation.

1.6.2.5   Other Federal Climate Regulations

The federal government has several regulatory measures either in place or under development aimed at reducing emissions in different ways. Canada was the first country to require coal-fired power plants to meet performance standards in 2012, and then in 2016 set a goal of eliminating coal-fired generation of electricity by 2030.[201] Alberta stopped using coal in electricity generation in 2024, which leaves only the provinces of Saskatchewan, Nova Scotia, and Manitoba using coal for electricity in Canada.[202]

The *Clean Fuel Regulations*, which took effect in 2023, set increasingly strict requirements to reduce the carbon intensity of transportation fuels such as gas and diesel. The standards will be ratcheted up over time, reaching full stringency in 2030.[203] Federal natural gas regulations impose performance standards on new natural gas electricity generators.[204] These regulations are complimented by a mandate to achieve 100% zero-emission vehicle sales for light duty vehicles by 2035.[205]

For methane, the federal government pledged to reduce methane emissions from oil and gas by 75 percent by 2030, compared to 2012 levels. The *Regulations Respecting Reduction in the Release of Methane and Certain Volatile Organic Compounds (Upstream Oil and Gas Sector)* require

---

201   *Reduction of Carbon Dioxide Emissions from Coal-Fired Generation of Electricity Regulations*, SOR/2012-167.

202   Government of Canada, *Powering Past Coal Alliance: phasing out coal,* online: canada.ca [perma.cc/S4YP-N95A].

203   *Clean Fuel Regulations*, SOR/2022-140.

204   *Regulation Limiting Carbon Dioxide Emissions from Natural Gas-fired Generation of Electricity*, SOR/2018-261.

205   Canada Gazette Part II, *Regulations Amending the Passenger and Light Truck Greenhouse Gas Emission Regulation*, Dec 20, 2023).

reductions of 40 to 45 percent below 2012 levels by 2025, with plans to ramp this up to 75 percent below 2012 levels by 2030.[206]

The *Clean Electricity Standards*, enacted in December 2024, aim to encourage electrification of energy intensive activities such as transportation, heating and cooling and different industrial processes. The standards would establish performance standards to reduce GHG emissions from fossil-fuel generated electricity as of 2035.[207]

There are also plans to cap and reduce emissions from the oil and gas sector at the scale and pace needed to achieve net-zero by 2050. The government released draft regulations in November 2024, aimed at setting a cap on emissions (not production) equivalent to 35 percent below 2019 levels. The regulations would be implemented using a cap-and-trade system.[208]

For buildings, the federal government plans to develop a net-zero by 2050 building strategy, which will include changes to the building code, regulatory standards for heating systems, and an EnerGuide labeling system for homes at the point of sale, in addition to considerable investments. The federal government maintained its stated commitment to reduce inefficient fossil fuel subsidies.[209] It has also implemented tax incentives, including accelerated capital cost allowances for zero-emissions technology.[210] The government also introduced an investment tax credit for carbon capture, utilization, and storage.[211] The Emissions Reduction Plan entails measures in other sectors as well, including agriculture (e.g., national fertilizer emission reductions target of 30 percent below

---

206 SOR/2018-66; Environment and Climate Change Canada, "Proposed Regulatory Framework for Reducing Oil and Gas Methane Emissions to Achieve 2030 target" (last modified 20 September 2023), online: canada.ca [perma.cc/3LGF-BMA2].

207 Government of Canada, "Canada's Clean Electricity Future" (last modified 20 December 2024), online: canada.ca [perma.cc/57CQ-AKU5].

208 Canada Gazette, Part 1, Volume 158, Number 45: *Oil and Gas Sector Greenhouse Gas Emissions Cap Regulations*, November 9, 2024, online: canadagazette.gc.ca [perma.cc/E23J-9RUW].

209 Government of Canada, "Inefficient Fossil Fuel Subsidies" (last modified 27 October 2023), online: canada.ca [perma.cc/Q9ZN-DCLA].

210 *Ibid* at 44.

211 See Government of Canada, "Clean Technology (CT) Investment Tax Credit (ITC)" (last modified 18 October 2024), online: canada.ca [perma.cc/49WB-AXGV].

2020 by 2030), waste (e.g., landfill methane regulations), and measures relating to aviation and marine shipping.[212]

Although not an explicit part of its national climate plan, the federal government incorporated climate considerations into its updated *Impact Assessment Act* in 2019. Specifically, the *IAA* required evaluations to factor in the GHG emissions of a proposed project and includes climate change as one of the key public interest factors that must be considered in the decision of whether to approve a project.[213] It also conducted a Strategic Assessment of Climate Change that requires project proponents to provide information related to GHG emissions and, for long-term projects (beyond 2050), a plan to achieve net-zero emissions.[214] The Strategic Assessment has been criticized for not establishing a climate test or some form of cumulative benchmark against which projects can be measured to ensure national emission targets can be met.[215] In 2023, the Supreme Court of Canada struck down much of the *Impact Assessment Act* on jurisdictional grounds. The federal government has since amended the legislation, removing the provision that allowed GHG emissions to be considered as effects within federal jurisdiction in the decision-making phase of assessment, thus limiting the extent to which climate-considerations can factor into decisions relating to major projects.[216] The case is discussed in Chapter 3.

1.6.2.6 Adaptation

While most of the regulatory and policy focus at the federal level has been on mitigation, the federal government passed a National Adaptation Strategy in 2023[217] and an updated Action Plan in 2024. The strat-

---

212 Environment and Climate Change Canada, *2030 Emissions Reduction* Plan, above note 198 at 51.

213 *Impact Assessment Act*, SC 2019, c 28, ss 22(i) and 63(e).

214 Environment and Natural Resources, "Strategic Assessment of Climate Change" (last modified 6 October 2022), online: canada.ca [perma.cc/9NUB-HUNJ].

215 *Ibid*.

216 Government of Canada, "Amended *Impact Assessment Act* Now in Force" (last modified 21 June 2024), online: canada.ca [perma.cc/397R-8KJD].

217 Environment and Natural Resources, "Canada's National Adaptation Strategy" (last modified 9 September 2024), online: canada.ca [perma.cc/ZB74-L6L8].

egy and plan focus on five components, including reducing the impacts of climate-related disasters, improving health, protecting and restoring nature and biodiversity, building and maintaining resilient infrastructure, and supporting workers and the economy. The federal government has added an Adaptation Platform of tools and resources, such as guidance on planned retreat from high-risk areas and information to help support adaptation by municipalities.

1.6.2.7 Provincial, Territorial, and Municipal Measures

There is a diverse range of climate-related policies at the provincial, territorial and municipal level. From provincial carbon pricing frameworks and clean fuel standards to technological standards and transportation policies, provinces and territories have enacted a wide range of regulations aimed at reducing GHG emissions.[218] Given the breadth and diversity of policies, they are not reviewed in detail here. The Canadian Climate Policy Partnership (C2P2) and 440 Megatonnes websites provide a database of measures that illustrate the range of policies.[219] The measures are comprised of three different policy instruments, including abatement support (incentives or "carrots"), mandatory policies requiring compliance ("the sticks") and indirect policies. Of the 327 climate policies measured by C2P2, the vast majority are of the "carrot" variety.[220]

1.6.2.8 GHG Projections for Canada

Even with the rapidly evolving legal and policy framework, Canada is not on track to meet its 2030 or 2035 targets, though the gap between

---

218 Centre for Climate and Energy Solutions, "Canadian Province Climate Policy Maps" (last accessed 14 January 2025), online: c2es.org [perma.cc/58E7-Q8ED].

219 University of Calgary, "The Canadian Climate Policy Partnership" (last accessed 14 January 2025), online: spp.ucalgary.ca [perma.cc/C4AZ-K57H]; Katherine Koch et al, "The State of the Canadian Climate Policy Landscape" (22 August 2024), online: 440megatonnes.ca [perma.cc/V4LZ-WZSN]. See also Federation of Canadian Municipalities, "Municipalities for Climate Innovation Program" (last accessed 14 January 2025), online: fcm.ca [perma.cc/3YEW-TGTQ] (describing initiatives at the municipal level).

220 Koch et al, above note 219.

emissions and targets is narrowing.[221] In its latest report on GHG projections, the federal government estimated that Canada will emit between 491 and 514 Mt in 2030 depending upon the different scenarios of policies and measures implemented and funded by all governments by 2024.[222] An earlier estimate based on policies and measures implemented as of 2022 projected emissions of 625 Mt.[223] While an overshoot of 66–77 Mt is better than one of 200 Mt, it is still another missed target. In its 2024 report on implementation of the *Canadian Net-Zero Emissions Accountability Act*, the Auditor-General's office noted that only nine of twenty federal measures audited were on track.[224] The Auditor General's report also noted that Canada has the worst record of the G7 nations on GHG emissions reductions.[225]

Canada's ability to meet its targets depends on the country effectively implementing the measures that have been announced and are under development. This includes methane regulations for the oil and gas sector and proposed landfill gas regulations. Federal clean fuel standards have also only recently been implemented.[226] Accounting for these additional measures, emissions are projected to fall to 458 Mt in 2035, overshooting the new 2035 target by approximately 18–56 Mt, and decline to 431Mt in 2040.[227] It is important to note that the estimates make certain assumptions about policies. For instance, the estimates account for contributions from land use, nature-based solutions. and credits purchased under the Western Climate Initiative.[228] If any of the policies assumed

---

221 Environment and Climate Change Canada, "Chapter 1: Introduction and Executive Summary" in *Canada's First Biennial Transparency Report under the Paris Agreement*, Developed in accordance with the Paris Agreement's Enhanced Transparency Framework (Gatineau: Environment and Climate Change Canada, 2024) [Biennial Transparency Report].

222 *Ibid* at 12.

223 Environment and Climate Change Canada, *Canada's Greenhouse Gas and Air Pollutant Emissions Projections 2021* (Gatineau: Environment and Climate Change Canada, 2022) at 9, online: <publications.gc.ca [perma.cc/8PVT-DUVU] [ECCC Projections 2021].

224 Office of the Auditor General of Canada, *Report 7*, above note 183.

225 *Ibid* at 7.3.

226 *Clean Fuel Regulations*, above note 206.

227 Biennial Transparency Report, above note 220.

228 *Ibid*.

to be in place under the projections are modified or abolished (following a change of government, for instance), the projections will necessarily change accordingly. With a minority government in place federally, divisive politics in the federation, and the ongoing global political instability fueled in part by the Trump administration's policies, including tariffs and a retreat from the *Paris Agreement*, there is much uncertainty. The house of cards is only as stable as the political will holding them in place.

1.7 CONCLUSION

As the IPCC has recognized, "(e)ffective climate action is enabled by political commitment, well-aligned multilevel governance, institutional frameworks, laws, policies and strategies and enhanced access to finance and technology. Clear goals, coordination across multiple policy domains, and inclusive governance processes facilitate effective climate action."[229] Canada faces some important challenges in this regard, especially in the context of the highly charged political context within which decarbonization is taking place in this country.

However, it is essential that Canada act responsibly. Canada is a wealthy, privileged country with the means to transition our energy systems while safeguarding core structures such as health care and social programs. In recognizing our global responsibility to address climate change, the Supreme Court of Canada noted that failing to act hinders our ability to push for international action to reduce GHGs.[230] Chapter 3 examines the way in which federalism and our division of legislative powers authorizes aspects of the Canadian legal framework for climate change. Before turning to that, Chapter 2 considers the climate issue from the perspective of Indigenous Peoples.

Additional Resources

Ben Clarke et al, "Extreme Weather Impacts of Climate Change: An Attribution Perspective" (2022) 012001 Environmental Research: Climate 1.

---

229 IPCC, "AR6," above note 11 at 32.
230 *GGPPA References*, above at 2 at para 190.

Nathalie J Chalifour & Jessica Earle, "Feeling the Heat: Climate Litigation under the Charter's Right to Life, Liberty and Security of the Person" (2018) 42:2 Vermont LR 690.

Commissioner for Environment and Sustainable Development, Auditor General's Office, *Canadian Net-Zero Emissions Accountability Act –2024 Report*, Report 7 (2024), online: www.oag-bvg.gc.ca/internet/English/parl _cesd_202411_07_e_44576.html

Meinhard Doelle & Sara L Seck, *Research Handbook on Climate Change Law and Loss & Damage* (UK: Edward Elgar Publishing Limited, 2021).

Meinhard Doelle "The Paris Agreement: Historic Breakthrough or High Stakes Experiment?" (2016) 6:1–2 Climate Law 1.

Sam Fankhauser et al, "The Meaning of Net Zero and How to Get it Right" (2022) 12 Nature Climate Change 15. Government of Canada, "Canada's First Biennial Transparency Report under the Paris Agreement" (December 31, 2024).

James Hansen et al, "Target Atmospheric CO_2: Where Should Humanity Aim?" (2008) 2 The Open Atmospheric Science Journal 217.

James Hansen et al, "Young People's Burden: Requirement of Negative CO_2 Emissions" (2017) 8:3 Earth System Dynamics 577.

Katharine Hayhoe, *Saving Us: A Climate Scientist's Case for Hope and Healing in a Divided World*. (Simon and Schuster, 2022).

Amy Janzwood & Kathryn Harrison, "The Political Economy of Fossil Fuel Production in the Post-Paris Era: Critically Evaluating Nationally Determined Contributions" (2023) Energy Research & Social Science 102.

Matthew W Jones et al, "Global and Regional Trends and Drivers of Fire Under Climate Change" (2022) 60:3 Rev of Geophysics 1.

Luke Kemp et al, "Climate Endgame: Exploring Catastrophic Climate Change Scenarios" (2022) 119:34 Proceedings of the National Academy of Sciences of the United States of America 1.

Richard J Lazarus "Super Wicked Problems and Climate Change: Restraining the Present to Liberate the Future" (2009) 94 Cornell L Rev 115.

Andrew Leach, *Between Doom & Denial: Facing Facts about Climate Change* (McGill Max Bell Lectures, 2023).

Timothy M Lenton et al, "Climate Tipping Points: Too Risky to Bet Against" (2019) 575 Nature (27 November 2019) 592.

Kelly Levin et al, "Overcoming the Tragedy of Super Wicked Problems: Constraining Our Future Selves to Ameliorate Global Climate Change" (2012) 45 Policy Sciences 123.

Johan Rockstrom et al, "Safe and just Earth System Boundaries" (2023) 619 Nature 102.

Marina Romanello et al, "The 2024 Report of the Lancet Countdown on Health and Climate Change: Facing Record-breaking Threats from Delayed Action"

(2024) 404 The Lancet 1847, online: www.thelancet.com/action/showPdf?pi
i=S0140-6736%2824%2901822-1.

Jocelyn Stacey, *The Constitution of the Environmental Emergency* (Hart Publishing, 2018).

Farhana Sultana, *Confronting Climate Coloniality: Decolonizing Pathways for
Climate Justice* (Taylor & Francis Group, 2024), online: www.routledge.com
/Confronting-Climate-Coloniality-Decolonizing-Pathways-for
-Climate-Justice/Sultana/p/book/9781032737850.

United Nations Environment Program, Emissions Gap Report 2024, online:
www.unep.org/resources/emissions-gap-report-2024.

John Vaillant, *Fire Weather: The Making of a Beast* (Penguin Random House Canada, 2023).

chapter 2

INDIGENOUS PEOPLES, CLIMATE CHANGE, AND CONSTITUTIONS[*]

"Colonialism caused climate change. Indigenous rights are
the solution."[1]
Indigenous Climate Action

"As an Indigenous person I see first-hand impacts that damaging
the land has on not only our day to day lives but also our spirit."
Jordan Brown, Inuk[2]

"Canadian law will remain problematic for Indigenous peoples
as long as it continues to assume away the underlying title and
overarching government powers that First Nations possess."
Professor John Borrows[3]

---

[*] The author wishes to thank colleagues who provided input on and/or reviewed drafts of this chapter. This includes Professor Aimée Craft, Professor Heather McLeod-Kilmurray, Professor Anne Levesque and Professor David Boyd. I also wish to thank Gérick Girard (JD/LLL Candidate, uOttawa) and Ronald Cheung (JD Candidate, uOttawa) for their research assistance. Any errors are the author's alone.

1 Indigenous Climate Action, *Decolonizing Climate Policy in Canada: Report from Phase One* (Indigenous Climate Action, 2021) at 5, online: indigenousclimateaction.com [perma.cc/K2FY-YVYJ].

2 Indigenous Climate Action, *Decolonizing Climate Policy in Canada: Report from Phase 2 | Part 1* (Indigenous Climate Action, 2023) at 21, online: indigenousclimateaction.com [perma.cc/NT7E-VR4J].

3 John Borrows, "The Durability of Terra Nullius: *Tsilhqot'in Nation v British Columbia*" (2015) 48:3 UBC L Rev 701.

Indigenous Peoples, who have played little role in creating the global problem of climate change,[4] bear a disproportionate burden of its impacts.[5] As Anishinaabe scholar and Professor Deborah McGregor explains, "[i]t is recognized, internationally and in Canada, that Indigenous peoples are more vulnerable to the impacts of climate change than other peoples due to distinct connections to the natural world."[6] Additionally, Indigenous Peoples "are confronted with disparities and disadvantages in every conceivable indicator of well-being" and climate change threatens to exacerbate these challenges."[7] Policies aimed at addressing climate change risk reinforcing existing racism and inequality unless they are carefully designed in collaboration with Indigenous Peoples and informed by their laws, governance systems, knowledge and leadership.[8] Against the backdrop of colonization — marked by dispossession, extractivism, exclusion, systemic racism, and oppression — climate change constitutes another layer of colonial violence inflicted upon Indigenous Peoples.[9]

---

4   The problem is caused predominantly by so-called developed nations with a worldview that treats nature as a commodity for human benefit (see Deborah McGregor, "Mino-Mnaamodzawin: Achieving Indigenous Environmental Justice in Canada" (2018) 9:1 Envt & Society at 7–24). For an exploration of critical, decolonizing approaches to addressing climate change, see also Kyle Powys Whyte, "Indigenous Climate Change Studies: Indigenizing Futures, Decolonizing the Anthropocene" (2017) 55:1/2 English Language Notes 153.

5   Deborah McGregor, "Reconciliation, Colonization, and Climate Futures" in Carolyn Tuohy et al, eds, *Policy Transformation in Canada: Is the Past Prologue?* (Toronto: University of Toronto Press, 2019) 139 at 139 [McGregor, "Reconciliation"], citing Center for Indigenous Environmental Resources, *Report 2: How Climate Change Uniquely Impacts the Physical, Social and Cultural Aspects of First Nations*, Report Prepared for the Assembly of the First Nations (Winnipeg: Center for Indigenous Environmental Resources, 2011).

6   McGregor, "Reconciliation," above note 5 at 139.

7   *Ibid*. See also Allyson K Menzies et al, "'I See My Culture Starting to Disappear': Anishinaabe Perspectives on the Socioecological Impacts of Climate Change and Future Research Needs" (2021) 7 FACETS 509 at 514.

8   The policies may raise the cost of food or fuel, for instance, which disproportionately impacts communities living in rural areas and those living in poverty (see generally Indigenous Climate Action, *Phase one*, above note 1).

9   See Leanne Betasamosake Simpson, *As We Have Always Done: Indigenous Freedom Through Radical Resistance* (Minneapolis: University of Minneapolis Press 2017). For an exploration of the relationship between Indigenous knowledge, environmental justice

Despite these profound challenges, Indigenous Peoples are at the forefront of climate resilience and advocacy. They are leaders whose knowledge systems and governance practices offer essential insights into sustainable resource management, ecosystem restoration, and adaptation to environmental change.[10] Indigenous Climate Action, for example, draws upon teachings from the Medicine Wheel within Anishinaabe, Métis, and Nehiyaw traditions to illustrate how colonialism drives climate change and how the assertion of Indigenous rights and sovereignty provide a promising path forward.[11] Initiatives such as Indigenous Protected and Conserved Areas (IPCAs) and Indigenous-led adaptation projects demonstrate how Indigenous leadership in conservation addresses climate change while affirming land rights and cultural renewal.[12]

As a settler scholar and descendant of white francophone settlers, I acknowledge my position within this colonial system. It is not my role to interpret or define Indigenous experiences or perspectives on climate change and the Constitution. Instead, my intention is to situate this book, which focuses on the settler Constitution, within the broader context of colonialism and its intersections with the climate crisis. With humility and respect, I seek to amplify Indigenous voices, knowledge systems, and lived experiences.

This chapter begins with a reminder of Canada's colonialist and racist past and present, including how Indigenous rights, laws, languages. and cultures were and are suppressed and how Indigenous Peoples were, and continue to be, dispossessed of their lands and homes.[13] It briefly exam-

---

and reconciliation, see Deborah McGregor, "Reconciliation and Environmental Justice: In Search of Intersectionality" (2018) 14:2 J Global Ethics 222.

10   See, e.g., Indigenous Climate Action, *Phase one*, above note 1. For more on this topic, see the series of reports by Indigenous Climate Action in the Decolonizing Climate Policy project.

11   Indigenous Climate Action, *Phase two*, above note 2.

12   Indigenous Circle of Experts, *We Rise Together: Achieving Pathway to Canada Target 1 Through the Creation of Indigenous Protected and Conserved Areas in the Spirit and Practice of Reconciliation* (Gatineau: Parks Canada, 2018).

13   Some authors have cautioned against allowing decolonization to become a metaphor that enables a "settler move to innocence" that problematically tries to reconcile settler guilt and complicity (see Eve Tuck & K Wayne Yang, "Decolonization is Not a Metaphor" (2012) 1:1 Decolonization: Indigeneity, Education & Society 1).

ines the devastating impacts of colonialization on Indigenous Peoples in Canada before turning to Indigenous scholarship to explore the unique impacts of climate change on Indigenous communities and the ways settler courts have begun to acknowledge these impacts. Recognizing the diversity of Indigenous Peoples, communities, and cultures, this chapter aims to avoid generalizing or essentializing Indigenous identities and experiences while acknowledging the harmful impacts of past and ongoing systemic discrimination and colonialism.

Next, the chapter highlights ongoing steps in reconciliation, including the work of the Truth and Reconciliation Commission, judicial recognition of Indigenous rights under section 35 of the *Constitution Act*, and the implementation of the *United Nations Declaration on the Rights of Indigenous Peoples (UNDRIP)*.[14] Importantly, none of the discussions in this chapter or book are intended to minimize questions about the legitimacy of the Canadian Constitution as it relates to Indigenous self-government and rights. As part of the ongoing process of repair, decolonization, reconciliation and trust-building, I encourage readers to engage with Indigenous lawyers, scholars, legal traditions, and community leadership to learn about how Indigenous Peoples live in reciprocity with the land and water and contend with the challenges of climate change.[15]

2.1 THE CANADIAN CONSTITUTION AND RECONCILIATION

"How is it possible that any Pope, King or Queen, or explorers from Europe could 'discover' lands in the New World if Indigenous Peoples were

---

14   *United Nations Declaration on the Rights of Indigenous Peoples*, UNGA, 61st Sess, UN Doc A/61/L.67/Add.1 (2007) GA Res 61/295 [*UNDRIP*].

15   See, for instance, the works of scholars such as Professor Aimée Craft, Professor Deborah McGregor, Professor Val Napoleon, Professor John Borrows, the work of organizations such as Indigenous Climate Action, the Assembly of First Nations, the Inuit Tapiriit Kanatami, the work of communities and Nations, such as the Athabasca Chipewyan First Nation and Lubicon Lake Cree Nation and the many individuals and organizations listed in the Additional Resources section at the end of this chapter.

already occupying such lands, according to (their) own laws and legal orders?"[16]

This question is at the root of colonialization and the ongoing challenge of reconciling Indigenous sovereignty with the British and French Crown's assertion of sovereignty and the eventual confederation of Canada. Indigenous Peoples have lived on the lands we call Canada for millennia, long before European settlement began in the 1700s. Yet, in 1763, King George III unilaterally claimed these territories for European settlement through the *Royal Proclamation of 1763*. First Nations were active participants in the formulation of the Proclamation as part of a larger treaty between the British Crown and First Nations that included the treaty of Niagara ratified in 1764.[17] Together, these agreements contained a positive guarantee of First Nations self-government but such a guarantee was not adequately reflected in the language of the Proclamation.[18] While the Proclamation introduced European treaty-making processes and enabled the Crown to purchase lands from Indigenous Peoples, it also entrenched colonial dominance.[19] Although some early treaties began as agreements of mutual recognition, they soon devolved into unilateral arrangements laden with broken promises. Following Confederation, nation-to-nation relationships were systematically dismantled as Canada pursued a colonial agenda that excluded Indigenous Peoples from meaningful treaty-making.[20]

Colonial legal systems were a tool for the dispossession and subjugation of Indigenous Peoples. Settler courts relied on the racist doctrine of discovery, as evidenced in cases like *Johnson v McIntosh*[21] in the United

---

16 Assembly of First Nations, *Dismantling the Doctrine of Discovery* (Ottawa: Assembly of First Nations, 2018) at 2 [Assembly of First Nations, *Doctrine of Discovery*].

17 See John Borrows, "Wampum at Niagara: The Royal Proclamation, Canadian Legal History, and Self-Government" in Michael Asch, ed, *Aboriginal and Treaty Rights in Canada: Essays on Law, Equality and Respect for Difference* (Vancouver: UBC Press, 1997) at 155.

18 *Ibid*.

19 *Ibid* at 159–62.

20 *Ibid*.

21 *Johnson v M'Intosh*, 21 US (8 Wheat) 543 at 573–74 (1823).

States and *St Catharines Milling and Lumber Co v The Queen*[22] in Canada, to extinguish Indigenous rights and legalize colonial exploitation. These decisions were made through settler law, without regard to Indigenous legal traditions, reinforcing systemic oppression.[23] In 1973, the Supreme Court of Canada shifted course in *Calder v Attorney-General of British Columbia*,[24] rejecting the doctrine of discovery and acknowledged that Aboriginal title predated colonization. This marked a turning point in recognizing the Crown's duty to uphold its treaty obligations.

The Supreme Court of Canada has since recognized that Indigenous Peoples "were here when Europeans came and were never conquered," and that treaties are instruments to reconcile inherent Indigenous sovereignty with assumed Crown sovereignty.[25] Although the Vatican formally repudiated the doctrine of discovery[26] and Canadian courts have stated that the related doctrine of *terra nullius* never applied in Canada,[27] the colonial frameworks they justified continue to harm Indigenous communities. As Indigenous legal scholar John Borrows observes, "Canadian law will remain problematic for Indigenous peoples as long as it continues to assume away the underlying title and overarching government powers that First Nations possess."[28]

Colonial assumptions are deeply embedded in Canada's constitutional framework. The *Constitution Act, 1867* allocates legislative authority over "Indians and lands reserved for Indians" to Parliament, disregarding pre-existing Indigenous jurisdiction.[29] This division of powers, which is

---

22   [1887] 13 SCR 577.

23   Assembly of First Nations, *Doctrine of Discovery* above note 16 at 2.

24   *Calder v British Columbia (Attorney General)*, [1973] SCR 313 at 328.

25   *Haida Nation v British Columbia (Minister of Forests)*, 2004 SCC 73 at paras 20 and 25 [*Haida Nation v BC*].

26   Holy See Press Office, Daily Bulletin, "Joint Statement of the Dicasteries for Culture and Education and for Promoting Integral Human Development on the 'Doctrine of Discover'" (30 March 2023), online: press.vatican.va [perma.cc/N7BA-PKX3]. In fact, the Royal Proclamation of 1763 confirms that the related doctrine of *terra nullius* never applied in Canada.

27   *Tsilhqot'in Nation v British Columbia*, 2014 SCC 44 at 69 [*Tsilhqot'in Nation*].

28   John Borrows, "Terra Nullius," above note 3.

29   See *Constitution Act, 1867* (UK), 30 & 31 Vict, c 3, ss 91–95, reprinted in RSC 1985, Appendix II, No 5.

the subject of Chapter 3, justified legislation such as the *Indian Act*.[30] This Act, imposed without Indigenous consent, sought to eradicate Indigenous governance and culture under the guise of assimilation. As Prime Minister John A. Macdonald stated, its purpose was to "do away with the tribal system and assimilate the Indian people in all respects."[31] The Act codified federal control over Indigenous lives, criminalizing, banning or regulating ven cultural practices and reducing Indigenous Peoples to wards of the state.

More recently, the Supreme Court of Canada has begun to recognize Indigenous Peoples' inherent rights to self-government. Through a broad interpretation of section 91(24), the Court in *Reference re an Act Respecting First Nations, Inuit and Métis Children, Youth and Families*[32] upheld federal legislation affirming Indigenous jurisdiction over child and family services. While a step in the right direction, offering a model for future recognition of Indigenous sovereignty, the decision maintains the framing of Indigenous governance as stemming from federal authority, rather than acknowledging it as inherent and pre-existing.

Settler recognition of Indigenous sovereignty after centuries of settler governance requires dismantling colonial assumptions and recognizing Canada as a pluralist state of overlapping Nations. The inclusion of section 35 in the *Constitution Act, 1982,* which recognizes and affirms the existing Aboriginal and treaty rights of First Nations, Inuit and Métis Peoples, was a step in the right direction but far from what is required. Indigenous scholars, including John Borrows and Val Napoleon, emphasize the need for Canada to respect plural constitutional narratives informed by Indigenous legal and political traditions.[33] As Feltes and others argue,

---

30 RSC 1985, c I-5.

31 Robert P C Joseph, *21 Things You May Not Know About the Indian Act: Helping Canadians Make Reconciliation with Indigenous Peoples a Reality* (Port Coquitlam: Indigenous Relations Press, 2018); See also Hayden King & Shiri Pasternak, *Canada's Emerging Indigenous Rights Framework: A Critical Analysis* (Toronto: Yellowhead Institute, 2018), online: yellowheadinstitute.org [perma.cc/GL24-UU5Y].

32 2024 SCC 5 [*Indigenous Child Welfare Reference*].

33 Jenson suggests that there are three collective projects in Canada that have legitimacy and authority: Canada as a whole, Indigenous Peoples and Quebec (see Jane Jenson, "Recognizing Differences: Distinct Societies, Citizenship Regimes and

these traditions "bust through Canada's legal foundation" and demand a reorientation of settler institutions.[34] The Supreme Court of Canada's use of the "braiding" metaphor to refer to the integration of Indigenous laws, Canadian statutes, and international standards like *UNDRIP* in child and family services governance offers a model for advancing Indigenous sovereignty. However, such efforts must be part of a broader commitment to upholding Indigenous legal systems and self-determination, across all areas of governance and recognizing inherent Indigenous legal traditions alongside common and civil law in a pluralistic nation.

2.2 THE HARMFUL LEGACY OF COLONIALISM ON INDIGENOUS PEOPLES IN CANADA

Colonialism has inflicted, and continues to inflict, profound and enduring harm on Indigenous Peoples in Canada, undermining Indigenous cultures, governance systems, economies, and well-being. Rooted in policies of dispossession and assimilation, colonialism has eroded, and continues to erode, Indigenous identities and ties to land, disrupting intergenerational transmission of knowledge and fostering widespread marginalization.

2.2.1 Residential School System

One of the atrocities committed by the federal government, in conjunction with Christian religious institutions, was the residential school

---

Partnerships" in Guy Laforest & Roger Gibbins, eds, *Beyond the Impasse: Towards Reconciliation* (Montreal: Institute for Research on Public Policy, 1998) 215 at 232–234. See also the Call for Papers for "The Supreme Court at 150: The Past, Present and Future" co-convened by Professors Noura Karazivan, Vanessa MacDonnell & Naiomi Metallic (on file with author).

34 Emma Feltes et al, "Crisis, Colonialism and Constitutional Habits: Indigenous Jurisdiction in times of Emergency" (2023) 38:1 CJLS 1 at 9, citing John Borrows, *Freedom and Indigenous Constitutionalism* (Toronto: University of Toronto Press, 2016); James Sákéj Youngblood Henderson et al, *Aboriginal Tenure in the Constitution of Canada* (Toronto: Thomson Reuters, 2022); Keira L Ladner & Michael McCrossan, "The Road Not Taken: Aboriginal Rights after the Re-Imagining of the Canadian Constitutional Order," in James B Kelly & Christopher P Manfredi, eds, *Contested Constitutionalism: Reflections on the Canadian Charter of Rights and Freedoms* (Vancouver: UBC Press, 2009) 263.

system in Canada.[35] For over 150 years, First Nations, Inuit, and Métis children were forcibly removed from their families and communities and placed in residential boarding schools, often far from their homes, families, and communities. These schools were deliberately designed to eradicate Indigenous languages and cultures in an attempt to assimilate Indigenous Peoples into settler society.[36] The residential school system inflicted profound physical, emotional, and cultural harm. It constitutes a form of cultural genocide, as children were prohibited from speaking their languages, practicing spiritual traditions, or maintaining family ties, leading to the loss of cultural knowledge and identity.[37] Many students endured severe physical, emotional, and sexual abuse at the hands of school staff. The trauma of these experiences continues to reverberate through generations, resulting in intergenerational suffering and erosion of language, culture, knowledge, and tradition.[38]

Thousands of Indigenous children in the residential school system died due to neglect, malnutrition, disease, and abuse.[39] The recent discovery of unmarked graves at former residential school sites, including those in Kamloops and Cowessess, highlights the system's devastating toll.[40] The

---

35 Truth and Reconciliation Commission of Canada, *Honouring the Truth, Reconciling for the Future: Summary of the Final Report of the Truth and Reconciliation Commission of Canada* (Winnipeg: The Truth and Reconciliation Commission of Canada, 2015) [Truth and Reconciliation Commission of Canada, *Summary*]. See also James Rodger Miller, *Shingwauk's Vision: A History of Native Residential Schools* (Toronto: University of Toronto Press, 1996).

36 Marie Ann Battiste, *Reclaiming Indigenous Voice and Vision* (Vancouver: UBC Press, 2000).

37 The Truth and Reconciliation Commission of Canada (TRC) characterized this as a form of cultural genocide (see National Center for Truth and Reconciliation, "Residential School History" (last visited 14 January 2025), online: nctr.ca [perma.cc/HTY4-DQWL].

38 *Ibid*; Theodore Fontaine, *Broken Circle: The Dark Legacy of Indian Residential Schools: A Memoir* (Victoria: Heritage House, 2010).

39 Truth and Reconciliation Commission, *Canada's Residential Schools: Missing Children and Unmarked Burials*, The Final Report of the Truth and Reconciliation Commission of Canada: Volume 4 (Montreal: McGill-Queen's University Press, 2015) [Truth and Reconciliation Commission, *Missing Children*].

40 Truth and Reconciliation Commission, *Missing Children* above note 39; Government of Canada, "Missing Children and Burial Information" (last modified, 24 Mai 2024), online: rcaanc-cirnac.gc.ca [perma.cc/YZC3-YACB].

deaths of over 4,000 Indigenous children in the residential school system have now been confirmed, though the true number is likely higher. The ongoing search for further unmarked graves and missing children across the country remains a deeply painful and unresolved process.[41] In 2022, the National Advisory Committee on Residential Schools Missing Children and Unmarked Burials was established to support Indigenous communities in this critical and heartbreaking work. At this time, Kimberly R Murray was also appointed as an Independent Special Interlocutor to ensure that the Indigenous children whose burial sites are being recovered are treated with honour, respect, and dignity.[42] The Interlocutor's powerful Final Report, published in 2024, identified an urgent need for an Indigenous-led search and recovery mechanism and an Indigenous-led reparations framework for truth, accountability, justice, and reconciliation.[43]

2.2.2 Missing and Murdered Indigenous Women and Girls

Colonialism in Canada has, and continues to have, devastating impacts on Indigenous women, girls, and Two-Spirit people.[44] The 2019 final report of the National Inquiry into Missing and Murdered Indigenous Women and Girls (MMIWG) revealed that the legacy and impacts of colonialism today is deeply gendered, with Indigenous women, girls, and Two-Spirit people disproportionately affected.[45] The Inquiry unequivo-

---

41 Truth and Reconciliation Commission, *Missing Children* above note 39.

42 National Advisory Committee on Residential Schools Missing Children and Unmarked Burials, "Navigating the Search for Missing Children and Unmarked Burials: An Overview for Indigenous Communities and Families" (last visited 15 January 2025), online: nac-cnn.ca [perma.cc/Z8FL-G4UW].

43 Office of the Independent Special Interlocutor for Missing Children and Unmarked Graves and Burial Sites associated with Indian Residential Schools, *Executive Summary: Final Report on the Missing and Disappeared Indigenous Children and Unmarked Burials in Canada*, Kimberley R Murray (Oshweken: Office of the Independent Special Interlocutor, 2024), online: osi-bis.ca [perma.cc/A9KC-UHX5].

44 Kim Anderson, *A Recognition of Being: Reconstructing Native Womanhood* (Toronto: Sumach Press, 2000); Betasamosake Simpson, above note 9.

45 National Inquiry into Missing and Murdered Indigenous Women and Girls, *Reclaiming Power and Place: The Final Report of the National Inquiry into Missing and Murdered Indigenous Women and Girls*, vol 1b (Ottawa: Privy Office Council, 2019).

cally described this crisis as a form of genocide, rooted in colonial policies and systemic structures that have devalued Indigenous lives and increased their vulnerability to violence. The report documented that Indigenous women, girls and Two-Spirit people are four times more likely than non-Indigenous women to be the victims of violence, twice as likely to experience intimate partner violence, and significantly more likely to experience physical and sexual assault. Between 2001 and 2014, the homicide rate for Indigenous female victims was four times higher than that of non-Indigenous females.[46]

The Inquiry highlighted intersectional violence, showing how Indigenous women face compounded risk due to systemic poverty, inadequate housing, and displacement from traditional lands and the intersecting impacts of racism and misogyny.[47] It also revealed systemic and interjurisdictional neglect, where colonial law enforcement and judicial systems have historically failed to protect Indigenous women, investigate cases of violence adequately, or pursue justice, reflecting entrenched institutional racism and sexism.[48] Furthermore, the report identified intergenerational impacts, including the connections between the residential school system, the Sixties Scoop — which forcibly removed Indigenous children into non-Indigenous foster homes — and the current overrepresentation of Indigenous children in the child welfare system.[49]

To address these issues, the MMIWG Inquiry issued 231 Calls for Justice, urging governments, institutions, and Canadians to implement systemic reforms.[50] These reforms aim to end the cycles of trauma, marginalization, and violence while ensuring safety, equality, and justice for

---

46 Assembly of First Nations, "Ending the Critical Situation of Violence, Disappearance, and Murder of First Nations Women, Girls, and Gender-Diverse People" (last visited 15 January 2025), online: afn.ca [perma.cc/HZ9X-JUPW] [Assembly of First Nations, "Ending violence"]. See also National Inquiry into Missing and Murdered Indigenous Women and Girls, above note 45.

47 National Inquiry into Missing and Murdered Indigenous Women and Girls, above note 45.

48 *Ibid.*

49 *Ibid.*

50 See also Truth and Reconciliation Commission of Canada, *Truth and Reconciliation Commission of Canada: Calls to Action*, (Winnipeg: Truth and Reconciliation Commission of Canada, 2015) [Truth and Reconciliation Commission of Canada, *Calls to Action*].

Indigenous women, girls, and Two-Spirit people.[51] The report emphasized the importance of centering Indigenous knowledge, leadership, and perspectives in creating lasting change.

2.2.3 Land Dispossession, Economic Marginalization, and Health Inequities

Colonialism has systematically dispossessed Indigenous Peoples of their lands in various ways including outright appropriation without compensation.[52] For First Nations Peoples, treaties were often unilaterally negotiated, misinterpreted, or outright violated, prioritizing settler access to resources over the rights and sovereignty of Indigenous Peoples.[53] Furthermore, the establishment of the reserve system and the imposition of the *Indian Act* further restricted access to traditional territories. The Métis, for their part, were dispossessed of their land through the script system — meaning the allocation of money in exchange for their rights to land — and persistent patterns of inattention and neglect in the land settlement scheme.[54] Government evictions and relocations forced many Inuit communities to abandon their homes and land in favour of newly established communities aimed at securing Canada's Arctic sovereignty, causing a loss of cultural and traditional ways of life.[55] All of these policies have resulted in undermining self-sufficiency and severing ties to ecological knowledge, cultural traditions, and subsistence practices such as hunting, fishing,

---

51 Assembly of First Nations, "Ending violence," above note 46. See also National Inquiry into Missing and Murdered Indigenous Women and Girls, above note 45.

52 Aimée Craft, *Breathing Life into the Stone Fort Treaty: An Anishinaabe Understanding of Treaty Making* (Saskatoon: Purich Publishing, 2013) [Craft, *Breathing Life*].

53 *Ibid.*

54 See House of Commons, *"We Belong to the Land": The Restitution of Land to Indigenous Nations: Report of the Standing Committee on Indigenous Rights and Recognition* (May 2024) (Chair: John Aldag). See also *Manitoba Metis Federation Inc v Canada (AG)*, 2013 SCC 14 at para 108. For a more complete history of the dispossession of the Métis land settlement, see Paul LAH Chartrand, "Aboriginal Rights: The Dispossession of the Métis" (1991) 29:3 *Osgoode Hall LJ* 7.

55 Carol Brice-Bennett, *Dispossessed: The Eviction of Inuit from Hebron, Labrador, Isberg Series* (Montréal: Imaginaire | Nord, 2017).

gathering, and agriculture. Land dispossession contributes to the ongoing violence against women: what we do to the land, we do to the women.[56]

Economic marginalization continues to impact Indigenous communities due to historical and ongoing colonial policies. Resource extraction authorized by governments on Indigenous lands often proceeds without free, prior, and informed consent, resulting in environmental degradation and loss of traditional livelihoods.[57] While judicial interpretation of the Honour of the Crown and the duty to consult and accommodate have brought some progress in recognizing Indigenous rights, conflicts over land and water rights persist. For instance, industrial developments, including pipelines and mining projects, threaten treaty lands and traditional territories despite growing efforts to resolve comprehensive land claims through negotiations and court cases; these industrial developments have extensive cumulative impacts.[58]

Indigenous communities also face disproportionately high rates of poverty, unemployment, and inadequate housing, outcomes rooted in policies that systematically restricted access to lands, resources, education, health care, and economic opportunities.[59] These inequities are compounded by profound health disparities, including higher rates of chronic diseases, mental health crises, and substance abuse.[60] These issues are driven by systemic inequities and intergenerational trauma.

---

56   See Native Women's Association of Canada, "Land Justice is Gender Justice" (last visited 15 January 2025), online: nwac.ca [perma.cc/9HHK-MSE4].

57   Deborah McGregor, "Traditional Ecological Knowledge and Sustainable Development: Towards Coexistence," in Mario Blaser, Harvey A Feit & Glenn McRae, eds, *In the Way of Development: Indigenous Peoples, Life Projects and Globalization* (London: Zed Books, 2004) 140.

58   As one example, the Wet'suwet'en Nation has been resisting pipeline construction through their traditional territory, resulting in the arrests of many Indigenous land defenders. See Shreya Shah, "Wet'suwet'en Explained" (last visited 15 January 2025), online: theindigenousfoundation.org [perma.cc/GT2L-86HK]. The discussion below on the *Yahey* decision is also relevant here (see Section 2.7, below in this chapter).

59   Toronto Metropolitan University, "First Nations Poverty in Canada," online: torontomu.ca [perma.cc/ED4A-VR8R].

60   See Government of Canada, "Suicide Prevention in Indigenous Communities" (last visited 15 January 2025), online: sac-isc.gc.ca [perma.cc/Q7F2-PYBG].

For example, the Athabasca Chipewyan First Nation has experienced increased cancer rates linked to pollution from nearby oilsands developments, while the Grassy Narrows First Nation continues to suffer severe health consequences from mercury poisoning caused by industrial dumping by a pulp and paper mill.[61]

In many cases, the imposition of Western medical systems has marginalized traditional healing practices, eroding holistic approaches to health and wellness. Infrastructure failures and jurisdictional disputes have left many First Nations reserves without consistent access to safe drinking water.[62] Long-standing boil water advisories, stemming from systemic underfunding, contamination, and inadequate maintenance, exemplify the harmful impacts of ongoing colonial policies.[63] As Professor Aimée Craft notes, water governance in Canada often ignores Indigenous legal traditions, perpetuating the colonial mindset of control rather than relationship.[64] These challenges underscore the need to uphold Indigenous rights and self-determination as essential components of addressing these systemic issues.

2.2.4 Child Welfare System and Children Services

Indigenous children in Canada have long faced systemic discrimination when seeking to access government services, a stark example of the

---

61 See, e.g., Ian Willms, "A Life – and Death – in Fort Chipewyan, Downstream from the Oilsands" (October 29, 2022) *The Narwhal* (29 October 2022) online: thenarwhal. ca [perma.cc/23YW-TB75]; Sarah Law, "Mercury Poisoning Near Grassy Narrows First Nation Worsened by Industrial Pollution, Study Suggests" *CBC News* (24 May 2024) online: cbc.ca [perma.cc/3BB2-ZE5U].

62 Constance MacIntosh, "Testing the Waters: Jurisdictional and Policy Aspects of the Continuing Failure to Remedy Drinking Water Quality on First Nations Reserves" (2007) 39:1 Ottawa L Rev 63 at 67–68, 70, 73 80–90.

63 See, e.g., Shoal Lake 40 First Nation, "Our History: Water," online at: shoallake40.ca / [perma.cc/M34X-7GZ8].

64 Aimée Craft, "Navigating Our Ongoing Sacred Legal Relationship with Nibi (Water)" in John Borrows et al, eds, *Braiding Legal Orders: Implementing the United Nations Declaration on the Rights of Indigenous Peoples* (Waterloo: Centre for International Governance Innovation, 2019) 101.

ongoing inequities stemming from colonial policies.[65] In 2005, Jordan River Anderson, a young boy from the Norway House Cree Nation, tragically passed away at the tender age of five after spending his entire life in hospital. The failure of federal and provincial governments to agree on who was financially responsibility for his medical care contributed to his death.[66] In his memory, *Jordan's Principle* was established to ensure that no First Nations child is ever denied access to the services they need due to jurisdictional disputes.[67]

In 2007, the First Nations Child and Family Caring Society, led by advocate Cindy Blackstock and the Assembly of First Nations, filed a landmark complaint with the Canadian Human Rights Commission. They claimed that the federal government had violated the *Canadian Human Rights Act* by failing to provide equitable funding for child welfare services for First Nations children, creating conditions of systemic discrimination.[68] In its 2016 decision, the Canadian Human Rights Tribunal (CHRT) ruled that the federal government's practices were discriminatory, depriving 165,000 First Nations children equitable access to essential services.[69] This underfunding not only exacerbated the overrepresentation of First Nations children in the child welfare system but also perpetuates intergenerational cycles of harm and trauma.[70]

Since the hearing on the merits, the CHRT has had to issue multiple orders directing the federal government to implement systemic reforms and to compensate affected children and families for the harm it has caused. Subsequent negotiations and rulings have addressed the scope of

---

65   First Nations Child & Family Caring Society, "Jordan's Principle: Ensuring First Nations Children Receive the Supports They Need When They Need them" (last visited 15 January 2025), online: fncaringsociety.com [perma.cc/9UFA-PXFA]. See also Truth and Reconciliation Commission of Canada, *Summary*, above note 35.

66   First Nations Child & Family Caring Society, "Honouring Jordan River Anderson" (last visited 15 January 2025), online: fncaringsociety.com [perma.cc/M87H-AHK7].

67   Cindy Blackstock, "Jordan's Principle: Canada's Broken Promise to First Nations Children" (2012) 17:7 Pediatrics & Child Health 368.

68   Cindy Blackstock, "The Complainant: The Canadian Human Rights Case on First Nations Child Welfare" (2016) 62:2 McGill LJ.

69   *First Nations Child and Family Caring Society of Canada v Attorney General of Canada (for the Minister of Indian and Northern Affairs Canada)*, 2016 CHRT 2.

70   *Ibid.*

eligibility and the amounts of compensation, reflecting the ongoing struggle to hold Canada accountable and to dismantle the systemic inequities that persist.[71] At the time of writing, however, negotiations relating to the long-term reform of child welfare services for First Nations children and families and Canada's implementation of Jordan's Principle aiming to end the discrimination remain ongoing. [72] This case remains a pivotal example of Indigenous advocacy and resilience in the fight for justice and equality for their children.

2.3 THE IMPACT OF CLIMATE CHANGE ON INDIGENOUS PEOPLES IN CANADA

"We are deeply connected to this Earth and witnessing the suffering impacts us – her trauma is our trauma which leads to a feeling of helplessness and hopelessness."[73]

Indigenous Climate Action

Climate change profoundly impacts Indigenous Peoples and their rights in Canada, intersecting with their cultural, environmental, and economic practices in complex and multi-faceted ways.[74] Indigenous communities experience these impacts disproportionately due to their deep relationships

---

71   For example, the Federal Court dismissed an application for judicial review of the Tribunal's decisions (see *Canada (AG) v First Nations Child and Family Caring Society of Canada*, 2021 FC 969). For an overview of the timeline, see Indigenous Services Canada, "Timeline: Jordan's Principle and First Nations Child and Family Services" (last modified 30 October 2023), online: sac-isc.gc.ca [perma.cc/S9FB-ZXTS]. See also Anne Levesque, "L'égalité réelle et la mise en œuvre intégrale du principe de Jordan" (2020) 36:1 Windsor YB Access Just 231.

72   First Nations Child and Family Caring Society of Canada, Press Release, First Nations Leadership Vote for a New Process to End Canada's Discrimination Against First Nations Children" (21 October 2024), online: fncaringsociety.com [perma. cc/2RQH-KSW2].

73   Indigenous Climate Action, *Phase two*, above note 2 at 21.

74   Pelin Kinay et al, "Reporting Evidence on the Environmental and Health Impacts of Climate Change on Indigenous Peoples of Atlantic Canada: A Systematic review" (2023) 2:2 Env Research: Climate, online: DOI: 10.1088/2752-5295/accb01 [perma.cc /QAG6-TKUR].

with the land, waters, and ecosystems.[75] As Menzies and others note, "for people and entire nations whose sense of community, identity, and well-being emerge from strong connections to the land and to place, these biophysical changes and disruptions to ways of life are linked to declining emotional, spiritual, and sociocultural well-being."[76] These challenges are compounded by colonial legacies, such as the reserve system and systemic inequities, which constrain adaptive responses.[77]

As the Mikisew Cree Nation have stated, Indigenous Peoples have a special connection to nature and the land, which means they are directly affected and more susceptible to even slight environmental changes caused by climate change.[78] The rapid pace of environmental changes, such as rising sea levels, poses significant threats to them.[79] In the Atlantic region, Mi'kmaq communities face increasing coastal erosion and flooding, as seen during Hurricane Fiona in 2022,[80] which devastated Lennox Island First Nation in Prince Edward Island, leaving the community submerged and without power.[81] Similarly, sea-level rise in Esgenoôpetitj First Nation, New Brunswick, has contaminated

---

75   Indigenous Climate Action, *Phase two*, above note 2 at ch 1.

76   Menzies et al, above note 7 at 510, citing Ashley Cunsolo Willox et al, "'From This Place and of This Place:' Climate Change, Sense of Place, and Health in Nunatsiavut, Canada" 75:3 Soc Science & Medicine.

77   Natural Resources Canada, *Canada in a Changing Climate: Regional Perspectives Report*, Chapter 5: British Columbia (Ottawa: Natural Resources Canada, 2022), online: changingclimate.ca [perma.cc/KY9F-R7GR]. See also Nancy J Turner & Helen Clifton, "'It's So Different Today': Climate Change and Indigenous Lifeways in British Columbia, Canada" (2009) 19:2 Global Envtl Change 180.

78   Mikisew Cree First Nation, "Written Brief of the Mikisew Cree First Nation to the Standing Committee on Environment and Sustainable Development" (15 November 2016), online: ourcommons.ca [perma.cc/32KK-JAQU].

79   *Ibid*; Stéphane M McLachlan & Clayton H Riddel, *"Water Is a Living Thing": Environmental and Human Health Implications of the Athabasca Oil Sands for the Mikisew Cree First Nation and Athabasca Chipewyan First Nation In Northern Alberta. Phase Two Report* (Winnipeg: Environmental Conservation Laboratory, 2014), online: landuse.alberta.ca [perma.cc/R5MM-72KV].

80   Oscar Baker III, "Mi'kmaw Communities Still Cleaning Up from Fiona Prepare for the Next Hurricane Season" *CBC News* (31 May 2023), online: cbc.ca [perma.cc/WZH2-63KX].

81   Logan MacLean, "First Nations in P.E.I. Recovering from Structural and Ecological Damage, Week Long Power Failures," *PNI Atlantic News* (30 September 2022), online: saltwire.com [perma.cc/W8FB-AZ75].

traditional food sources like clam beds and mussels, directly affecting cultural and subsistence practices.[82]

In the Pacific coastal region, Indigenous communities such as the Haida and Gitga'at Nations confront distinct challenges from climate change, including forest fires, drought, and rising ocean levels.[83] There has been a massive decline in Pacific salmon populations, upon which many western Nations rely. Traditional resources like the western red cedar, crucial for cultural artifacts and housing, are in decline due to environmental stressors.[84] These changes disrupt long-standing practices, such as canoe-making and salmon fishing, that sustain cultural identity and community well-being.

The Arctic is one of the most vulnerable regions to climate change, with warming occurring nearly three times faster than the global average.[85] The Inuit face some of the most acute effects of climate change. Thinning sea ice disrupts hunting practices for species like seals and polar bears, while permafrost thaw damages infrastructure, displacing communities and exacerbating economic instability.[86] Meanwhile, boreal forest communities, including Métis and First Nations in Alberta and Saskatchewan, face increasing wildfires, pest outbreaks, and shifts in wildlife migration patterns. These changes undermine traditional livelihoods and threaten culture and subsistence.[87]

Beyond physical and cultural impacts, climate change amplifies existing inequities in housing, health, and infrastructure and introduces new stressors

---

82   Oscar Baker III, "Indigenous Communities Along Atlantic Coast at Risk from Rising Sea Levels" *CBC News* (8 January 2022), online: cbc.ca [perma.cc/7L5B-34H3]; Natural Resources Canada, above note 77 at ch 1, 41.

83   Natural Resources Canada, above note 77 at ch 5, 11–23.

84   Richard J Hebda & Rolf W Mathewes, "Holocene History of Cedar and Native Indian Cultures of the North American Pacific Coast" (1984) 225:4663 Science 711.

85   Inuit Tapiriit Kanatami, *National Inuit Climate Change Strategy* (Ottawa: Inuit Tapiriit Kanatami, 2019) at 11–13; *References re Greenhouse Gas Pollution Pricing Act*, 2021 SCC 11 at para 11 [*GGPPA References*].

86   Inuit Tapiriit Kanatami, *Unikkaaqatigiit – Putting the Human Face on Climate Change: Perspectives from Inuit in Canada* (Ottawa: Inuit Tapiriit Kanatami, 2005), online: itk.ca [perma.cc/5JS4-BWS5].

87   See, e.g., National Collaborating Centre for Indigenous Health, *Climate Change and Indigenous Peoples' Health in Canada* (Prince George: NCCIH, 2022), online: nccih.ca [perma.cc/D876-9QQ2].

for Indigenous Peoples. Communities like the Kashechewan First Nation in Ontario endure recurring floods, forcing evacuations and destabilizing social structures.[88] Climate-induced displacement, coupled with the mental health impacts of witnessing environmental degradation, often intersects with historical trauma from colonial policies, creating compounded vulnerabilities.[89]

Despite these disruptions and challenges, Indigenous Peoples have demonstrated resilience through their traditional knowledge systems and adaptive practices. Indigenous knowledge systems remain critical for navigating changing environmental conditions and for challenging settler approaches to climate law and policy.[90]

2.4   INDIGENOUS LEADERSHIP AND CLIMATE RESILIENCE

> "Indigenous Peoples are the original stewards. To that end, we must still be the leaders on climate policy, not a government driven by money."[91]
>
> *Sandra Boucher, Seine River First Nation*

Indigenous Peoples have adapted to changing environments for millennia without the influence of European settlers.[92] While climate change is changing the environment at a far faster rate than in the past, deep Indigenous knowledge and connection to the land are essential for mitigation and for adapting to the changes. Not only has a lot of Indigenous knowledge been erased by settlers, but Indigenous knowledge systems and

---

88 Muhammad-Arshad K Khalafzai et al, "Flooding in the James Bay Region of Northern Ontario, Canada: Learning from Traditional Knowledge of Kashechewan First Nation" (2019) 36 Intl J Disaster Risk Reduction 1.

89 Jacqueline Middleton et al, "Indigenous Mental Health in a Changing Climate: A Systematic Scoping Review of the Global Literature" (2020) 15:5 Envtl Research Letters 1.

90 Deborah McGregor, "Anishnaabe-kwe, Traditional Knowledge and Water Protection" (2008) 26:3 Can Woman Studies 26; Turner & Clifton, above note 77.

91 Indigenous Climate Action, *Phase two*, above note 2 at 23.

92 Turner & Clifton, above note 77; Shawn Singh & James Gacek, "Erasure and Erosion: Exploring Federal Government Efforts to Complicate Socio-Legal and Environmental Obligations Owed to Indigenous People" (2022) 45:4 Man LJ 121.

experiences continue to be highly underrepresented in discussions about climate change in Canada.[93]

Indigenous perspectives emphasize the interconnectedness of humans and the environment, a worldview that underpins many traditional knowledge systems and Indigenous laws. Indigenous-led organizations and leaders, such as the Assembly of First Nations and Inuit Tapiriit Kanatami (ITK), have advocated for the inclusion of indigenous ecological science and knowledge in climate adaptation and mitigation strategies.[94] This knowledge is critical for understanding local environmental changes and developing sustainable solutions. For instance, the Gwich'in in the Northwest Territories are actively monitoring caribou populations and advocating for climate policies that protect these key species.[95] Similarly, Indigenous Protected and Conserved Areas (IPCAs), such as Thaidene Nëné in the Northwest Territories, exemplify how Indigenous governance can contribute to climate resilience.[96]

By identifying and addressing barriers to Indigenous-led climate policy, the work of Indigenous Climate Action (ICA) is helping to identify ways in which Indigenous communities are already taking action to address the impacts of climate change.[97] For example, the ICA shared seven pathways for restoring balance that include truth and transparency, centering Indigenous knowledge systems, honouring consent, abolishing the carceral system, decentralizing and lateralizing governance systems, respecting Indigenous sovereignty, and considering perspectives from the Global South.[98]

---

93   James D Ford et al, "Including Indigenous Knowledge and Experience in IPCC Assessment Reports" (2016) 6:4 Nature Climate Change 349.

94   The Manitoba Chiefs' Factum in the carbon pricing reference is a good example of this (see *References re Greenhouse Gas Pollution Pricing Act*, 2021 SCC 11 (Factum, Assembly of Manitoba Chiefs)).

95   Gwich'in Council International, "The Gwich'in," online: gwichincouncil.com [perma. cc/4F4F-RYPQ].

96   See Indigenous Circle of Experts, above note 12; Conservation through Reconciliation Partnership, *Indigenous Protected and Conserved Areas*, online: conservation through reconciliation partnership [perma.cc/Z8L5-P7KA].

97   Indigenous Climate Action, *Phase two*, above note 2 at 13.

98   *Ibid* at 13-14.

2.5 INDIGENOUS CONSTITUTIONS AND SELF-DETERMINATION: INTERSECTIONS WITH THE CANADIAN CONSTITUTION

Indigenous Peoples had and continue to have their own constitutions, including laws, governance systems, and principles developed and maintained by Indigenous Peoples based on their traditions, worldviews, and inherent rights as self-determining nations. These constitutions predate European colonization and form the foundation for Indigenous governance structures, encompassing decision-making processes, justice systems, and social organization. The concept of Indigenous self-determination is rooted in the inherent right of Indigenous Peoples to govern themselves and their lands in accordance with their own laws and customs.

The 1996 *Royal Commission on Aboriginal Peoples* recognized that Indigenous Nations have the right to govern themselves as part of the inherent sovereignty they possess as distinct Peoples. By recognizing and affirming the existence of these Aboriginal rights, section 35 of the *Constitution Act, 1982* established a settler constitutional foundation that has the potential to acknowledge and recognize Indigenous self-determination within Canada (though it has failed to deliver fully on its promise). Today, *UNDRIP* recognizes that "Indigenous peoples have the right to self-determination" and that this entails the "right to autonomy or self-government in matters relating to their internal and local affairs."[99] Section 2.6 below addresses section 35 and *UNDRIP* in more detail.

Despite these legal foundations, respect for Indigenous self-determination often faces practical and systemic challenges. Efforts to address these include negotiated self-government agreements, modern treaties, and legislative frameworks. For instance, the *Nisga'a Final Agreement* (2000) established a constitutionally protected self-government regime for the Nisga'a Nation, recognizing its authority over areas such as education, health, and land management.[100] The *Framework Agreement on*

---

99   *UNDRIP*, above note 14, arts 3, 4.
100 Nisga'a Lisims Government, *Nisga'a Treaty*, online: Nisg'a Treaty [perma.cc/NE75-8CM9].

First Nation Land Management is a government-to-government agreement signed between an original group of thirteen First Nations and the federal Crown that provides pathways for First Nations to exercise greater control over governance and land-use decisions.[101] The framework today has many more signatories.[102] One of the best and most recent examples of nation-to-nation understandings is the agreement between the Haida Nation and the province of British Columbia, and with the federal government, that recognizes Aboriginal title over the lands that form Haida Gwaii.[103] Provincial, territorial, and federal laws incorporating *UNDRIP* are important steps, since the Declaration reinforces the legal obligation to align Canadian laws with Indigenous rights, including the right to self-determination.[104]

The coexistence of Indigenous Constitutions and the Canadian Constitution reflects a pluralistic approach to governance, requiring meaningful dialogue and collaboration. This process entails respecting Indigenous legal traditions as equal partners in shaping Canada's constitutional framework. Initiatives such as Indigenous-led legal education and the establishment of Indigenous-led governance structures offer examples of progress toward reconciliation.[105] Ultimately, fully realizing Indigenous self-government within Canada necessitates ongoing legal, political, and social efforts to harmonize Indigenous and Canadian constitutional and legal orders. By

---

101 Lands Advisory Board and First Nations Land Management Resource Centre, *Framework Agreement*, online: Lands Advisory Board [perma.cc/P4TU-RVYE].

102 *Ibid.*

103 Haida Nation, "Haida Nation Approves the Gaayhllxid/Giihlagalgang 'Rising Tide' Haida Title Lands Agreement (April 6, 2024), online: haidanation.ca [perma.cc/JWB2-BT4K]; APTN News, "Haida celebrate right to control their own destiny" (February 18, 2025), online: aptnnews.ca [perma.cc/3GD5-YY7S].

104 See, e.g., *United Nations Declaration on the Rights of Indigenous Peoples Act*, SC 2021, c 14 [*UNDRIP* Act]; British Columbia, *Declaration on the Rights of Indigenous Peoples Act*, SBC 2019, c 44; The Northwest Territories, *UNDRIP Implementation Act*, SNWT 2023, c 36.

105 The National Centre for Indigenous Laws at the University of Victoria, for instance, offers a joint degree program in Indigenous Legal Orders (JID) and Canadian Common Law (JD). See University of Victoria, "A New Home for Indigenous Legal Resurgence," online: uvic.ca [perma.cc/F26A-2VDU]; See also University of Victoria, "World's First Indigenous Law Program Launches with Historic and Emotional Ceremony" (Oct 22, 2018), online: uvic.ca [perma.cc/MSS3-RQJQ].

addressing the divergences between these systems, Canada can move closer to a relationship based on mutual recognition, respect, and partnership.

2.6 TOWARD REPAIR, BUILDING TRUST, AND RECONCILIATION

> "We've had a very abusive relationship with policy. Since we've had this lived experience of the abuses we face through the Indian Act and different ways of colonization, that means we're actual hardcore policy experts that deserve to be in those spaces to actually make them better and less colonial. There are already experts everywhere within the territories that know what needs to be done."
>
> *Katelynn Herchak, Inuk*[106]

2.6.1 The Truth and Reconciliation Commission (TRC)[107]

The Truth and Reconciliation Commission (TRC) was officially established in 2008 to document the historical and lasting impacts of the Canadian residential school system on Indigenous students, families, and future generations. In 2015, the TRC released 94 Calls to Action which are actionable items that Canadians can take to reconcile with Indigenous Peoples. While the TRC did not make recommendations specific to climate change, the recommendations stress the importance of Indigenous sovereignty, land stewardship, and knowledge systems, all of which are integral to addressing climate change.[108] Indigenous approaches to environmental management often emphasize sustainability, relationality, and resilience, making them invaluable in addressing the climate emergency.

Call to Action 45 calls for the adoption and implementation of *UNDRIP*, as well as to renew and establish treaty relationships, and reconcile past hurts in the *Royal Proclamation*. Call to Action 47 urges

---

106 Indigenous Climate Action, *Phase two*, above note 2.
107 Truth and Reconciliation Commission of Canada, *Calls to Action*, above note 50.
108 See McGregor, "Reconciliation" above note 5.

governments to repudiate doctrines that disregard Indigenous sovereignty, such as the "doctrine of discovery," replacing them with policies rooted in mutual respect. This includes supporting Indigenous stewardship of land and water, which is crucial for climate mitigation and adaptation. Calls to Action 6 to 12 relate to education and the importance of integrating Indigenous knowledge and perspectives, which is relevant to climate policy.

Some have criticized the TRC for failing to emphasize the ongoing and perpetual colonization of Indigenous peoples.[109] Critics point to the slow progress on the Calls to Action, noting to challenges such as paternalism, structural discrimination, insufficient resources, and reconciliation.[110] There have also been concerns about a lack of transparency when it comes to reporting the TRC outcomes.[111]

2.6.2   Section 35 of the *Constitution Act, 1982*

In 1982, the *Constitution Act* was amended to include section 35, which states in subsection 1 that "[t]he existing aboriginal and treaty rights of the Aboriginal Peoples of Canada are hereby recognized and affirmed."[112] By

---

109   See, e.g., Jeff Corntassel et al, "Indigenous Storytelling, Truth-Telling, and Community Approaches to Reconciliation" (2009) 35:1 *English Studies in Canada* 137; Eva Jewell & Ian Mosby, *Calls to Action Accountability: A 2023 Status Update on Reconciliation* (2023, Yellowhead Institute), online: yellowheadinstitute.org [perma.cc/F2GK-22C4].

110   *Ibid* at 11.

111   See, e.g., Eva Jewell, "A Decade of Disappointment: Reconciliation and the System of a Crown" (2024), YellowHead Institute, online: yellowheadinstitute.org [perma. cc/5TQR-8XED]; Ma Junxiang, TRC Calls to Action - Critical Review, Thompson Rivers University, online: junxlearning.eddl.tru.ca [perma.cc/X2DD-CTAS].

112   The journey leading to the addition of section 35 was long and tumultuous. The federal government had attempted in 1969 to abolish the *Indian Act* to ostensibly achieve equality for all Canadians. However, Indigenous leaders protested this as another form of forced assimilation. A revised proposal in 1980 included a general Aboriginal rights clause that was the precursor to section 35. Harry Daniels, a Metis leader from Saskatchewan, is credited for helping ensue that Metis would be recognized as one of the three Aboriginal Peoples in the amended Constitution. See Canada, Indian and Northern Affairs, *Statement of the Government of Canada on Indian Policy* (Ottawa: Department of Indian and Northern Affairs, 1969); Singh and Gacek, above note 92; Jean Teillet, "The Métis and Thirty Years of Section 35: How Constitutional Protection for Métis Rights has led to the Loss of the Rule of Law" (2012) SCLR

situating section 35 outside of the *Canadian Charter of Rights and Freedoms* (also added to the Constitution in 1982), the provision is shielded from override by the notwithstanding clause, reflecting its importance in the constitutional order.[113]

Section 35 establishes a constitutional framework for the recognition and protection of Aboriginal rights, which include ancestral and treaty rights, land title, cultural practices, and self-determination. However, the Supreme Court of Canada has interpreted the provision to allow Crown infringements under specific conditions, provided such actions serve a compelling and substantial public objective, such as resource conservation, and align with the Honour of the Crown.[114] This interpretation reveals a persistent tension: while the provision acknowledges Aboriginal rights, it also situates them within a settler legal framework which often subordinates these rights to state interests.

As interpreted to date, the provision falls short of recognizing Indigenous self-determination and related rights. The Supreme Court of Canada interpreted "existing" rights as excluding those extinguished prior to 1982.[115] This limitation reinforces the colonial legal hierarchy, privileging settler constitutional authority over Indigenous legal traditions and sovereignty, rather than advancing a nation-to-nation vision.[116] Critics, including Bruce McIvor, argue that this approach reduces Indigenous rights to categories defined and validated by non-Indigenous judges, thereby often fragmenting Indigenous cultures and marginalizing their dynamic legal orders.[117] Despite these limitations, Professor John Bor-

---

58; Daum Shanks, "The Wastelander Life: Living Before and After the Release of *Daniels v Canada*" (2017) 54:4 OHLJ 1341.

113   The notwithstanding clause is discussed in Chapter 4.

114   *R v Sparrow*, [1990] 1 SCR 1075 (ancestral rights); *R v Badger*, [1996] 1 SCR 771; and *Mikisew Cree First Nation v Canada (Minister of Canadian Heritage)*, 2005 SCC 69 (relating to treaty rights). To respect the Honour of the Crown, the infringement must be minimal, not unnecessarily restrictive, and accompanied by appropriate consultation and accommodation. *R v Sparrow*, above note 114.

115   *Ibid*.

116   See, e.g., Lee Maracle, "The Operation Was Successful, But the Patient Died" online: ojs.library.ubc.ca [perma.cc/UU5K-P6LR].

117   Bruce McIvor, "Aboriginal Rights as a Tool of Colonialism" September 29, 2023, online: firstpeopleslaw.com [perma.cc/8X6S-6Q5Z].

rows notes that section 35 provided Indigenous communities with an institutional mechanism to resist rights violations, highlighting the duality of its potential as both a tool and a constraint.[118]

Section 35 has been interpreted to confirm the obligation on the Crown to act honourably in recognizing and respecting the rights of Indigenous Peoples.[119] This obligation entails a duty to consult and accommodate, which obliges the federal government to meaningfully engage with Indigenous communities when proposed government actions might affect their rights.[120] The Court has provided guidance about whether and how certain aspects of the obligation may be delegated to administrative tribunals or statutory bodies or in the context of a regulatory process, such as an impact assessment.[121] If there is a strong potential claim and threat of severe potential infringement, the Crown will be required to accommodate.[122] While the Supreme Court of Canada has held that this duty does not amount to an Indigenous veto, it has recognized that consent is required for certain uses of land subject to Aboriginal title.[123] In *Tsilhqot'in Nation*, the Court held that governments or individuals wishing to use land upon which Aboriginal title is claimed can avoid a charge of rights infringement or failure to consult if they obtain the consent of the Nation first.[124]

---

118 John Borrows, "Measuring a Work in Progress: Canada, Constitutionalism, Citizenship and Aboriginal Peoples" in Ardith Walkem & Halie Bruce, eds, *Box of Treasures or Empty Box? Twenty Years of Section 35* (Vancouver: Theytus, 2003) at 225.

119 See *Haida Nation v BC*, above note 25 at para 32. The obligation to act honourably is not new under section 35. It was recognized judicially in *Province of Ontario v The Dominion of Canada and Province of Quebec* and *In re Indian Claims*, (1895) 25 SCR 434. The duty to consult also applies to Crown corporations. See *Rio Tinto Alcan v Carrier Sekani Tribal Council*, [2010] 2 SCR 650 at para 81 [*Rio Tinto*].

120 See, e.g., *Delgamuukw v British Columbia*, [1997] 3 SCR 1010 at para 168; *Haida Nation v BC*, above note 25 at paras 43–45; *Rio Tinto*, above note 119; *Tsleil-Waututh Nation v Canada (AG)*, 2018 FCA 153. In *Mikisew Cree First Nation v Canada (Governor General in Council)*, [2018] 2 SCR 765).

121 See, e.g., *Chippewas of the Thames First Nation v Enbridge Pipelines Inc*, 2017 SCC 41 [*Chippewas*] at para 5; *Clyde River (Hamlet) v Petroleum Geo-Services Inc*, 2017 SCC 40 at paras 1 and 22.

122 *Haida Nation v BC*, above note 25 at para 47.

123 *Ibid* at para 48; *Chippewas*, above note 121 at para 59.

124 *Tsilhqot'in Nation*, above note 27 at para 97.

Although section 35 has achieved significant milestones, such as recognizing title in *Tsilhqot'in Nation* and recognizing Indigenous legal traditions as integral to Canadian law, it has fallen short of fulfilling its transformative potential. The courts' reluctance to fully embrace Indigenous self-determination perpetuates a colonial framework. Moreover, limits on economic rights, such as restricting commercial fishing to "moderate livelihoods,"[125] and the imposition of limits on the ways in which lands under Aboriginal title can be used,[126] continue to marginalize Indigenous economic autonomy. These constraints have raised critical questions about equity, particularly regarding sustainable land use. As Bruce McIvor aptly critiques, instead of recognizing "Indigenous Peoples as nations with legal rights rooted in their own legal orders and lifeworlds," settler courts have interpreted section 35 as something recognizing "Aboriginal rights," the content of which is to be determined by Canadian judges who are almost exclusively non-Indigenous.[127]

Despite these challenges, Indigenous leaders and legal scholars emphasize the ongoing importance of section 35 as a platform for asserting rights, reclaiming self-governance, and challenging systemic inequities. Its potential lies not only in the recognition of rights but also in fostering a dialogue that advances reconciliation and respects Indigenous legal traditions and sovereignty.

2.6.3 United Nations Declaration on the Rights of Indigenous Peoples (UNDRIP)

UNDRIP, adopted by the UN General Assembly in 2007, is a universal framework of minimum standards for the survival, dignity, and well-being of Indigenous peoples globally. The Declaration includes 46 articles addressing self-determination, cultural rights, and access to health, education, and other fundamental rights, while also supporting the maintenance of Indigenous political, legal, and social systems.[128]

---

125 See *R v Marshall*, [1999] 3 SCR 456; McIvor, above note 117.
126 *Tsilhqot'in Nation*, above note 27.
127 McIvor, above note 117.
128 *UNDRIP*, above note 14.

Canada originally rejected UNDRIP but reversed its position and endorsed UNDRIP without qualification in 2016. In 2021, the federal government passed the *United Nations Declaration on the Rights of Indigenous Peoples Act (UNDRIP Act)*.[129] This legislation requires the alignment of federal laws with UNDRIP principles, the creation of an action plan, and annual progress reports.[130] Several provinces and territories have also begun implementing UNDRIP through legislation.[131]

Indigenous leaders and organizations, including the Assembly of First Nations (AFN), have welcomed these laws as a step toward affirming rights such as self-determination, and free, prior, and informed consent (FPIC). However, concerns remain about the pace and depth of implementation. Ongoing conflicts, such as disputes over resource development and land rights (e.g., Wet'suwet'en opposition to pipeline projects on their territory) highlight challenges in applying FPIC and achieving substantive Crown-Indigenous reconciliation. Scholars and advocates emphasize the importance of co-developing processes to ensure Indigenous leadership in interpreting and applying UNDRIP, warning that unilateral approaches risk perpetuating colonial dynamics.[132]

In June 2023, Canada released the 2023–2028 Action Plan for implementing UNDRIP.[133] The plan was developed in consultation with First Nations, Inuit, and Métis peoples and outlines priorities for aligning federal laws with UNDRIP, including reviewing and proposing amendments to ensure the respect of Indigenous rights.[134] Key priorities include respecting Aboriginal titles and rights, as well as supporting Indigenous participation in resource development. The government commits in the

---

129 SC 2021, c 14 [*UNDRIP Act*].

130 *Ibid.*

131 See, e.g., British Columbia's *Declaration on the Rights of Indigenous Peoples Act*, SBC 2019, c 44 and the NWT's *UNDRIP Implementation Act*, SNWT 2023, c 36.

132 Brenda L Gunn et al, *UNDRIP Implementation: Braiding International, Domestic and Indigenous Laws* (Waterloo: Centre for International Governance Innovation, 2017); CIGI, *UNDRIP Implementation: More Reflections on the Braiding of International, Domestic and Indigenous Laws* (Waterloo: Centre for International Governance Innovation, 2018).

133 Government of Canada, *The United Nations Declaration on the Rights of Indigenous Peoples Act: Action Plan*, online (PDF): justice.gc.ca [perma.cc/E3QT-SBJQ].

134 *Ibid* at 23.

Plan to consult with Indigenous Peoples on projects affecting their lands and territories.[135] Environmental commitments focus on incorporating Indigenous knowledge into climate action, conservation, and policy development, as outlined in Article 29 of UNDRIP. The Action Plan also seeks to remove barriers to Indigenous leadership in climate change and to integrate Indigenous science into environmental policymaking. It emphasizes Indigenous involvement in the *Federal Sustainable Development Strategy*[136] as a core principle moving forward.[137]

While the *UNDRIP Act* and Action Plan demonstrate potential, their impact will depend on meaningful collaboration, comprehensive legal reform, and transformative policy changes.

In 2024, the Supreme Court of Canada issued its first decision interpreting *UNDRIP* since the Declaration was signed in *Reference re An Act respecting First Nations, Inuit and Métis children, youth and families*.[138] The Court had not mentioned *UNDRIP* in the Indigenous rights cases it heard between 2007 and 2023, though the Declaration would have been relevant and might have helped frame the issues before the Court.[139] The decision to engage with the Declaration in this recent case is likely due to the fact that the federal government had, by the time the *Child Welfare Reference* was heard, enacted the implementation Act. The heightened attention given to the Declaration was also likely influenced by several factors including the submissions made by multiple intervening parties regarding its relevance and the historic presence of the first Indigenous judge on the Supreme Court of Canada.

---

135   *Ibid* at 31.

136   Government of Canada, *Federal Sustainable Development Strategy*, online: canada.ca [perma.cc/BB3R-UC7K].

137   *Ibid* at 34.

138   2024 SCC 5.

139   A concurring opinion by Binnie and Major JJ in *Mitchell v MNR*, 2001 SCC 33, mentioned the Declaration while it was in draft form but did not offer much helpful guidance. See Nigel Bankes and Robert Hamilton, "What Did the Court Mean When It Said that UNDRIP 'Has Been Incorporated into the Country's Positive Law'? Appellate Guidance or Rhetorical Flourish?" (28 February 2024), online: ablawg.ca [perma.cc/7QUV-NMH8]. Note that British Columbia's law incorporating UNDRIP has been addressed by the BC Supreme Court in *Gitxaala v British Columbia (Chief Gold Commissioner)*, 2023 BCSC 1680 [*Gixaala*].

The Supreme Court of Canada's interpretation of the legal significance of the Declaration stands in contrast to the BC Supreme Court's interpretation of provincial legislation implementing the Declaration in *Gixaala*. There, the court interpreted the relevant provision of the provincial Act as serving an interpretative function but not creating substantive rights.[140] In contrast, the Supreme Court of Canada in the *Child Welfare Reference* found that the Declaration "has been incorporated into the country's positive law" through the *UNDRIP Act*.[141] This suggests a greater recognition of legal pluralism and a step in the direction of reconciliation.

The federal law at issue in the *Child Welfare Reference* included provisions that referentially incorporate Indigenous rights and give effect to Indigenous self-government in relation to child and family services. By interpreting the legislation as furthering the goals of the *UNDRIP Act*, the Supreme Court of Canada offered a model for how federal laws incorporating UNDRIP can facilitate the assumption of jurisdiction by Indigenous nations in other areas.[142]

2.7 JUDICIAL ENGAGEMENT WITH CLIMATE CHANGE AND INDIGENOUS RIGHTS

Settler courts in Canada have yet to fully engage with the complex intersections between climate change and Indigenous rights, though there has been incremental progress in recognizing the disproportionate impacts of climate change on Indigenous Peoples. A significant step was taken in the *References re Greenhouse Gas Pollution Pricing Act*, where the Supreme Court of Canada acknowledged the severe and unequal burdens climate change imposes on Indigenous communities.[143] Chief Justice Wagner stated that "climate change has had a particularly serious effect on Indigenous peoples,

---

140 *Gixaala*, above note 139 at paras 461–62.

141 *Indigenous Child Welfare Reference*, above note 32 at para 4.

142 See Robert Hamilton, "Legislative Reconciliation and Indigenous Rights of Self-Government: Reference re An Act respecting First Nations, Inuit and Métis Children, Youth and Families" (20 February 2024), online: ablawg.ca [perma.cc/JG8Q-8AQE]. See also Bruce McIvor, First Peoples Law, "The Troubling Basis for the Supreme Court's Child Welfare Law Decision" (February 12, 2024), online: firstpeopleslaw.com [perma.cc/EP9L-PEBJ].

143 *GGPPA References*, above note 85 at paras 11 and 187.

threatening the ability of Indigenous communities in Canada to sustain themselves and maintain their traditional ways of life."[144] The Court recognized that climate change has led to "significant environmental, economic, and human harm nationally and internationally, with especially high impacts in the Canadian Arctic, in coastal regions, and on Indigenous peoples."[145] These statements underscore the Court's recognition of the interconnectedness of Indigenous laws, livelihoods, traditional practices, and environmental sustainability and the existential threat posed by climate change to these systems and Indigenous Peoples themselves.[146]

Other Canadian courts have also acknowledged the disproportionate vulnerabilities of Indigenous Peoples to climate change. For example, in 2017 the Federal Court recognized that the interests and concerns of Indigenous Peoples are "often uniquely at risk in ways that the interests of non-[i]ndigenous people are not."[147] In 2023, the Federal Court of Appeal acknowledged the dramatic and rapidly unfolding effect that climate change is having on northern and Indigenous communities in particular.[148] In 2024, the Ontario Court of Appeal in *Mathur v Ontario* noted that climate change has disproportionate impacts on Indigenous Peoples and that interveners had raised important issues, such as the *Charter* rights of Indigenous Peoples and their section 35 rights, that were not addressed by the Superior Court.[149] By remitting the case back for rehearing, the Court of Appeal left open the possibility of these issues being addressed.[150]

The road to reconciliation in Canada remains fraught with numerous obstacles, including legal and policy decisions that undermine efforts to uphold and protect Indigenous rights. One such decision concerns the *Impact Assessment Act* (IAA),[151] enacted in 2019 to replace the *Can-*

---

144 *Ibid* at para 11.

145 *Ibid* at para 187.

146 *Ibid* at para 206.

147 *Taseko Mines Ltd*, 2017 FC 1100 at para 153.

148 *La Rose v Canada*, 2023 FCA 241 at para 76.

149 *Mathur v Ontario*, 2024 ONCA 762 at para 6.

150 *Ibid*. The Supreme Court of Canada denied the request by Ontario for leave to appeal the ONCE decision, which means the decision will be reheard at the Superior Court likely in the fall of 2025.

151 SC 2019, c 28, s 1 [*IAA*].

adian Environmental Assessment Act (2012).[152] The IAA sought to enhance environmental assessments, embedding provisions that recognized Indigenous rights. These included early engagement with Indigenous communities, integration of Indigenous knowledge, explicit consideration of impacts on Indigenous rights during decision-making, and support for Indigenous-led assessments.[153] These provisions aligned with principles enshrined in the *United Nations Declaration on the Rights of Indigenous Peoples* (UNDRIP), such as free, prior, and informed consent (FPIC).[154] Furthermore, the IAA aimed to evaluate the extent to which proposed projects could undermine Canada's climate change commitments — an essential factor for Indigenous Peoples, who bear disproportionate burdens of climate-related harms.

However, in 2023, the Supreme Court of Canada struck down significant portions of the IAA, including decision-making provisions related to climate change and the protection of Indigenous rights.[155] The *Reference re Impact Assessment Act* drew over thirty intervenors, including Mikisew Cree First Nation, Athabasca Chipewyan First Nation, Woodland Cree First Nation, and the Lummi Nation.[156] Indigenous intervenors underscored the federal government's constitutional responsibility to regulate activities affecting their rights, grounded in section 91(24) of the *Constitution Act, 1867* and section 35 of the *Constitution Act, 1982*.[157] For instance,

---

152 SC 2012, c 19 [Repealed 2019, c 28, s 9].

153 *IAA*, ss 60-63 and 84.

154 Despite improvements to the assessment process in the *IAA*, many Indigenous leaders remain concerned about whether the IAA adequately ensures meaningful participation or grants Indigenous communities' sufficient authority to influence decisions about resource development on their lands. For instance, some argue that while the Act requires the consideration of Indigenous knowledge, it does not mandate decision-makers to act on it. See, e.g., Aaron Bruce et al, "Indigenous Peoples' Authority, Rights and Engagement in Impact Assessment: Experiences and Perspectives from Canada and Aotearoa New Zealand," in Tanya Burdett & A John Sinclair, *Handbook of Public Participation in Impact Assessment* (Edward Elgar, 2024) 286.

155 *Reference re Impact Assessment Act*, 2023 SCC 23 [*IAA Reference*]. The Court held that sections 81-91 of the *IAA* dealing with projects on federal lands are *intra vires* Parliament.

156 *Ibid.*

157 See, e.g., Factum of the Intervener, Woodland Cree First Nation, SCC File Number: 40195, online (PDF): scc-csc.ca [perma.cc/86X5-NKDH].

the Athabasca Chipewyan First Nation emphasized their reliance on federal oversight of critical environmental impacts, including navigation, shipping, and species protection, to safeguard their Treaty 8 rights to hunt, fish, and trap.[158]

In a spilt 5:2 decision, the Supreme Court of Canada interpreted the IAA's provisions as overreaching Parliament's jurisdiction.[159] The majority deemed that the federal government's authority to block projects based on potential "changes to Indigenous Peoples' health, social, or economic conditions" without a clear threshold exceeded its constitutional remit.[160] The dissenting judges, however, argued for a purposive and contextual reading of the Act, consistent with precedents upholding environmental legislation that uses broad language.[161] Following the decision, Parliament introduced amendments narrowing the Act's language to focus on significant and adverse impacts.[162]

Another avenue for advancing Indigenous rights and environmental accountability has emerged through judicial recognition of cumulative effects. Cumulative effects — resulting from layered environmental harms over time — pose significant challenges for Indigenous communities, particularly in resource-rich territories. These impacts are amplified by climate change, which exacerbates ecological degradation on lands and waters already fragmented by industrial development.[163]

The groundbreaking *Blueberry River First Nation (Yahey)* case marked a watershed moment in addressing cumulative effects on treaty rights.[164] In this case, the British Columbia Supreme Court determined

---

158   Factum of the Intervener, Athabasca Chipewyan First Nation, In the Matter of *Reference re Impact Assessment Act*, at para 26, online (PDF): scc-csc.ca [perma.cc /M4TN-YKTW].

159   *IAA Reference*, above note 155.

160   *Ibid* at para 196.

161   *Ibid* at paras 223, 230, and 234.

162   Bill C-69, *An Act to implement certain provisions of the budget tabled in Parliament on April 16, 2024*, c 17, Division 28, Part 4 (sections 269 to 319).

163   Climate change is a cumulative effects problem since it is caused by the accumulation of GHGs in the atmosphere over time.

164   *Yahey v British Columbia*, 2021 BCSC 1287 [*Yahey*]. For an Indigenous perspective on the interpretation of treaty rights and environmental impacts, see also Craft, *Breathing Life*, above note 52.

that the cumulative effects of industrial activities had severely undermined the Blueberry River First Nation's ability to exercise Treaty 8 rights to hunt, fish, and trap.[165] Justice Burke's lengthy decision concluded that the province's failure to manage these impacts constituted an infringement of the Nation's rights.[166] By choosing not to appeal the decision, the provincial government signaled a willingness to recalibrate its approach to resource management.[167] This precedent is expected to influence the regulation of industrial development in British Columbia and beyond.

A similar lawsuit, brought by the Beaver Lake Cree Nation in Alberta under Treaty 6 underscores similar concerns regarding cumulative effects.[168] The Nation alleges that resource extraction has irreparably harmed water quality, biodiversity, and forests essential to their traditional practices. The King's Bench of Alberta recognized the financial barriers to pursuing the claim awarding Beaver Lake Cree Nation advanced costs to continue litigation, reflecting the gravity of their claims and the systemic importance of addressing cumulative impacts.[169]

2.8 CONCLUSION

Indigenous Peoples in Canada and globally have been stewards of lands, waters, and ecosystems for millennia, embodying worldviews rooted in interconnectedness, respect, relationality, and sustainability. These

---

165   Treaty 7 (or 8) protects their right to hunt, trap, and fish in their Treaty area, but there were many questions about how these promises should be upheld in the course of the 120 years since.

166   See *Yahey*, above note 164. Martin Olszynski, "Counting Straws: Yahey v British Columbia and the Future of Cumulative Effects Management in Canada" (13 July 2023), online: ablawg.ca [perma.cc/LLB9-6Q9Y].

167   Government of British Columbia, *BC, Blueberry River First Nations Reach Agreement on Existing Permits, Restoration Funding*, online: news.gov.bc.ca [perma.cc/8DMC-VYX8].

168   *Beaver Lake Cree Nation v Alberta*, 2020 ABCA 192 (on advanced costs). See also Beaver Lake Cree Nation, "Defend the Treaties," online: beaverlakecreenation.ca [perma.cc/5YEC-LY5W].

169   See *Anderson v Alberta*, 2024 ABKB 524. The award applies to Alberta and requires the Nation to contribute $150,000 annually. The Nation has an agreement with Canada to pay $2.6 million in costs.

practices, integral to Indigenous self-determination, have been profoundly disrupted by colonialism through systemic land dispossession, resource exploitation, cultural suppression, and ongoing racism and oppression. Climate change compounds these historical and contemporary harms, threatening the ecological and cultural foundations that underpin Indigenous rights, including the rights to land, self-determination, and cultural continuity.

The climate crisis not only endangers the physical environment but also exacerbates conflicts over land and resource governance, disproportionately burdening Indigenous communities. It acts as a multiplier of systemic inequities, creating profound social, economic, and cultural challenges while serving as a source of grief and ecological anxiety. Without Indigenous leadership and meaningful participation, policies addressing climate change risk perpetuating these inequities or creating new forms of colonial oppression.

Settler governments must urgently address climate change in ways that upholds Indigenous rights and respects principles of equity and justice in a manner that is respectful of Indigenous laws and legal traditions.[170] This includes fully adopting the UNDRIP into domestic law and implementing its principles, such as free, prior, and informed consent (FPIC), as a foundation for climate governance. Recognizing Indigenous legal systems and integrating them into Canada's constitutional framework is essential for fostering genuine reconciliation and shared responsibility in the face of the climate crisis.

Indigenous Peoples are not passive victims of climate change but vital leaders with knowledge systems that offer transformative approaches to sustainability. Their governance practices and relational philosophies provide pathways to recalibrate humanity's relationship with the natural world toward harmony, respect, hope, healing and accountability. Addressing the climate emergency requires embracing Indigenous leadership and

---

170   Heather Castleden et al, "Hishuk Tsawak"(Everything is One/Connected): A Huu-ay-aht Worldview for Seeing Forestry in British Columbia, Canada" (2009) Society & Natural Resources 22(9), 789–804; John L Lewis & SRJ Sheppard, "Ancient Values, New Challenges: Indigenous Spiritual Perceptions of Landscapes and Forest Management" (2005) Society and Natural Resources, 18(10), 907–20.

knowledge as central to achieving climate justice. The inseparability of Indigenous rights and climate justice demands that policymakers and settlers prioritize and uphold the former as a prerequisite for addressing the latter.

Sandra Lamouche of the Bigstone Cree Nation offers a powerful vision for decolonizing climate policy in *The Power of Acimowin (Storytelling)*, with actionable recommendations that reflect the necessity of Indigenous leadership and holistic governance.[171] Her vision includes Indigenous leadership and co-creation with federal and provincial governments of climate policies that integrate Indigenous governance models and approaches. She recommends applying an anti-Indigenous racism process across all sectors of Canadian society but especially policy makers whose decisions impact Indigenous peoples and the lands and waters to which they are inextricably connected. She encourages the inclusion of storytellers, artists, spiritual Elders and cultural knowledge holders in climate policy development and efforts to support the intergenerational transfer of stories and knowledge, which should be respected and incorporated into climate policy without the need for validation through Western scientific frameworks.[172]

Indigenous knowledge, science, leadership and culture should be integral to environmental and climate deliberations. As Professor McGregor eloquently states, "[r]econciliation, if it is to achieve its stated goals, must not only be concerned with healing relationships among peoples, but also with the land itself, and must occur at a societal level to be truly transformative and secure a sustainable future."[173]

Additional Resources

Michael Asch, *On Being Here to Stay: Treaties and Aboriginal Rights in Canada* (Toronto: University of Toronto Press, 2014).

---

171   See Sandra Lamouche, "The power of Acimowin (Storytelling) for Climate Change Policy" (2023), online: climateinstitute.ca [perma.cc/42XF-YTW2].

172   *Ibid*. Note that this is not the complete set of recommendations provided by the author.

173   McGregor, "Reconciliation" above note 5 at 145.

John Borrows, *Canada's Indigenous Constitution* (University of Toronto Press, 2016).

John Borrows, *Freedom and Indigenous Constitutionalism* (Toronto: University of Toronto Press, 2016).

John Borrows et al, eds, *Braiding Legal Orders: Implementing the United Nations Declaration on the Rights of Indigenous Peoples* (McGill-Queen's University Press, 2023).

Shaugn Coggins et al, "Indigenous Peoples and Climate Justice in the Arctic" (2021) Georgetown J of Int'l Affairs, online: https://gjia.georgetown.edu/2021/02/23/indigenous-peoples-and-climate-justice-in-the-arctic/

Aimée Craft, *Breathing Life into the Stone Fort Treaty: An Anishnabe Understanding of Treaty One* (Vancouver: UBC Press, 2013).

Aimée Craft, "Living Treaties, Breathing Research" (2014) 26:1 Cdn J of Women & Law 1–22.

Aimée Craft, "Navigating Our Ongoing Sacred Legal Relationship with Nibi (Water)" in John Borrows et al, eds, *Braiding Legal Orders: Implementing the United Nations Declaration on the Rights of Indigenous Peoples* (Centre for International Governance Innovation, 2019) pp. 101–110.

Aimée Craft, "Neither Infringement nor Justification – the SCC's Mistaken Approach to Reconciliation" in B Gunn & K Drake, eds, *Renewing Relationships: Indigenous Peoples and Canada* (University of Saskatchewan Native Law Centre, 2019) Chapter 3, pp 59–82.

Aimée Craft & Jill Blakley, *In Our Backyard: Keeyask and the Legacy of Hydroelectric* (University of Manitoba Press, 2022).

Aimée Craft & Rachel Plotkin, *Governance Back: Exploring Indigenous Approaches to Reclaiming Relationships with Land* (David Suzuki Foundation, 2022).

Aimée Craft & Rachel Plotkin, *Shared Governance* (David Suzuki Foundation, 2023).

Aimée Craft et al, "Decolonizing Anishiinabe nibi inaakonigewin and gikendaasowin Research" in Sujith Xavier et al, eds, *Decolonizing Law: Indigenous, Third World and Settler Perspectives*(New York: Routledge, 2021) 17–33. https://doi.org/10.4324/9781003161387

E Deranger et al, "Decolonizing Climate Research and Policy: Making Space to Tell Our Own Stories, in Our Own Ways" in *Community Development Journal. Special Issue: Environmental Community Development in the Climate Emergency* (Oxford University Press and Community Development Journal, 2022). https://doi.org/10.1093/cdj/bsab050, online: www.academia.edu/78749805/Decolonizing_Climate_Research_and_Policy_making_space_to_tell_our_own_stories_in_our_own_ways

Emma Feltes & Sharon H Venne, "Decolonization, Not Patriation: The Constitution Express at the Russell Tribunal" (2022) 212 BC Studies 65–102.

James (Sákéj) Youngblood Henderson, Marjorie Benson & Isobel Findlay, *Aboriginal Tenure in the Constitution of Canada* (Saskatoon: Purich Publishing, 2000)

Carol Anne Hilton, *Indigenomics* (New Society Publishers, 2021).

Indigenous Climate Action, Decolonizing Climate Policy in Canada, Report from Phase One (March 2021), online: www.indigenousclimateaction.com /programs/decolonizing-climate-policy.

Bob Joseph, *21 Things You May Not Know about the Indian Act: Helping Canadians Make Reconciliation with Indigenous Peoples a Reality* (Indigenous Relations Press, 2018).

Robin Wall Kimmerer, *Braiding Sweetgrass: Indigenous Wisdom, Scientific Knowledge, and the Teachings of Plants* (Minneapolis, Milkweed Publishing, 2013) https://milkweed.org/book/braiding-sweetgrass.

Keira L Ladner & Michael McCrossan, "The Road Not Taken: Aboriginal Rights after the Re-Imagining of the Canadian Constitutional Order" in James B Kelly & Christopher P Manfredi, eds, *Contested Constitutionalism: Reflections on the Canadian Charter of Rights and Freedoms* (Vancouver: UBC Press, 2010) 263–283.

Darcy Lindberg, "Mediated Relations: The Indian Act and the Politics of Ignorance" in Ryan Beaton, Robert Hamilton & Josh Nichols, eds, *Wise-practices: Exploring Indigenous Economic Justice and Self -Determination* (University of Toronto Press.

Deborah McGregor, "Reconciliation, Colonization, and Climate Futures" in Carolyn Tuohy et al, eds, *Policy Transformation in Canada: Is the Past Prologue?* (University of Toronto Press: Toronto, 2019) 139.

Deborah McGregor, "An Indigenous Peoples' Approach to Climate Justice." CarbonBrief: Clear on Climate. Guest post. Carbon Brief Ltd. online: www .carbonbrief.org/guest-post-an-indigenous-peoples-approach-to-climate -justice. Released October 8th, 2021.

Deborah McGregor & Aimée Craft, Special Issue Co-Editor 2021. "Sustainable Water Governance through Indigenous Research Approaches." Water Journal. online: www.mdpi.com/journal/water/special_issues /Indigenous_Water_Governance

K Menzies et al, "I See My Culture Starting to Disappear": Anishinaabe Perspectives on the Socioecological Impacts of Climate Change and Future Research Needs" (2021) 7 Facets 509–527.

Sarah Morales & Joshua Nicols, *Reconcilation Beyond the Box: The UN Declaration and Plurinational Federalism in Canada* (CIGI, 2018).

Nicole Redvers et al, "Indigenous Peoples: Traditional Knowledges, Climate Change, and Health" (2023) 3:10 PLOS Global Public Health.

G Reed et al, "Toward Indigenous Visions of Nature-Based Solutions: An Exploration into Canadian Federal Climate Policy." Climate Policy, DOI:10.1080/1 4693062.2022.2047585, online: www.tandfonline.com/doi/pdf/10.1080/1469 3062.2022.2047585.

G Reed et al, "Indigenizing Climate Policy in Canada: A Critical Examination of the Pan-Canadian Framework and the ZéN RoadMap" (2021) 12(3) Frontiers in Research www.frontiersin.org/articles/10.3389/frsc.2021.644675/full

Heidi Kiiwetinepinesiik Stark, "Changing the Treaty Question: Remedying the Right(s) Relationship" in John Borrows & Michael Coyle, eds, *The Right Relationship; Reimagining the Implementation of Historical Treaties* (Toronto: University of Toronto Press, 2017) 248–276.

Farhana Sultana, *Confronting Climate Coloniality: Decolonizing Pathways for Climate Justice* (Taylor & Francis Group, 2024), online: www.routledge.com /Confronting-Climate-Coloniality-Decolonizing-Pathways-for-Climate -Justice/Sultana/p/book/9781032737850.

CLIMATE FEDERALISM[*]

> "The undisputed existence of a threat to the future of humanity
> cannot be ignored."[1]
> *Chief Justice Wagner, GGPPA References*

3.1 INTRODUCTION

Responding effectively to climate change requires a complex, multi-faceted, systemic transition that engages many areas of law. In the Canadian federation, those laws must be enacted by the government with

---

[*]   The author wishes to thank the many people who have had a positive impact on this chapter, from those with whom I have engaged in discussions and those with whom I have co-authored previous pieces of related research, to those who reviewed and commented upon portions of, or the whole, chapter. These include Professor Martin Olszynski, Professor David Boyd, and Joshua Ginsberg, who reviewed draft material, and Professor Peter Oliver and Taylor Wormington, with whom I have co-authored related work. I also wish to thank the research assistants who helped with editing and footnotes on this chapter, notably Alexandria Peacock (JD Candidate, uOttawa) and Gérick Girard (JD/LLL Candidate, uOttawa).

[1]   *References re Greenhouse Gas Pollution Pricing Act*, 2021 SCC 11 [*GGPPA References*].

jurisdictional authority to do so.[2] Determining which order of government has authority to implement climate laws in Canada has proven to be a subject of political controversy and legal debate.

Jurisdictional complexities are nothing new; they are an unwavering feature of the Canadian federation. Jurisdictional wrangles are woven into the fabric of our federation and an inherent part of finding the right balance between maintaining provincial autonomy and flexibility and national goals and interests. The courts play an important role in clarifying the bounds of jurisdictional authority in areas of public policy, including the environment and climate change. In fact, many important division of powers cases by the Supreme Court of Canada over the last fifty years have been about environmental legislation.[3]

Climate change is proving to be no different, with the division of powers over climate policy emerging as a central theme in federal-provincial relations in recent years. Parliament's first major climate law — the *Greenhouse Gas Pollution Pricing Act* (*GGPPA*) — was immediately subject to a constitutional challenge by the provinces of Saskatchewan, Ontario, and Alberta.[4] While the *GGPPA* was ultimately upheld by the Supreme Court of Canada, Alberta's challenge to the federal *Impact Assessment Act* — which allowed Parliament to consider the extent to which a project's GHG emissions could impact Canada's ability to meet its national climate targets in decision-making — was largely successful.[5] Challenges to federal climate policy promise to continue, with the provinces of Alberta and Saskatchewan having both signalled their plans to challenge emerging federal climate initiatives, including the federal clean electricity standards, methane regulations, and the proposed cap on greenhouse gas (GHG)

---

2 See Chapter 2 of this book for a discussion about Indigenous self-determination and jurisdiction.

3 See, e.g., *R v Hydro-Quebec*, 1997 CanLII 318 (Supreme Court of Canada) [*Hydro-Quebec*]; *R v Crown Zellerbach Canada Ltd*, 1988 CanLII 63 (Supreme Court of Canada) [*Crown Zellerbach*].

4 *GGPPA References*, above note 1.

5 *Reference re Impact Assessment Act*, 2023 Supreme Court of Canada 23 [*IAA Reference*]. See section 3.3.2.3 below for further discussion of the *IAA Reference*.

emissions from oil and gas.[6] In 2024, Alberta launched a second challenge to the *Impact Assessment Act* and Saskatchewan filed for judicial review and injunctive relief to avoid paying the carbon levy under the *GGPPA* on natural gas for home heating.[7]

This chapter examines the way in which jurisdictional authority over matters related to climate change, specifically GHG emissions mitigation, is allocated in Canada. While the legal framework for climate change necessarily involves an array of measures that go beyond mitigating emissions (such as adapting to the impacts of climate change and addressing loss and damage), this chapter focuses on mitigation for reasons of scope. The chapter outlines what can be stated with confidence about provincial and federal jurisdiction over GHG emissions and identifies areas of uncertainty and tension, considering the implications of the most recent judicial decisions, especially *GGPPA References* and *IAA Reference*. The introduction includes some commentary about constitutionalism and the rule of law, then introduces jurisdictional issues in the environmental and climate space. Next, the chapter reviews the division of powers analysis and several key principles that are relevant to jurisdictional challenges of climate laws. Third, the chapter discusses the division of powers over the mitigation of GHG emissions, surveying key provincial powers (including property and civil rights and the natural resources amendment) and then the most relevant federal powers: criminal law, both the national concern and emergency doctrines of the peace, order, and good government clause (POGG), trade and commerce, taxation, and the declaratory power.

---

6   See, e.g., Danielle Smith, "Proposed Clean Electricity Regulations: Premier Smith" (10 August 2023), online: alberta.ca [perma.cc/D4Z2-YHWR]; Government of Saskatchewan, "Government of Saskatchewan Rejects Federal Oil and Gas Emissions Cap and Methane 75 Regulations" (24 September 2024), online: saskatchewan.ca [perma. cc/5HSB-3EEU].

7   Province of Alberta, Order in Council, (November 20, 2024), online: kings-printer. alberta.ca [perma.cc/NZ4K-2XUD]; Alexander Quon, "Saskatchewan Files for Injunction as CRA Attempts to Take $28M its Owed Under Federal Carbon Tax," *CBC News* (July 4, 2024), online: cbc.ca [perma.cc/A6RT-PAPL].

3.1.1 A Precondition – Constitutionalism and the Rule of Law

Constitutionalism and the rule of law are hallmarks of a functional, liberal democratic society and preconditions to peace and prosperity.[8] These precepts mean that "law is supreme over the acts of both government and private persons" and "the exercise of all public power must find its ultimate source in a legal rule."[9] The provinces and federal governments derive their authority from the powers allocated to them in the Constitution.[10] Because the Constitution is the supreme law of Canada, "any law inconsistent with the Constitution is, to the extent of the inconsistency, of no force or effect."[11]

These notions are so fundamental to our society that it may seem unnecessary to state them here. However, there have been recent signs of small cracks in the foundation of our legal institutions in this country (as described in the next paragraph), which could lead to a constitutional crisis. Combined with the need for decolonialization and reconciliation, as elaborated in Chapter 2, this is a tenuous time for the Canadian Constitution. It is not the first time Canada has faced tension in the federation; Canada has weathered several constitutional crises.[12] However, the global rise of populism and recent events in the United States under the Trump administration warrant a reminder that these vital principles should never be taken for granted.

---

8   As the Supreme Court of Canada has noted, the "principles of constitutionalism and the rule of law lie at the root of our system of government" (see *Reference re Secession of Quebec*, 1998 CanLII 793 at para 70 (Supreme Court of Canada) [*Secession Reference*]).

9   *Ibid* at para 71.

10   *Ibid* at paras 32, 52, and 72. Note that the Constitution includes unwritten principles that both empower and limit government authority. The Crown Prerogative, for instance, is briefly mentioned in the constitutional text but is elaborated in common law rules to provide authority for foreign affairs and treaty-making, *inter alia* (see *ibid*).

11   The Constitution binds all governments, including the executive branch. This principle is embodied in s 52(1) of the *Constitution Act, 1982*, being Schedule B to the *Canada Act 1982* (UK), 1982, c 11 [*Constitution Act, 1982*]. See also *Secession Reference*, above note 8 at para 72.

12   Consider the oil crisis of the 1970s and 1980s and the Quebec's separatist movement, for instance.

When Danielle Smith was elected Premier of Alberta in October 2022, her government passed unprecedented legislation — the *Alberta Sovereignty Within a United Canada Act* — which purports to grant the province authority to ignore any federal legislation or regulatory decision that intrudes on provincial jurisdiction, regardless of what the judiciary decides.[13] Saskatchewan followed suit, enacting the *Saskatchewan First Act* in March 2023 which aims to "defend the province's economic autonomy and potential from federal overreach."[14] These laws threaten to destabilize federalism by attempting to unilaterally grant the provinces the power to shield themselves from federal legislation and attempting to bypass the judiciary's role in interpreting the division of powers.[15] This is unchartered territory and threatens the constitutional stability of the country.[16]

In a similarly unprecedented move, the Quebec legislature in 2022 led by Premier Francois Legault enacted legislation purporting to amend section 90 of the *Constitution Act, 1867*.[17] Since the legislation declares what is already understood about Quebec (that people from Quebec form a nation and that French is the common language of the province), it was not substantively controversial. Procedurally, however, it represents an attempt to modify the *Constitution* unilaterally, something that is not permitted, even when the amendment pertains to only one province. Section 43 provides the formula for constitutional amendment that involves only

---

13   The law, for instance, attempts to grant the provincial cabinet power to direct institutions such as the police force and municipalities to refuse to enforce federal legislation and creates a bypass of the courts, as explained by constitutional law professor Eric Adams (see Eric M Adams, "Danielle Smith Didn't Give Us a Watered-Down Version of Alberta's Sovereignty Act," *CBC News* (29 November 2022), online: cbc.ca [perma.cc/UC52-VGCH]) See also Martin Olszynski & Nigel Bankes, "Running Afoul the Separation, Division, and Delegation of Powers: The Alberta Sovereignty Within a United Canada Act" (December 6, 2022), online: ablawg.ca [perma.cc/QG5Y-NLH5].

14   *Saskatchewan First Act*, SS 2023, c 9.

15   See Mark P Mancini, "Foxes, Henhouses, and the Constitutional Guarantee of Judicial Review: Re-Evaluating *Crevier*" (2024) 102:2 Can Bar Rev 315 at 350–52.

16   See Emmett Macfarlane, "Provincial Constitutions, the Amending Formula, and Unilateral Amendments to the Constitution of Canada: An Analysis of Quebec's Bill 96" (2023) 60:3 Osgoode Hall LJ 655 at 695–97.

17   Bill 96, *An Act Respecting French, the Official and Common Language of Quebec*, 2nd Sess, 42nd Leg, Quebec, 2022 (assented to 1 June 2022), SQ 2022, c 14.

one province, and that formula requires the federal government's approval as well.[18]

Although using the notwithstanding clause does not counter the rule of law, since it is explicitly permitted within the *Constitution Act*, its increased invocation in policy discussions and practice in the last several years is also a sign of intergovernmental tensions in the country.[19] For instance, Ontario invoked the clause for the first time in 2021 to limit third-party election financing and Saskatchewan recently invoked the clause in a bill requiring parental consent for name and gender pronoun changes for students under sixteen years of age.[20]

While key principles, such as constitutionalism and the rule of law, remain intact in Canada, it is important to bear this context in mind when discussing jurisdiction over laws that have great significance for the provinces where unity is politically tenuous. Judges will be cognizant of the fragile state of the federation in interpreting division of powers cases.[21] However, judges will also need to bear in mind the climate emergency, which threatens to undermine geopolitical and socioeconomic stability. A

18　See section 43 of the *Constitution Act, 1982*, above note 11. See also Tanzim Rashid, "The Slow-Moving, Silent, and Creeping Constitutional Crises Facing Canada" (1 November 2022), online: thecourt.ca [perma.cc/BX6D-PH6X].

19　The clause, embodied in section 33 of the *Charter*, has been used by the governments of Quebec, Alberta, the Yukon, and most recently Saskatchewan and Ontario. For instance, Ontario Premier Doug Ford attempted to use the notwithstanding clause to rewrite Toronto's municipal election laws part way through the campaign, to allow rules to limit third party advertising during provincial elections, and to prevent education workers from going on strike. Quebec Premier Francois Legault has applied the clause twice in Quebec, and Saskatchewan Premier Scott Moe invoked it recently in the context of gender identity and parental rights (see Government of Saskatchewan, "'Parents' Bill of Rights' Passed and Enshrined in Legislation" (20 October 2023), online: saskatchewan.ca [perma.cc/E3V3-UKU4]).

20　Dave Snow, "Saskatchewan Bill Shows Why the Notwithstanding Clause Exists," *The Hill Times* (30 November 2023), online: hilltimes.com [perma.cc/55QP-WM3A].

21　It is possible that the restrictive interpretation of jurisdiction in the *IAA Reference* was influenced by a desire to reassure Alberta that federal power is appropriately constrained. See Nathalie J Chalifour, "Missing the Forest for the Trees: The Supreme Court of Canada's Formalistic Approach in the *IAA Reference* a Setback for Environmental and Climate Law in Canada." (2025) SCLR (3d) 31 [Chalifour, "Missing the Forest for the Trees"].

robust legal framework founded on the rule of law and constitutionalism is critical in the face of the climate crisis and divisive politics.

3.1.2 Environmental Federalism

It is unsurprising that neither climate change nor GHG emissions are featured in the allocation of legislative authority in the Constitution. Like the environment more generally, the subject was not in the consciousness of those who drafted the Constitution. While control over resources was a central part of the debate surrounding the constitutional amendments of 1982, no one was contemplating the context we would find ourselves in forty years later. We are thus left with the task of applying and interpreting the division of powers as drafted a century and a half ago, with only modest amendments along the way, to the very unique and formidable realities presented by the changing climate.

As Canada's environmental regulatory framework evolved over the last fifty years, key federal laws — ranging from environmental impact assessment to pollution control, fisheries management, and species at risk protections — have faced jurisdictional challenges by industry and provincial governments. While largely unsuccessful, these lawsuits cause delays, consume resources, and often lead to the weakening or narrowing of federal legislation to mitigate the risk of constitutional challenges.

In 1992, the Supreme Court of Canada famously observed that "the environment is a constitutionally abstruse matter,"[22] yet it has repeatedly affirmed the importance of environmental protection.[23] Until recently, the Court consistently upheld environmental legislation as *intra vires* except in one case.[24] These decisions have acknowledged the complexity of environmental matters, noting their multi-faceted nature and shared jurisdiction

---

22　*Friends of the Oldman River Society v Canada (Minister of Transport)*, 1992 CanLII 100 at 64 (Supreme Court of Canada) [*Oldman River*].

23　See, e.g., *Ontario v Canadian Pacific Ltd*, 1995 CanLII 112 at para 51 (Supreme Court of Canada) [*Canadian Pacific*]; *Hydro-Quebec*, above note 3; *114957 Canada Ltée (Spraytech, Société d'arrosage) v Hudson (Town)*, 2001 Supreme Court of Canada at para 1 [*Spraytech*]; *GGPPA References*, above note 1.

24　See *Fowler v The Queen*, 1980 CanLII 201 (Supreme Court of Canada).

between federal and provincial governments. By favouring a cooperative federalism approach, the jurisprudence has created ample jurisdictional space for both provinces and federal governments to regulate aspects of the environment.

Guided by this jurisprudence, the Canadian federation has largely sorted out jurisdictional authority over environmental matters, including air and water pollution, toxic substances, fisheries, waste management, and species protection.[25] Provinces exercise broad authority over environmental issues within their borders, from industrial emissions to water quality and building codes, while the federal government has authority over matters such as fisheries, navigable waters, and toxic substances.[26] These are not enclaves; considerable jurisdictional overlap exists, and ongoing debates persist regarding the effectiveness of environmental regulation across the country, even in areas where jurisdiction appears settled.

A variety of governance mechanisms have been developed to support intergovernmental coordination and cooperation. Meetings of the Canadian Council of First Ministers, for instance, happen annually though they take place behind closed doors with a convention of consensus decision-making.[27] Federal, provincial, and territorial environment ministers also meet annually through the Canadian Council of Ministers of the Environment (CCME) to discuss a range of issues and, at times, arrive at intergovernmental agreement. Indigenous governments are not invited

---

25 See *Hydro-Quebec*, above note 3; *Crown Zellerbach*, above note 3; *Oldman River*, above note 22; *Spraytech*, above note 23; *Interprovincial Co-operatives Ltd v R*, [1976] 1 SCR 477; *Groupe Maison Candiac Inc v Attorney General of Canada*, 2020 FCA 88, leave to appeal to Supreme Court of Canada dismissed). Note that the *IAA Reference* and subsequent amendments to the legislation raise question about federal authority over interprovincial air pollution — something that is discussed below in Section 3.3.2.3

26 Jamie Benidickson, *Environmental Law*, 5th ed (Toronto: Irwin Law, 2019) at 33–54. While the Constitution allocates law-making authority to Parliament over Indigenous matters to the federal government, this ignores pre-existing sovereignty and inherent Indigenous jurisdiction. This is addressed in Chapter 2 of this book. See also *ibid* at 45–48).

27 Kathyrn Harrison, "Climate Governance and Federalism in Canada" in Alana Fenna, Sébastien Jodoin & Joana Setzer, eds, *Climate Governance and Federalism: A Forum of Federations Comparative Policy Analysis* (Cambridge: Cambridge University Press, 2023) at 69.

to the First Ministers' meetings, nor do they participate in meetings of the CCME even though they have treaty rights, title, and claims to large swaths of unceded ancestral lands and waters across the country.[28] There has been some effort to consult with Indigenous leaders ahead of such ministerial meetings, but these efforts are insufficient.

Overall, the courts have provided a mostly enabling interpretation of environmental jurisdiction balancing provincial autonomy with a meaningful federal role. While far from perfect, the relatively settled state of environmental federalism in Canada has allowed the development of a comprehensive, albeit consistently weak set of environmental laws across the country. Enter climate change.

3.1.3   Climate Federalism

Even though the Canadian climate legal framework is nascent, several climate-related laws have already been subject to jurisdictional challenges. The first dispute was initiated in 2011 by Syncrude Canada which challenged the federal *Renewable Fuel Regulations* enacted under the *Canadian Environmental Protection Act, 1999*.[29] These regulations require a minimum content of renewable fuel in gas and diesel, something Syncrude argued was outside the scope of federal authority. The Federal Court dismissed the challenge, upholding the regulations as valid under the criminal law power (a decision that was confirmed in 2016 by Rennie J, Ryer J, and Boivin J of the Federal Court of Appeal).[30] The Federal Court of Appeal confirmed that GHG emissions contribute to what they characterized as the evil of climate change and that behaviour-modifying incentives like minimum content requirements are valid criminal provisions.[31]

---

28   *Ibid.*

29   SC 1999, c 33 [*CEPA*].

30   *Syncrude Canada Ltd v Canada (Attorney General)*, 2014 FC 776 [*Syncrude* FC], upheld in *Syncrude Canada Ltd v Canada (Attorney General)*, 2016 FCA 160 [*Syncrude* FCA]. The scope of the criminal law power is discussed in detail later in this chapter in section 3.2.2.

31   *Syncrude* FC, above note 30 upheld in *Syncrude* FCA, above note 30 at paras 24, 62, and 68–69. The case is discussed in further detail in Section 3.3.2.2.

In 2018, the province of Saskatchewan challenged federal carbon pricing legislation, with Ontario and Alberta filing similar challenges shortly thereafter.[32] In 2021, the Supreme Court of Canada heard appeals from all three cases and issued one ruling upholding the *Greenhouse Gas Pollution Pricing Act* under the national concern branch of POGG.[33] The Court held that the goal of the legislation was to establish minimum national standards of GHG price stringency to reduce GHG emissions.[34] Writing for a six judge majority, Wagner CJ held that this objective was a matter of concern to Canada as a whole and that the test for single-ness, distinctiveness and indivisibility was met. The Court also held that upholding the legislation would have an acceptable scale of impact on provincial jurisdiction.[35] The Court's findings were influenced by the backstop design of the *GGPPA*, the application of the double aspect doctrine, and the gravity of climate change if Parliament was constitutionally unable to enact this law.[36]

In 2019, the province of Alberta challenged the federal *Impact Assessment Act* (IAA). The Supreme Court of Canada ruled in 2023 that core parts of that legislation were outside of federal jurisdiction.[37] Although the IAA is not climate legislation *per se*, impact assessment plays an important role in evaluating the risks and benefits of projects before they go ahead, allowing governments to consider these factors (and impose conditions) before approving a project.[38] One of Alberta's concerns was that the IAA

---

32 *Reference re Greenhouse Gas Pollution Pricing Act*, 2019 SKCA 40 [*GGPPA References* SKCA]; *Reference re Greenhouse Gas Pollution Pricing Act*, 2019 ONCA 544 [*GGPPA References* ONCA]; *Reference re Greenhouse Gas Pollution Pricing Act*, 2020 ABCA 74 [*GGPPA References* ABCA].

33 *GGPPA References*, above note 1. This case is discussed in greater detail further in this chapter (section 3.3.2.3).

34 *GGPPA References*, above note 1 at para 80.

35 *Ibid* at para 207.

36 *GGPPA References*, above note 1 at para 206. Interpretation of the national concern branch and *GGPPA References* is discussed in more detail in section 3.3.2.3 of this chapter.

37 *IAA Reference*, above note 5.

38 Impact assessment can be helpful in meeting GHG reduction goals, for instance. See Meinhard Doelle, "Integrating Climate Change into Environmental Impact Assessments: Key Design Elements" in Francesco Sindico, Stephanie Switzer & Tianbao Qin, eds, *The Transformation of Environmental Law and Governance: Risk, Innovation and*

authorized federal decision-making based on a range of criteria, including considering the impact of a proposed project on the government's ability to meet climate change goals.[39] A 5:2 majority decision penned by Wagner CJ agreed that the law strayed out of Parliament's constitutional lane. The majority expressed concern about the broad way in which "effects within federal jurisdiction" was defined in the law. With respect to GHG emissions, the Court noted that broad language in the legislation could allow federal jurisdiction over projects based on any threshold of emissions crossing provincial boundaries.[40] The majority noted that Canada made no attempt to apply the clarified national concern test elaborated in *GGPPA References* to justify including language broad enough to capture interprovincial GHG emissions in the definition of interprovincial effects.[41] The *IAA Reference* did not hold that Parliament has no jurisdiction over aspects of national GHG emissions; it reiterated its findings in *GGPPA References* that jurisdiction over GHG emissions is shared. However, the Court held that Parliament would have had to make the case to have appropriately circumscribed jurisdiction to consider GHG emissions in its decision-making — something it did not do.[42] The decision signals a shift toward a more cautious interpretation of jurisdiction whereby Parliament should ground provisions in uncontested federal spheres of authority (or be ready to defend reliance upon less established sources). In this case, conferring jurisdiction to consider a project's interprovincial GHGs emissions without any significance threshold was too broad.[43]

Resilience (Cheltenham, Edward Elgar Publishing, 2021) 111. See also David V Wright, "Supreme Court of Canada Will Soon Rule on the Constitutionality of the Federal *Impact Assessment Act.* Here's What to Watch for…" (3 October 2023), online: ablawg. ca [perma.cc/75LW-2U9W]; David V Wright, "Constitutional Caution, Correction, and Abdication: The Proposed Amendments to the *Impact Assessment Act*" (May 10, 2024), online: ablawg.ca [perma.cc/4XK6-5LMB].

39   See, e.g., *Impact Assessment Act*, SC 2019, c 28, s 1, s 63.

40   *IAA Reference*, above note 5 at para 183. See also Chalifour, "Missing the Forest for the Trees," above note 21 for a critique of the majority's approach in the *IAA Reference*.

41   *IAA Reference*, above note 5 at para 189.

42   *Ibid* at para 189.

43   *Ibid* at para 184.

Following the decision, Parliament amended the *IAA* in accordance with the majority's reasons.[44] Alberta has already launched a second challenge to the amended *Impact Assessment Act*.[45] Alberta has also challenged the constitutionality of the federal Clean Electricity Standards[46] and announced its intention to challenge regulations to limit methane emissions[47] and the proposed cap[48] on GHG emissions from oil and gas.[49] Saskatchewan Premier Scott Moe has also indicated a constitutional challenge to the oil and gas emissions cap is likely.[50] It is likely that most of the provisions slated to be challenged will be enacted under the *Canadian Environmental Protection Act* and justified as a matter of criminal law. This probably seemed like a reasonably safe approach constitutionally, in light of the *Hydro-Quebec* and *Syncrude* decisions (discussed below). However, a recent decision of the Federal Court finding certain provisions of *CEPA* to be *ultra vires* the criminal law power raises some questions about the scope of this power for environmental regulations.[51] In *Responsible Plastic Use Coalition v Canada (Environment and Climate Change)*, the Federal Court held that the inclusion of all plastic manufactured items on the list

---

44 See *Budget Implementation Act, 2024, No 1*, SC 2024, c 17, ss 269–319.

45 See Quon, above note 7.

46 CPAC, *Alberta Govt to Challenge Federal Clean Energy Regulations*, May 1, 2025, online: https://www.cpac.ca/headline-politics/episode/alberta-govt-to-challenge-federal-clean-energy-regulations--may-1-2025?id=aa08e168-2ae0-4b44-b68a-4ac409fbe244. Environment and Climate Change Canada, "Draft Clean Electricity Regulations" (10 August 2023), online: canada.ca [perma.cc/TVG3-7CH8].

47 The methane regulations aim to reduce oil and gas methane emissions by a minimum of 75 percent by 2030 (see Environment and Climate Change Canada, "Canada Confirms its Support for the Global Methane Pledge and Announces Ambitious Domestic Actions to Slash Methane Emissions" (11 October 2021), online: canada.ca [perma.cc/GA2W-4XY4].

48 Environment and Climate Change Canada, "Canada Introduces Framework to Cap Greenhouse Gas Pollution from Oil and Gas Sector" (7 December 2023), online: canada.ca [perma.cc/U9CM-79RF].

49 See Jason Markusoff, "Danielle Smith Would Never Accept Ottawa's Oil Emissions Rules No Matter How Flexible," *CBC News* (7 December 2023), online: cbc.ca [perma.cc/88Z3-A3AJ]. See also Smith, above note 6.

50 Will McLernon, "Sask Premier Says Ottawa's New Emissions Framework for Oil and Gas Will Hurt Province's Energy Sector," *CBC News* (8 December 2023), online: cbc.ca [perma.cc/34SG-RM5Z].

51 *Responsible Plastic Use Coalition v Canada (Environment and Climate Change)*, 2023 FC 1511 [*Responsible Plastic Use Coalition*].

of toxic substances in an unqualified manner — regardless of the potential for each of those items to cause harm — exceeded Parliament's authority.[52] The decision was under appeal at the time of writing.

3.1.4 Looking Ahead

The fact that there have already been several constitutional challenges to federal climate-related legislation and promises of a challenge to every major federal climate policy announced is a sign that we are in an era of constitutional controversy. While most agree that climate change is important and must be addressed, there are deep-seated differences about how that should happen, how quickly, and who has the jurisdictional authority to do what. The politics of decision-making are especially charged in the context of polarization and powerful fossil fuel interests anchoring aspects of provincial economies, such as in Alberta and Saskatchewan. Climate policies are especially threatening for provinces whose economies are dependent on fossil fuel extraction.[53] The provinces of Alberta and Saskatchewan account for 91 percent of Canada's oil production, whereas British Columbia produces the most coal of the Canadian provinces.[54] As long as Canada remains committed to meeting its GHG reduction goals under the *Paris Agreement*, we can expect the rapidly evolving legal framework for climate change to continue to be the subject of jurisdictional challenges. The rest of this chapter discusses the way in which jurisdictional disputes are resolved by Canadian courts and examines the contours of the key jurisdictional powers that are likely to be at play in mitigating GHG emissions.

---

52 *Ibid* at para 119.

53 Harrison, above note 27 at 69. See also Angela V Carter, *Fossilized: Environmental Policy in Canada's Petro-Provinces* (Vancouver: UBC Press, 2020).

54 Natural Resources Canada, "Oil Supply and Demand" (16 December 2019), online: natural-resources.canada.ca [perma.cc/GNK6-YAWP]; Natural Resources Canada, "Coal Facts", online: natural-resources.canada.ca [perma.cc/ZJ76-XTJG].

3.2 THE DIVISION OF POWERS IN THE CANADIAN CONSTITUTION

Authority to enact laws in Canada is allocated within the Constitution among Parliament and the provincial legislatures. The key provisions are sections 91, 92, 92A, 93, 94(a), and 95 of the *Constitution Act*, 1982. Section 91 allocates authority to the federal Parliament, while section 92 outlines the powers granted to provincial legislatures. Section 92A was added to the Constitution in 1982, clarifying provincial authority over the exploration, development, conservation, and management of non-renewable resources in the province[55] but also granting to provinces the authority to make laws related to the export of natural resources from the province to other parts of Canada[56] and to levy indirect taxes on natural resources and their production.[57] Section 93 grants provinces jurisdiction to make laws about education, and sections 94(a) and 95 establish shared jurisdiction on a limited set of subjects including agriculture and immigration. Section 52(1) (sometimes referred to as the supremacy clause) explains that any provision inconsistent with the Constitution is of no force or effect.

Federalism in the messy real world – upholding unity and national goals while respecting autonomy and diversity

The Canadian Constitution was drafted in the wake of the US civil war, so it aimed to be more centralized than its US counterpart. The federal government was given unlimited taxation powers and residual power to make laws for Peace, Order and Good Government (POGG). To make up for their more limited taxation powers, provinces were given control of provincial crown lands as a source of income.[58] Despite an initial aim at centralization, broad judicial interpretation of provincial jurisdiction

---

55 *Constitution Act, 1982*, above note 11, s 92A(1). Section 92A(1) largely confirms jurisdiction over natural resources under property and civil rights and matters of a local or private nature.

56 *Ibid*, s 92A(2).

57 *Ibid*, s 92A(4). See also Carissima Mathen & Patrick Macklem, *Canadian Constitutional Law*, 6th ed (Toronto: Emond Publishing, 2022) at 391; Nigel Bankes & Andrew Leach, "Preparing for a Mid-Life Crisis: Section 92A at 40" (2023) 60:4 Alta L Rev 853.

58 Harrison, above note 27 at 68.

over property and civil rights and the cautious approach to POGG have resulted in one of the most decentralized federations in the world.[59]

When you consider the scale, scope and diversity of our federation, it is a marvel that it functions. Faced with the need to regulate areas of amorphous jurisdiction such as health, transportation, food, and social policy, not to mention the environment and climate change, the Canadian federation has managed remarkably well.[60] It has survived numerous touchpoints of tension, including the energy crisis of the 1970s and early 1980s and the independence movement in Quebec. The country's ability to withstand a variety of challenges is a tribute to the political prowess of key players, the good faith of stakeholders of all stripes, and a commitment to Canada as a country, not to mention the key role of the judiciary in addressing jurisdictional disputes throughout difficult times.

We are once again in a period of transition stemming from the imperative to decolonize, address the tragic harms of colonialization, and pursue reconciliation with Indigenous Peoples, as well as navigating federal-provincial tensions amidst ideological differences on a range of issues, including federal transfers and social security, natural resources, energy policy, and climate change. As Professors Nigel Bankes and Andrew Leach have recently noted, "Canada finds itself in the early stages of another federation-defining conflict over resource-related issues including the construction of new pipelines, legislative and policy responses to greenhouse gas emissions, and the reach of federal impact assessment legislation."[61] If the Canadian state is to survive, we will need to find a path that navigates the delicate balance between provincial autonomy and national unity. The imposition of tariffs, threats of annexation, and withdrawal from the Paris

---

59 *Ibid*. See also Paolo Dardanelli et al, "Conceptualizing, Measuring, and Theorizing Dynamic De/Centralization in Federations" (2019) 49:1 Publius: J Federalism 1.

60 The country's survival is not without grave repercussions, given the country was built on a legacy of genocide and violence against Indigenous Peoples (see, e.g., National Inquiry into Missing and Murdered Indigenous Women and Girls, *A Legal Analysis of Genocide: Supplementary Report of the National Inquiry into Missing and Murdered Indigenous Women and Girls* (National Inquiry into Missing and Murdered Indigenous Women and Girls, 2019), online: mmiwg-ffada.ca [perma.cc/BQZ5-KTXG]; see also Chapter 2 of this book.

61 See Bankes & Leach, above note 57 at 882.

Agreement by the United States under the Trump administration have added a layer of uncertainty and complexity that will undoubtedly influence Canadian policy choices. Interestingly, the United States' actions have had the effect of fostering Canadian unity and patriotism. How this new context translates to energy and climate policy remains to be seen. If anything, Canada has an even greater responsibility to show leadership in doing its part to respond to the climate crisis. Fulfilling that responsibility will require ongoing, ambitious GHG emissions reductions by all governments across the country compatible with what the science dictates is necessary. The only way to achieve these climate goals in Canada is with a cooperative, dynamic approach to federalism that is enabling rather than restrictive and that accommodates necessarily overlapping, shared and concurrent responsibilities.

Exclusive powers

A fundamental tension that has permeated division of powers discourse and jurisprudence since confederation is the tension between exclusivity and overlap, a strain point that Professor Wade Wright refers to as one of federalism's underlying questions.[62] This jurisdictional push-pull is at the heart of federalism and has been approached in different ways over the years. While there are many nuances, two central approaches to federalism permeate the jurisprudence: the classical watertight compartment approach of early days characterized by an emphasis on exclusivity and the more modern approach of cooperative federalism characterized by flexibility and tolerance for overlap.[63] The modern approach favours allowing both orders of government to regulate aspects of society's complex activities within their respective spheres of authority, something that is necessary to accommodate today's complex, multi-faceted public policy challenges. As Mathen and others state, "[a] vast range of government functions are now concurrent, *de facto* if not *de jure*" and "concurrency,

---

62 Wade K Wright, "Canadian Federalism's Underlying Question: What it is and Why it Matters" (2020) 53:2 UBC L Rev 531 at 532 [Wright, "Canadian Federalism"].
63 *GGPPA References*, above note 1 at para 50.

overlapping, and shared responsibilities are fundamental features of Canadian federalism as in all other federations."[64]

In *GGPPA References*, the challenging provinces urged the courts to interpret the division of powers in a way that aligned with the watertight approach, essentially arguing that jurisdiction over GHG emissions was either federal or provincial. They argued that upholding the *GGPPA* would mean Parliament occupied the whole field of GHG regulation and would — as per the watertight compartment view they advocated — eviscerate provincial jurisdiction in this area. The Supreme Court of Canada rejected this argument, favouring cooperative federalism, a narrow definition of the law's pith and substance, and application of the double aspect doctrine to find that upholding the federal law did not negate provincial jurisdiction to legislate GHG emissions.[65]

Even though the modern approach favours concurrent application of laws in similar fact situations, the courts have repeatedly cautioned that cooperative federalism does not upend the division of powers, and that care must be taken to safeguard provincial jurisdictional autonomy.[66] The devil is often in the details, as seen in the jurisdictional wrangling over climate policy in the Canadian federation.

3.2.1 The Division of Powers Analysis

The well-established two-part test used by courts to determine whether legislation is within jurisdictional constitutional authority (*intra vires*) is deceptively simple. First, the court identifies the law's "pith and substance," or the dominant purpose of the matter being legislated (the "characterization" stage).[67] To determine the pith and substance of a law, courts will consider its legislative purpose as well as its legal and practical effects.

---

64   Mathen & Macklem, above note 57 at 166–67.

65   *GGPPA References*, above note 1 at para 206.

66   See, e.g., *ibid* at para 50.

67   *Ibid* at para 47; *Reference re Genetic Non-Discrimination Act*, 2020 Supreme Court of Canada 17 at paras 28–30 [*Genetic Non-Discrimination*]; *Reference re Pan-Canadian Securities Regulation*, 2018 Supreme Court of Canada 48 at para 86 [*Pan-Canadian Securities*]; *Reference re Firearms Act (Can)*, 2000 Supreme Court of Canada 31 at para 18 [*Firearms Reference*]; *GGPPA References*, above note 1 at para 51.

Together, these help to identify a law's essential character.[68] Second, the court must classify the law as characterized by reference to provincial and federal heads of power (the "classification" stage).[69] The second step is distinct from the first, in that the pith and substance determination should not be shaped to suit a particular class of subject matters.[70]

The courts have developed and applied a variety of constitutional principles and doctrines to adjudicate an array of *vires* cases over the years that clarify and rationalize the division of powers over such issues as securities regulation, marriage, gun control, secession, various aspects of health care, senate reform, and — of note — the environment and climate change.[71] I describe some of the key principles below before turning to discuss jurisdictional authority over GHG emissions reductions.

3.2.2 Important Principles That Underpin Jurisdictional Analysis

While this section highlights some important principles that underpin the division of powers analysis, these are by no means the only relevant principles in this space. The ones discussed below are the ones most likely to be at play in resolving climate federalism disputes.

3.2.2.1 Presumption of Constitutionality

An important starting point in constitutional challenges to jurisdictional authority is the presumption of constitutionality. The Supreme Court of Canada recently described this as a "cardinal principle of our division of

---

68 *Genetic Non-Discrimination*, above note 67 at para 30.

69 *Ibid* at para 26.

70 *Ibid* at para 31.

71 See, e.g., *Crown Zellerbach*, above note 3; *RJR-MacDonald v Canada (Attorney General)*, 1995 CanLII 64 (Supreme Court of Canada) [*RJR-MacDonald*]; *Hydro-Quebec*, above note 3; *Canada (Attorney General) v PHS Community Services Society*, 2011 Supreme Court of Canada 44 [*PHS*]; *Secession Reference*, above note 8; *Firearms Reference*, above note 67; *Reference re Same-Sex Marriage*, 2004 Supreme Court of Canada 79 [*Same-Sex Marriage Reference*]; *Reference re Assisted Human Reproduction Act*, 2010 Supreme Court of Canada 61 [*Assisted Human Reproduction*]; *Quebec (Attorney General) v Canada (Attorney General)*, 2014 Supreme Court of Canada 14 [*Quebec v Canada*]; *GGPPA References*, above note 1; *IAA Reference*, above note 5.

powers jurisprudence."[72] This presumption has three important, related components. First, and most obviously, laws are presumed to be constitutionally valid.[73] In the context of the division of powers, this means there is a presumption that a legislative body has the *bona fide* intention of enacting laws within its jurisdictional sphere.[74] Courts may fairly presume that laws are enacted in good faith and are intended to conform to the government's jurisdictional authority.

Second, this presumption has the practical effect of placing the burden of proving unconstitutionality on the challenging party. It is a fundamental principle of law, with limited exceptions, that the burden of proof lies with the challenging party.[75] In the context of the division of powers, and in light of the presumption of constitutionality, this means that if a party believes that a law or some part of it is *ultra vires*, the onus is on that party to bring a challenge and provide adequate evidentiary proof of such.[76]

Third, the presumption of constitutionality has implications for statutory interpretation.[77] When a law is sufficiently vague that it can be interpreted in more than one way, courts must lean toward the interpretation that conforms to the Constitution.[78] This is important in the context of laws that impart discretion on the executive branch. When the discretion

---

72   *IAA Reference*, above note 5 at para 69, citing *Murray-Hall v Quebec (Attorney General)*, 2023 Supreme Court of Canada 10 at para 79 [*Murray-Hall*].

73   *Murray-Hall*, above note 72 at para 79. See also *IAA Reference*, above note 5 at para 70.

74   *Nova Scotia Board of Censors v McNeil*, 1978 CanLII 6 at 687–88 (Supreme Court of Canada). See also *Re The Farm Products Marketing Act*, 1957 CanLII at 255 (Supreme Court of Canada).

75   See *Kruger v The Queen*, 1977 CanLII 3 at 112 (Supreme Court of Canada); *Canadian Industrial Gas & Oil Ltd v Government of Saskatchewan*, 1977 CanLII 210 at 573–74 (Supreme Court of Canada). See also Paul Daly, "Presumptions of Constitutionality in Canada" (2 December 2021), online: administrativelawmatters.com [perma. cc/655X-BSVC].

76   Note that in the context of alleged *Charter* violations, the initial burden is on the challenging party to establish that a *Charter* right has been infringed, followed by a shift in burden to the state to demonstrate that the infringement is saved by section 1 of the *Charter*.

77   *Osborne v Canada (Treasury Board)*, 1991 CanLII 60 at 103 (Supreme Court of Canada); *McKay v The Queen*, 1965 CanLII at 803–4 (Supreme Court of Canada) [*McKay*].

78   *IAA Reference*, above note 5 at para 314. *McKay, above note 77* at 803–4.

is exercised in a way that produces a violation, the violation is created by the administrative decision itself, rather than the provision that imparts the discretion (assuming there is an interpretation that the provision is *intra vires*).[79] As such, the matter is one to be resolved by administrative law rather than a jurisdictional challenge.[80]

The Supreme Court of Canada addressed the presumption of constitutionality in *IAA Reference*, calling it a cardinal principle. The Court referred to the first and third principles above and the corresponding need to select an interpretation of a challenged law that would allow it to stand.[81] However, the Supreme Court of Canada emphasized that the presumption does not create an "impenetrable shield" to jurisdictional challenge, and that questions of constitutionality and administrative application should not be conflated. As per the majority, an otherwise invalid (*ultra vires*) law cannot be saved by the prospect of future judicial review if the law is applied in a way that strays out of jurisdictional bounds.[82]

In contrast, in their dissent in the *IAA Reference*, Karakatsanis J and Jamal J would have relied upon the presumption of constitutionality to support a finding that the law was within Parliament's jurisdiction. They found that the presumption of constitutionality "requires a court to interpret the discretion granted under the legislation as being exercised in good faith and within constitutional bounds."[83] They argued that the broad, diffuse nature of environmental challenges requires the court to maintain a flexible, cooperative approach to interpreting jurisdictional powers which includes presuming that the legislature did not intend to exceed the constitutional limits on its authority.[84] By presuming constitutionality and good faith among legislatures, Karakatsanis J and Jamal J avoided the narrow, more literal, and restrictive interpretation applied by the majority that captured trivial or *de minimis* environmental effects.[85] The dissenting view is

---

79   See Daly, above note 75.
80   *Ibid.*
81   See *IAA Reference*, above note 5 at paras 69–73.
82   *IAA Reference*, above note 5 at paras 73–74. See also *Hydro-Quebec*, above note 3 at para 73.
83   *IAA Reference*, above note 5 at para 314.
84   *Ibid* at para 222.
85   *Ibid* at para 280. See also Chalifour, "Missing the Forest for the Trees," above note 21.

one that ascribes more trust and good faith to the legislature and resonates with the tone of most environmental federalism cases.

3.2.2.2 No Gaps

Another core principle in federalism is that the Canadian constitution fully distributes legislative authority among the provincial legislatures and Parliament.[86] The distribution is exhaustive, meaning there is no legislative subject matter that does not belong to either order of government.[87] Setting aside the critical need to reconcile jurisdictional authority with inherent Indigenous jurisdiction, this means that within the Constitutional allocation of powers, authority to enact any climate law must belong either to Parliament, the provincial legislatures, or (within the confines of constitutional doctrines such as the double aspect) both. It cannot belong to neither.

3.2.2.3 Federalism

Federalism is the guiding principle for how legislative authority is allocated in the country.[88] It is the mechanism by which the country aims to reconcile diversity with unity and foster intergovernmental cooperation for the common good.[89] The courts consistently emphasize the need to strike this balance in its division of powers jurisprudence: "A view of federalism that disregards regional autonomy is in fact as problematic as one that underestimates the scope of Parliament's jurisdiction."[90] While the overarching purpose of federalism might be to provide a mechanism to bind the jurisdictions into a nation, a nation is built to achieve a set of common goals. The federalist structure then "facilitates democratic

---

86 See, e.g., *Genetic Non-Discrimination*, above note 67 at para 21. Note that this exhaustive distribution does not account for inherent Indigenous sovereignty and jurisdiction. See John Borrows, *Canada's Indigenous Constitution* (Toronto: University of Toronto Press, 2010). See also Chapter 2 of this book.

87 See, e.g., *Same-Sex Marriage Reference*, above note 71; *Quebec v Canada*, above note 72; *GGPPA References*, above note 1 at para 461, Rowe J dissenting but not on this point.

88 See *GGPPA References*, above note 1 at paras 3 and 48–50.

89 *Ibid* at para 48; *Secession Reference*, above note 8 at para 43.

90 *R v Comeau*, 2018 Supreme Court of Canada 15 at para 82 [*Comeau*], cited in *GGPPA References*, above note 1 at para 49.

participation by distributing power to the government thought to be most suited to achieving the particular societal objective having regard to this diversity."[91] Interesting debates ensue about which order of government is best suited to addressing aspects of climate policy. While local jurisdictions may be best suited to build resilience and adapt to the impacts, efforts to mitigate national levels of GHG emissions in the context of international commitments require efforts by all orders of government, including federal. Federalism remains a central animating feature in both political discourse over energy and climate policy and the accompanying push and pull of jurisdictional disputes over regulation in this space.

3.2.2.4 Cooperative Federalism

As already noted, in the early days of the Constitution, the allocation of powers was viewed as a set of watertight compartments for which there was little room for overlap or interplay.[92] The Supreme Court of Canada has moved away from this classical approach to a more flexible and cooperative style of federalism that accommodates the complexity of modern society.[93] Cooperative federalism is an approach to federalism that recognizes that there will be "overlap between valid exercises of provincial and federal authority."[94] It is a response to the watertight components approach that would characterize jurisdictional spheres as zero-sum categories that are in either one column or the other. The more flexible or cooperative approach to federalism that has dominated Supreme Court of Canada jurisprudence for decades allows "for a fair amount of interplay and indeed overlap between federal and provincial powers."[95] The move toward greater flexibility and accommodation has been necessary to

---

91 *Secession Reference*, above note 8 at para 58.

92 See notes 62–66 and accompanying text on Exclusive Powers.

93 *Genetic Non-Discrimination*, above note 67 at para 22. See also Nathalie J Chalifour, Peter Oliver & Taylor Wormington, "Clarifying the Matter" (2020) 40:2 NJCL 153 at 164. See also Wright, "Canadian Federalism," above note 62 at 443–44, discussing three forms of federalism, classical, modern, and flexible, roughly corresponding to different periods in Canada's constitutional history.

94 *Genetic Non-Discrimination*, above note 67 at para 22; *Pan-Canadian Securities*, above note 67 at para 18.

95 *Ontario (Attorney General) v OPSEU*, 1987 CanLII at 18 (Supreme Court of Canada).

"account for the increasing complexity of modern society"[96] and to "avoid unnecessary constraints on provincial legislative action."[97]

Cooperative federalism is an enabling interpretation that facilitates each order of government enacting laws within its sphere of authority. The preference for an approach to federalism that accommodates cooperation and overlap has "often played a role in upholding the validity or constitutional operability of provincial legislation, particularly when the legislature has acted in an area in which Parliament has also legislated or over which there is a federal aspect."[98] Cooperative federalism has also been used to support federal legislation and interlocking federal/provincial legislative schemes.[99] While the Supreme Court of Canada has recognized the role of cooperative federalism in accommodating overlapping jurisdiction and encouraging intergovernmental cooperation, it has similarly warned that the principle cannot override nor modify the fundamental distribution of powers.[100]

3.2.2.5 Pith and Substance

The pith and substance doctrine is core to the characterization stage. It is key to ensuring that legislation remains anchored by a respective class of jurisdictional authority.[101] To identify the pith and substance of a provision or statute, courts may consider both intrinsic (such as the preamble and purpose clauses of legislation) and extrinsic (such as the Hansard or minutes of Parliamentary committees) evidence.[102] The pith and substance must be identified as precisely as possible to avoid the risk of exaggerating the extent to which the legislation or provision extends into the other level of government's jurisdictional space.[103] However, precise does not

---

96   *Genetic Non-Discrimination*, above note 67 at para 22.

97   *Ibid* at para 23.

98   *Genetic Non-Discrimination*, above note 67 at para 23.

99   See, e.g., *Pan-Canadian Securities*, above note 67 at para 18; *Genetic Non-Discrimination*, above note 67 at para 24.

100   *IAA Reference*, above note 5 at para 122. See also *GGPPA References*, above note 1.

101   *GGPPA References*, above note 1 at para 51; *Genetic Non-Discrimination*, above note 67 at paras 28 and 166.

102   *GGPPA References*, above note 1 at para 51.

103   *Ibid* at para 52.

necessarily mean narrow — the ultimate goal is to characterize what the law is ultimately "all about" so it may be properly classified at the second stage of analysis.[104]

Whether the means by which a statute's purpose is achieved is part of the pith and substance will depend on the circumstances. Although the means used to carry out the law's purpose is not determinative of the pith and substance, it may be appropriate to include, especially when the main thrust of a statute is closely tied to its means (as was the case with the *GGPPA*).[105]

3.2.2.6   Double Aspect Doctrine

Closely related to cooperative federalism is the double aspect doctrine, which recognizes that some subject areas or fact situations have both provincial and federal aspects and can thus validly co-exist. As noted by the Supreme Court of Canada, there is no need to "kill one (law) and let the other live"[106]. By engaging in a flexible approach to constitutional interpretation, the courts can allow each jurisdiction maximum flexibility to enact laws within their respective jurisdictional spheres without constraining the other order of government's scope of legislative authority. In *Tsilhqo'tin Nation v British Columbia*, the Supreme Court of Canada offered forestry on Aboriginal title land as an example of a domain that, for constitutional purposes, "possesses a double aspect, with both orders of government enjoying concurrent jurisdiction."[107] Indeed, courts need to recognize a triple aspect to acknowledge inherent Indigenous jurisdiction in addition to provincial and federal authority.[108] In *GGPPA References*, the Supreme Court of Canada confirmed the application of the double aspect doctrine to matters justified federally under the national concern branch of POGG, resolving uncertainty regarding the application of the doctrine

---

104   *Ibid* at para 52.

105   *Ibid* at para 53.

106   *Multiple Access Ltd v McCutcheon*, 1982 CanLII 55 at 182 (Supreme Court of Canada) [*Multiple Access*].

107   *Tsilhqot'in Nation v British Columbia*, 2014 Supreme Court of Canada 44 at para 129.

108   Dayna Nadine Scott et al, "Constitutional Cases 2024 (Pt 2), Environmental Regulation and the Constitution" (Osgoode's Annual Constitutional Cases Conference, 12 April 2024), online (video): digitalcommons.osgoode.yorku.ca [perma.cc/JLP2-Q2S3].

to the national concern branch.[109] This careful balancing of double aspect with the potential for intrusion on provincial autonomy is an ongoing theme in division of powers jurisprudence and central to climate policy given how systemically integrated GHG emissions are to almost every aspect of the economy and the lives of Canadians.

3.2.2.7 Ancillary Powers

What if the pith and substance of a piece of legislation is justifiable under one head of power but certain provisions of that law have effects on a matter that would normally fall under the other order of government's jurisdiction? The ancillary powers doctrine allows some incidental provisions to be upheld even if they fall outside of the enacting government's jurisdiction if they are sufficiently connected to a valid legislative scheme and further the legislative purpose (the rational and functional connection test).[110] The otherwise invalid provisions can be constitutionally upheld if they are found to be "ancillary," "integral," or "necessarily incidental" to the legislative scheme that comes under a valid head of power.[111] There is a form of sliding scale applied to the analysis: if the intrusion on the other jurisdiction is substantial, it will only be justified if it is truly necessary to the effectiveness of the legislative scheme; if the intrusion is more minor, it can be justified if it can be shown to be functionally connected.[112]

The test for determining whether a provision can be validated under the ancillary powers doctrine has been considered many times by the

---

109 *GGPPA References*, above note 1 at para 130.

110 *Assisted Human Reproduction*, above note 71 at para 126. See also *Walter v Attorney General of Alberta*, 1969 CanLII 64 (Supreme Court of Canada), in which Alberta passed a law that forbade the new acquisition of communally owned land without provincial permission. The Court found that the law was passed in order to prevent the Hutterites, a religious group, from acquiring more communal land, which the group did as part of its religious practice. However, the Court found that the law was valid because it was "in relation to" property, a provincial power under s 92(13) of the *Constitution Act, 1982*, see above note 11, and it did not single out the Hutterites but rather prevented any group from acquiring land in common.

111 *Whitbread v Walley*, 1990 CanLII 33 at para 13 (Supreme Court of Canada).

112 *Assisted Human Reproduction*, above note 71 at para 127.

Supreme Court. In *Re GST*,[113] the Court was asked to consider the constitutional validity of the federal value-added goods and services tax (GST), including its system of input tax credits. The Court determined that the purpose of the GST was clearly to raise revenue, which was authorized by the federal taxation power in section 91(3). However, the Court agreed with the province of Alberta's argument that the GST affects matters which fall within provincial jurisdiction under section 92(13) to legislate in relation to property and civil rights in the province.[114] The Court therefore needed to consider the extent of the "overflow" into provincial matters and whether the incursion was "necessarily incidental" to the exercise of the federal taxation power and well-integrated into the scheme of the Act. The Court ultimately concluded that the GST, including the input tax credits, was a valid exercise of the federal taxation power, despite the incursion into provincial domain. The Court returned to the functional or necessary connection test in the *Assisted Human Reproduction Reference*, confirming that the more necessary ancillary provisions are to the effectiveness of the valid provisions, the more overflow into the other jurisdictions will be tolerated.[115] Viewed the other way, the more serious the overflow, the "closer the relationship between the impugned provisions and the otherwise valid statute must be."[116] In the *Assisted Human Reproduction* case, the Court was divided in its application of the doctrine. Chief Justice McLachlin and others determined that although the administrative, organizational, and enforcement provisions of the Act intruded in varying degrees into provincial matters, they were integrated with the prohibitive parts of the Act, functioning merely to assist in enforcing the Act and therefore valid. Justices LeBel and Deschamps held that some of the ancillary provisions were not necessary to the prohibitory provisions and thus not justified.[117]

---

113   *Reference re Goods and Services Tax*, 1992 CanLII 69 (Supreme Court of Canada) [*Re GST*].

114   *Ibid* at 29.

115   *Assisted Human Reproduction*, above note 71 at para 274.

116   *Ibid* at para 275.

117   *Ibid* at paras 276–79.

The ancillary powers doctrine could be quite important in a constitutional analysis of GHG policy, since a federal climate law regulating GHG emissions will almost inevitably have some incidence on provincial matters. The assessment will turn on the design of the federal measure and the degree of impact on provincial matters, as well as whether there is a strong rational connection between the law and any ancillary provisions that flow into provincial matters.

3.2.2.8 Paramountcy

The doctrines of paramountcy and interjurisdictional immunity (discussed next) establish the rules in the event of a conflict between federal and provincial laws. With minor exceptions, the Constitution offers no mechanisms for dealing with conflicts between validly enacted provincial and federal laws.[118] Perhaps the drafters of the Constitution did not anticipate the complexity of our future society and regulation, or assumed Parliament's declaratory, reservation and disallowance powers would be used to address conflict.[119] In the absence of clear rules, the courts developed the paramountcy doctrine. As per the paramountcy doctrine, in the event of a conflict, the provincial provision or law is suspended, becoming inoperative to the extent of the conflict and only as long as the conflict exists.[120]

---

118 Section 95 of the *Constitution Act, 1982* assigns concurrent jurisdiction over immigration and agriculture, stating that provincial laws are inoperative if they are repugnant to federal laws; section 92A grants provincial legislatures jurisdiction to enact laws related to natural resources exports to other provinces subject to federal paramountcy in the event of conflict; section 94A creates concurrency for old-age pensions and additional benefits but confers provincial paramountcy in the event that federal laws could affect the operation of such provincial laws. See *Constitution Act, 1982*, above note 11, ss 92A, 94A, and 95. See also Mathen & Macklem, above note 57 at 225.

119 See *Constitution Act, 1982*, above note 11, ss 90 and 92(10)(c). See also Mathen & Macklem, above note 57 at 225.

120 Mathen & Macklem, above note 57 at 226. See also *Multiple Access*, above note 106 at 191; *Canadian Western Bank v Alberta*, 2007 Supreme Court of Canada 22 at para 124 [*Canadian Western Bank*].

In *Rothmans, Benson & Hedges Inc v Saskatchewan*,[121] the court summarized the state of the law on paramountcy as dictating that "where there is an inconsistency between validly enacted but overlapping provincial and federal legislation, the provincial legislation is inoperative to the extent of the inconsistency"[122] The courts have been careful to interpret the paramountcy doctrine narrowly to limit its potential to bias jurisdiction in Parliament's favour. In *Multiple Access Ltd v McCutcheon*,[123] for instance, the Supreme Court of Canada held that federal paramountcy applies only where there is impossibility of dual compliance (in other words, a direct or genuine operational conflict between the laws where one cannot comply with both at the same time) or where the purpose of the federal measure would be frustrated by allowing the provincial measure to persist.[124] In other words, there are essentially two branches. The first involves an actual conflict in operation in the sense that one law says "yes" and the other says "no."[125] The second branch applies if the federal purpose is frustrated. The scope for frustration is fairly narrow. Legislation that is simply duplicative, for instance, will not generally be considered a conflict. A more restrictive provincial law that provides an entitlement rather than being permissive, however, could frustrate the federal purpose.[126]

There are several examples of the Supreme Court of Canada finding there to be no incompatibility between overlapping provincial and federal laws.[127] For instance, the Supreme Court of Canada held that there was no

---

121   2005 Supreme Court of Canada 13 [*Rothmans*].

122   *Ibid* at para 11.

123   *Multiple Access*, above note 106.

124   *Ibid* at 163.

125   Nigel Bankes & Martin Olszynski, "Going Through the Motions to Trigger the Sovereignty Act: Another Paper Tiger?" (16 December 2024), online: ablawg.ca [perma.cc/64XY-LYR6]. Duplicative federal and provincial enactments are not generally considered conflicting. *Ibid*.

126   *Alberta (AG) v Moloney*, 2015 Supreme Court of Canada 51 at paras 22–29 [*Moloney*]. See also Martin Z. Olszynski, "Testing the Jurisdictional Waters: The Provincial Regulation of Interprovincial Pipelines" (2018) 23:1 Rev Const Stud at 95 – 96.

127   See, e.g., Rothmans above note 121 at paras 22–23 (where the Supreme Court of Canada held that a retailer could easily comply with both the provincial and federal statutes in question in that case).

incompatibility between provincial farm securities legislation and the federal *Bankruptcy and Insolvency Act*, referencing cooperative federalism as an approach that favours a harmonious interpretation of provincial and federal laws.[128] It is only in the narrow and rare circumstances of a true direct conflict that makes dual compliance impossible or a genuine frustration of federal purpose that the doctrine of paramountcy will apply to render the provincial portion of the legislation that engages the conflict inoperative.[129] The burden of proving an operational conflict or frustration of purpose (which is on the party alleging such) is a high one, meaning that the paramountcy principle will be applied with restraint.[130] In the context of cooperative federalism, such restraint is necessary to allow "for interplay and overlap between federal and provincial legislation."[131]

The author is aware of no major environmental or climate-related laws that have been rendered inoperative due to paramountcy.[132] In the *Spraytech* decision, Justice L'Heureux-Dubé characterized federal legislation governing pesticides as permissive legislation, thereby avoiding the potential to characterize a municipal bylaw restricting pesticides as coming into conflict with the federal law.[133] No one was placed in the impossible situation of not being able to comply with both laws, and the bylaw did not frustrate Parliament's purpose in regulating pesticides. In the *Orphan Wells* case, the court was invited to find the federal *Bankruptcy and Insolvency Act* paramount over provisions of Alberta's *Environmental Protection and Enhancement Act*, which imposes obligations on companies to clean up abandoned oil wells. The Supreme Court of Canada held there was no conflict or frustration of purpose between the two regimes.[134]

---

128 *Saskatchewan (AG) v Lemare Lake Logging Ltd*, 2015 Supreme Court of Canada 53 at para 21 [*Lemare*].

129 *Ibid* at paras 21–22.

130 *Orphan Well Association v Grant Thornton Ltd*, 2019 Supreme Court of Canada 5 at para 66 [*Orphan Well*].

131 *Ibid* at para 66.

132 However, note that the National Energy Board (as it then was) declared several of Burnaby's bylaws inoperative and inapplicable in the context of the Transmountain Pipeline. *See* Martin A Olszynski, "Testing the Jurisdictional Waters: The Provincial Regulation of Interprovincial Pipelines" (2018) 23:1 Rev Const Stud.

133 *Spraytech*, above note 23 at para 35.

134 *Orphan Well*, above note 130.

It is possible that this doctrine may become more important in the context of conflicting policy objectives related to climate change, especially rules governing the stringency and timing of GHG emissions reductions. If provincial laws are stricter, and the federal law is not permissive (as in Spraytech) but rather confers an entitlement, the prospect of impossibility of dual compliance arises. The key in such a case would be to identify what exactly is prohibited and assess the degree of conflict and/or frustration.[135]

3.2.2.9 Interjurisdictional Immunity

The doctrine of interjurisdictional immunity serves to insulate one level of government from regulation by the other in certain narrowly defined core areas of competence. It is a vestige of the classical approach to federalism that valued exclusivity and watertight compartments of authority over the flexibility and overlap that is now a central feature of cooperative federalism.[136] When it applies, courts will read down the provisions of broad legislation in one jurisdiction to protect the core areas of legislation in the other jurisdiction from encroachment. Although the doctrine can apply to federal or provincial legislation, it has been successfully invoked only to protect core areas of federal jurisdiction.[137]

In *Bell Canada v Quebec (Commission de la Santé et de la Sécurité du Travail)*,[138] Beetz J applied the doctrine to hold that a provincial statute regulating health and safety in the workplace did not apply to a federal undertaking, in that case a trucking business with exclusively inter-provincial and international operations.[139] However, he held that the regulation had to affect the basic, minimum, and unassailable content or core of legislative power in question and a vital core of the federal undertaking to apply.[140] A decade later, the Supreme Court of Canada narrowed the doctrine in several ways, holding it applies only to laws

---

135 See dissenting reasons by Côté J in *Moloney*, above note 126 at para 110.

136 *Canadian Western Bank v Alberta*, above note 120 at para 34.

137 See, e.g., *PHS*, above note 71 at paras 57–70; *Carter v Canada (AG)*, 2015 Supreme Court of Canada 5 at paras 49–53.

138 1988 CanLII 81 (Supreme Court of Canada) [*Bell Canada*].

139 *Ibid* at 750.

140 *Ibid* at para 254.

that *impair* (not just affect) an essential part of an undertaking[141] and limited its application to areas of jurisdiction in which the doctrine has been invoked in the past. This includes federal jurisdiction over things, persons, or undertakings versus activities (e.g., federal works and undertakings including aviation, ports, interprovincial rail and trucking, and federal telecommunications).[142]

While the doctrine has not been repudiated completely, it stands in tension with the tide of cooperative federalism that relies upon the doctrines of pith and substance, double aspect, and paramountcy to rationalize jurisdiction in an increasingly complex society,[143] and where courts "should favour, where possible, the ordinary operation of statutes enacted by both levels of government."[144] The Supreme Court of Canada has also indicated that the interjurisdictional immunity doctrine should be considered only after applying the pith and substance and paramountcy doctrines.[145]

In *Insite*, the Supreme Court of Canada noted that the interjurisdictional immunity doctrine "has never been applied to a broad and amorphous area of jurisdiction," such as health, noting that the doctrine is "in tension with the dominant approach that permits concurrent federal and provincial legislation with respect to a matter, provided the legislation is directed at a legitimate federal or provincial aspect, as the case may be" and cooperative federalism.[146] The Supreme Court of Canada underscored the risks of applying the doctrine, noting that it:

> may overshoot the federal or provincial power in which it is grounded and create legislative "no go" zones where neither level of government regulates. Since it is not necessary for the government benefiting from

---

141 *Canadian Western Bank*, above note 120 at para 48.
142 See *PHS*, above note 71 at para 60; *Canadian Western Bank*, above note 120 at para 34. See also *Rogers Communications Inc v Châteauguay (City)*, 2016 Supreme Court of Canada 23 at paras 61–63; Mathen & Macklem, above note 57 at 256.
143 *Canadian Western Bank*, above note 120 at para 42.
144 *Ibid* at para 37.
145 *Ibid* at para 77.
146 See *PHS*, above note 71 at paras 62–63 and 70.

the immunity to actually regulate in the field in question, extension of the doctrine of interjurisdictional immunity risks creating "legal vacuums."[147]

The issue of whether the doctrine applies to shield federal undertakings from provincial environmental regulations has arisen in environmental cases. In *Ontario v Canadian Pacific Ltd*,[148] a federally regulated railway was charged under Ontario's environmental legislation for its discharge of a contaminant (smoke). The Supreme Court of Canada ruled that the provincial law was applicable to the federal railway. In two subsequent Supreme Court of Canada cases involving provincial legislation, the Court opted not to apply the doctrine, holding that the provincial legislation did not impact the core of federal legislative jurisdiction.[149] Without eliminating the doctrine altogether, the Supreme Court of Canada nonetheless further restricted its relevance, noting a preference for the paramountcy doctrine instead of interjurisdictional immunity to resolve conflicts.

In *Reference re Environmental Management Act (British Columbia)*,[150] the Court considered the application of provincial environmental rules requiring permits for heavy fuel oil, which would have applied principally to oil transported by the federally regulated Transmountain Pipeline. The Court decided the case without having recourse to the interjurisdictional immunity doctrine, as it determined the pith and substance of the environmental regulation was to regulate a substance to be carried only by a federal pipeline, rendering it *ultra vires* provincial authority.[151] As such, interjurisdictional immunity will likely have limited application to climate policy, since presumably the pith and substance and paramountcy doctrines will be used to resolve jurisdictional questions.

---

147 *PHS*, above note 71 at para 64, citing *Canadian Western Bank*, above note 120 at para 44.

148 *Canadian Pacific*, above note 23.

149 See, e.g., *Canadian Western Bank*, above note 120; *British Columbia (Attorney General) v Lafarge Canada Inc*, 2007 Supreme Court of Canada 23 [*Lafarge*].

150 2019 BCCA 181 [*EMA Reference*], leave to appeal to Supreme Court of Canada refused, 2020 Supreme Court of Canada 1.

151 *Ibid* at para 101.

3.3 THE DIVISION OF POWERS OVER CLIMATE LAWS

Since virtually all laws have implications for climate change and could therefore be described as climate law, the analysis here is limited to consideration of laws aimed at mitigating GHG emissions. The findings and tensions discussed in the chapter are likely relevant to other areas of climate policy. Readers are invited to reflect on how jurisdictional authority will be interpreted as we move further into our climate-impacted future.

3.3.1 Provincial Jurisdiction

3.3.1.1 Scope and Breadth of Provincial Jurisdiction

Provinces have a broad range of jurisdictional authority to regulate activities inside the province. Much of this jurisdiction stems from the provinces' authority to regulate property and civil rights under section 92(13) of the *Constitution Act, 1867*. This power is broad and far-reaching, enabling the provinces to govern matters of property, succession, families, contracts, torts, labour relations, business and industrial activity, subject to aspects of such matters that fall under federal jurisdiction, such as marriage, banking, interest, navigation and shipping, and interprovincial and international transportation and telecommunications.[152] Additionally, section 92(16) grants provinces the authority to regulate over "all matters of a merely local or private nature in the province." This residuary power has played a lesser role in light of the broad interpretation given to property and civil rights.[153] Provinces also have authority to legislate in relation to the management and sale of public lands in the province, including timber (s 92(5)). As compared to the federal government (and not accounting for Indigenous ownership and jurisdiction), provinces own an average of 80 percent of the land within their borders and retain the rights to most of the minerals beneath the land, including the 20 percent that

---

152 See Peter W Hogg & Wade Wright, *Constitutional Law of Canada*, 5th ed (Thomas Reuters, 2024) at 21.3.

153 *Ibid* at 21.4. See, e.g., *Seimens v Manitoba (Attorney General)* 2003 Supreme Court of Canada 3 at para 22.

is privately held.[154] Provinces also have jurisdiction over direct taxation within the province (s 92(2)), municipal institutions (s 92(8)), raising revenue by licenses (s 92(9), local works and undertakings other than interprovincial works (s 92(10)), and non-renewable, forestry, and electrical energy resources (s 92A).[155] The province's power over the administration of justice (section 92(14)) was recently interpreted to allow it to pass a law enabling a class action — led by the province and including a class consisting of all federal, provincial, and territorial governments and agencies that paid opioid-related healthcare costs — to recovery those costs from opioid manufacturers, marketers, and distributors.[156]

Known as the resources amendment, section 92A was added to the Constitution in 1982 at the same time as the *Charter of Rights and Freedoms*. The addition of section 92A was intended to confirm, clarify, and enhance the provinces' jurisdictional authority over resource exploitation within their borders. The amendment was motivated in part by the oil crisis of the 1970s, which led to high oil prices for Canadians and sparked debate in Canada about jurisdiction over resources and, importantly, their rents.[157]

Section 92A confirms provincial authority to legislate over natural resources (specifically forestry and electricity generating facilities) and enhances jurisdiction in relation to interprovincial trade in natural resources and taxation of natural resources and production of electricity.[158] The section includes provisions clarifying that provincial jurisdiction over interprovincial trade in natural resources does not derogate from federal authority, and similarly that nothing in section 92A derogates any authority from provinces.[159] Subsection 1 of section 92A is focused primarily on internal resource control whereas subsections 2 and 3 address

---

154 Harrison, above note 27 at 68.
155 For a detailed discussion of the history and judicial interpretation of section 92A, see Bankes & Leach, above note 57.
156 *Sanis Health Inc. v British Columbia*, 2024 Supreme Court of Canada 40.
157 Bankes & Leach, above note 57 at 856–57.
158 *Constitution Act, 1982*, above note 11, ss 92A(1), 92A(2), and 92A(4).
159 *Ibid*, ss 92A(3) and 92A(5). See also Bankes & Leach, above note 57 at 861.

export or external aspects of resource control.[160] Section 92A(2) enhances the jurisdictional authority of provinces by granting them authority to regulate exports of non-renewable and forestry resources from the province to another part of Canada, without price or supply discrimination. The provision does not authorize the province to regulate international trade, including exports. Section 92A(4) extends provincial authority to tax natural resources and electricity generation using indirect, as well as direct, taxation.

Judicial interpretation of section 92A has confirmed that the provision does not, unsurprisingly, extend the reach of provincial jurisdiction beyond provincial borders.[161] The Supreme Court of Canada has also confirmed that jurisdiction conferred in section 92A(1) over electrical generation in the provinces does not constrain federal authority, whether through POGG or the declaratory power.[162] The Supreme Court of Canada has underscored, however, that the bargain struck in section 92A must be recognized and respected.[163]

Provinces have enacted a wide range of environmental laws under this collection of powers. Indeed, Canadian provinces (and municipalities) have enacted laws regulating discharges to air and water, along with rules relating to waste management, land-use, pesticides, and drinking water.[164] There have been relatively few challenges to provincial authority to enact environmental laws. In contrast, there have been several challenges to federal environmental legislation that have ended up before the Supreme Court of Canada (for example, *Crown Zellerbach*, *Hydro-Quebec*, and *Oldman River*).[165]

---

160 Bankes & Leach, above note 57 at 871, citing William D Moull, "Section 92A of the Constitution Act, 1867" (1983) 61:4 Can Bar Rev 715.

161 *Reference re Newfoundland Continental Shelf*, 1984 CanLII 132 (Supreme Court of Canada) [*Continental Shelf*].

162 *Ontario Hydro v Ontario (Labour Relations Board)*, 1993 CanLII 72 (Supreme Court of Canada) [*Ontario Hydro*].

163 *IAA Reference*, above note 5 at 233.

164 See, e.g., *Environmental Protection Act*, RSO 1990, c E19; *Water Protection Act*, RSBC 1996, c 484; *Environmental Management and Protection Act, 2010*, SS 2010, c E-10.22.

165 *Crown Zellerbach*, above note 3; *Hydro-Quebec*, above note 3; *Oldman River*, above note 22.

Canadian provinces began enacting climate-related policies in the mid-2000s and provincial regulatory frameworks for climate change continue to rapidly evolve.[166] The suite of provincial policies includes a mix of regulation (reducing industrial emissions, increasing efficiency, and phasing out coal), province-wide targets, accountability legislation, spending initiatives,, carbon pricing, and rules requiring adaptation measures from municipalities.

The provinces of Alberta and Quebec were early actors on carbon pricing, each establishing a modest carbon price in 2007, through the *Specified Gas Emitters Regulation* (establishing the first output-based carbon price for oil sands producers) and the *Redevance Annuelle* respectively.[167] In 2008, British Columbia established a classically designed carbon tax, with an initial ten dollars per ton price that increased by gradual increments (subject to one freeze for a few years under Premier Clark) to eighty dollars today.[168] Quebec was the first jurisdiction to launch a cap and trade system in 2013, linked with the Californian market through the Western

---

166   These are described in detail in Annex II of the Pan-Canadian Framework on Clean Growth and Climate Change (see Canada, Environment and Climate Change Canada, *Pan-Canadian Framework on Clean Growth and Climate Change* (Gatineau: Environment and Climate Change Canada, 2016), online: publications.gc.ca [perma.cc /Q4MF-CPLA]). This period is known as a second phase of climate federalism. The first phase of climate federalism, which ran from 1990 to 2007, was marked by inaction, where the norm of consensus decision-making led some provinces to block action at a national scale (see Harrison, above note 27 at 69–70). A third phase of climate federalism started in 2015 with a change in federal government. While the phase began with a fragile consensus, represented in the Pan-Canadian framework on Climate Change, that consensus degraded quickly with the enactment of the national carbon pricing system, even though the four most populous provinces had a pricing system in place when the backstop was enacted (see Harrison, above note 27 at 72, 78).

167   *Specified Gas Emitters Regulation*, Alta Reg 139/2007; *Règlement sur la redevance annuelle payable à la Régie de l'énergie*, CQLR c R-6.01, r 7.

168   See Government of British Columbia, "Motor Fuel Tax and Carbon Tax" (last visited 15 January 2025), online: gov.bc.ca [perma.cc/N3LV-22CP]. A classically designed carbon tax taxes the negative externality of CO_2 and is revenue neutral, using the revenue to reduce other distortionary taxes. Revenues from BC's carbon tax are returned to taxpayers in the province through tax measures, such as reductions in personal and corporate income tax rates and a refundable low-income tax credit (see *Carbon Tax Act*, SBC 2008, c 40).

Climate Initiative.[169] Ontario followed suit with a cap and trade program launched in 2017 but cancelled it one year later after a change in government.[170] With the advent of the federal *Greenhouse Gas Pollution Pricing Act* establishing minimum national standards for carbon pricing, BC, Alberta, Saskatchewan, Ontario, Quebec, New Brunswick, Nova Scotia, and Newfoundland and Labrador have all enacted some form of carbon price.[171] When Prime Minister Mark Carney replaced Prime Minister Justin Trudeau in March 2025, he eliminated the federal fuel charge backstop in Part 1 of the Act. While Part 2 of the *GGPPA* remains in place, Saskatchewan opted to eliminate its Output-Based Performance Standards which means it will no longer be in compliance with the federal benchmark.[172]

In addition to carbon pricing, provinces have enacted a variety of other climate policies aimed at mitigating GHG emissions or adapting to climate change.[173] Ontario was the first Canadian jurisdiction to eliminate coal-fired electricity generation between 2003 and 2015, contributing an estimated twenty-five Mt of GHG emissions reductions.[174] Nova

---

169 For Quebec regulation on cap-and-trade see *Règlement concernant le système de plafonnement et d'échange de droits d'émission de gaz à effet de serre*, CQLR c Q-2, r 46.1; on the harmonization of California and Quebec cap-and-trade systems see OC 1181-2013, (2013) GOQ II, 3389. See generally International Carbon Action Partnership, "Canada - Quebec Cap-and-Trade System," online: icapcarbonaction.com [perma.cc /QG4V-C22Q].

170 Government of Ontario, "Archived - Cap and Trade in Ontario" (last modified 12 July 2021), online: ontario.ca [perma.cc/7TE7-EHZ6]; See also *Climate Change Mitigation and Low-carbon Economy Act, 2016*, SO 2016, c 7 (creating the program) and *Cap and Trade Cancellation Act, 2018*, SO 2018, c 13 (cancelling the program).

171 Government of Canada, "Carbon Pollution Pricing Systems Across Canada" (last modified 3 May 2024), online: canada.ca [perma.cc/F9YQ-9ZBW]. See also Government of Canada, "Update to the Pan-Canadian Approach to Carbon Pollution Pricing 2023-2030" (last modified 5 August 2021), online: canada.ca [perma.cc/NM6W-7LND].

172 Alexandar Quon, "Saskatchewan is Now Carton Tax Free, but Some Wonder About the Cost" (April 1, 2025) online: cbc.ca [perma.cc/X99R-CXTY].

173 An audit of climate policies across the country in 2018 concluded that while all but two jurisdictions (Saskatchewan and the Northwest Territories) had high-level mitigation strategies with some actions planned to achieve GHG emissions reductions, but most lacked details such as timelines and implementation plans (see Office of the Auditor General of Canada, *Perspectives on Climate Change Action in Canada - A Collaborative Report from Auditors General* (2018), online: oag-bvg.gc.ca [perma. cc/776S-3N33].

174 Harrison, above note 27 at 70.

Scotia was the first province to require local climate action plans by municipalities in 2014 as a condition of receiving gas tax funding.[175] BC was first out of the gates implementing climate accountability legislation that requires specific emissions reduction targets and reporting on progress toward meeting those targets (though it failed to meet the 2020 target set under that law).[176]

To date, no provincial GHG emissions reduction law has been subject to a jurisdictional challenge of which the author is aware.[177] If they were, they would likely be found to be safely anchored in one or more of the provincial heads of authority, assuming the pith and substance of the legislation were aimed at the source of the emissions.[178] For instance, regulations governing emissions from industrial actors within the province would be justifiable under the property and civil rights power, which authorizes provinces to "regulate manufacturing, farming, mining and related activities that generate GHGs."[179] Provincial carbon pricing policies, which come in a variety of forms, would likely be justifiable under the property and civil rights power but also the authority over direct taxation

---

175   Commissioner for Environment and Sustainable Development, *Report 5 – Lessons Learned from Canada's Record on Climate Change* (Officer of the Auditor General of Canada, 2021), online: oag-bvg.gc.ca [perma.cc/9C9Z-JU3N] at 22. See also Government of Nova Scotia, *Municipal Climate Change Action Plan Guidebook* (Halifax: Service Nova Scotia and Municipal Relations, Canada-Nova Scotia Infrastructure Secretariat, 2011), online: beta.novascotia.ca [perma.cc/RF86-V2TL].

176   *Climate Change Accountability Act*, SBC 2007, c 42.

177   One possible exception to this may be BC's effort to amend the *Environmental Management Act* to require permits for the release of heavy oil in the province was not technically a GHG emissions reduction law, though it would have had implications for the Trans Mountain Pipeline expansion (see *EMA Reference*, above note 150).

178   If the pith and substance were aimed at something intangible, the court needs to consider the "relationships among the enacting territory, the subject matter of the legislation and the persons made subject to it to determine whether the legislation, if allowed to stand, would respect the dual purpose of the territorial limitations in s. 92," referring to the need for provincial legislation to have a meaningful connecting to the enacting province and respects the legislative authority of other jurisdictions. See *British Columbia v Imperial Tobacco, Canada Ltd*, 2005 Supreme Court of Canada 49 at para 36.

179   *GGPPA References* ONCA, above note 32 at para 193. The Alberta Court of Appeal emphasized s 92A as a source of jurisdiction for regulations on GHG emissions (see *GGPPA References* ABCA, above note 32 at para 265).

(s 92(2)), taxation of natural resources (s 92A(4)), and even the power to impose regulatory fees and levies (s 92(9)), depending on their design.[180] Justice Strathy in the Ontario Court of Appeal judgment (which was upheld by the Supreme Court of Canada) stated that there "is no question that the provinces can address climate change by imposing fuel and excess emission charges."[181] In his dissent in the *GGPPA References* at the Supreme Court of Canada, Brown J commented that the various provincial carbon pricing policies would likely be authorized by sections 92(13) or 92(16), and in the case of Part 2 of the Act, section 92A.[182]

As the Supreme Court of Canada explained in *GGPPA References*, there is a double aspect to carbon pricing where both Parliament and the provinces can regulate to impose carbon pricing under distinct heads of power.[183] In the event of genuine conflict or frustration of purpose, the federal law would prevail under the doctrine of paramountcy.[184] This, of course, is where the tension lies. What happens when jurisdictions disagree with respect to GHG emissions requirements, specifically the mechanisms used (such as carbon pricing or regulation) and/or the stringency of that mechanism? We have an answer from the Supreme Court of Canada with respect to minimum standards of price stringency for GHG emissions — the federal system prevails. If a province wishes to impose a higher price, they are not precluded from doing so since the federal law imposes a floor but not a ceiling. If a province wishes to impose a lower

---

180 See Nathalie J Chalifour, "Making Federalism Work for Climate Change: Canada's Division of Powers over Carbon Taxes" (2008) 22:2 NJCL 119 at 154–199 [Chalifour, "Making Federalism Work"].

181 *GGPPA References* ONCA, above note 32 at para 194.

182 See *GGPPA References*, above note 1 at paras 343–47. Justice Brown held that the identification of areas of provincial authority for the subject matter of the *GGPPA* settled the matter of the federal law being *ultra vires*, barring application of the double aspect or ancillary powers doctrines which he held did not apply here (see *GGPPA References*, above note 1 at para 373).

183 *GGPPA References*, above note 1 at para 129. The Supreme Court of Canada clarified that the double aspect doctrine applies even when the federal power is the national concern doctrine (see *GGPPA References*, above note 1 at para 125).

184 *Ibid* at para 129.

price, the jurisdictionally affirmed federal law applies through the backstop mechanism.[185]

What if, however, Parliament (perhaps after a change in government) were to reframe the *GGPPA* so it imposes a price ceiling? Would a provincial law that surpasses that price be in direct conflict or frustrate the purpose of the federal legislation, triggering either branch of the paramountcy doctrine? It is unlikely that a direct conflict would be found. If a private sector actor challenged the constitutionality of the provincial law (such as BC's carbon tax), a court might hold that the measure is within the province's jurisdiction under sections 92(2), 92(9), or 92(13) and that the company can easily comply with both policies by paying the higher price in BC. Dual compliance is possible since — barring issues related to the design of the system and collection — paying the higher BC tax would necessarily complete payment of the lower federal charge.[186] It is not impossible to pay both.

However, if the federal provision was aimed at maintaining a low carbon price (perhaps to facilitate market access for fossil fuel companies located in Canada), could paying the higher provincial price be considered in defiance of the federal ceiling, frustrating its purpose and triggering the second branch of the paramountcy doctrine?[187] Whether a more stringent provincial carbon price would frustrate the federal purpose of a carbon pricing ceiling would depend on how the purpose of the federal measure was construed. If the federal carbon price was aimed at addressing climate change, reducing GHG emissions or some such related purpose, there would be no frustration of having a higher price in a province. However, if the purpose of a price ceiling was explicitly to maintain a lower price (to promote or protect the fossil fuel industry or other market players, for instance), providing a positive entitlement to those regulated, there could be a valid argument that the purpose would be frustrated by mandating a higher price elsewhere (depending, once again, on how the price was

---

185 *Ibid.*
186 See *Multiple Access*, above note 106 at 191.
187 *Bank of Montreal v Hall*, 1990 CanLII 157 at 130 (Supreme Court of Canada). See also *Rothmans*, above note 121 at paras 13–14.

administered). If a price ceiling was aimed at protecting a given industry, its jurisdictional validity might be called into question as encroaching upon provincial jurisdiction.

The Supreme Court of Canada has been clear in rejecting the idea that merely legislating in relation to a matter means Parliament occupies a field rendering provincial action in respect of that subject inoperative due to paramountcy.[188] As the Supreme Court of Canada has noted, "[t]he standard for invalidating provincial legislation on the basis of frustration of federal purpose is high; permissive federal legislation, without more, will not establish that a federal purpose is frustrated when provincial legislation restricts the scope of the federal permission."[189] The Supreme Court of Canada has noted that a conflict will not arise "where a provincial law is more restrictive than a federal law," though it could "frustrate the federal purpose if the federal law, instead of being merely permissive, provides for a positive entitlement."[190] The court has also left open the possibility that a federal intention to occupy the field could be a basis for paramountcy but only where "very clear statutory language to that effect" was used.[191]

In *Saskatchewan (AG) v Lemare Lake Logging Ltd*,[192] the Supreme Court of Canada cited the principle of cooperative federalism in support of the idea that "harmonious interpretations of federal and provincial legislation should be favoured over interpretations that result in incompatibility."[193] Relatedly, the Supreme Court of Canada has stated:

> In keeping with co-operative federalism, the doctrine of paramountcy is applied with restraint. It is presumed that Parliament intends its laws to co-exist with provincial laws. Absent a genuine inconsistency, courts will favour an interpretation of the federal legislation that allows the

---

188 *Ibid* at para 21.

189 *Quebec (Attorney General) v Canadian Owners and Pilots Association*, 2010 Supreme Court of Canada 39 at para 66.

190 *Moloney*, above note 126 at para 26.

191 *Canadian Western Bank*, above note 120 at para 74. See also the critique of the court's application of the paramountcy doctrine in *Lafarge*, above note 149. See also Carissima Mathen & Michael Plaxton, "Developments in Constitutional Law: The 2006–2007 Term" (2007) 38 SCLR (2d) 111 at 134–36.

192 *Lemare*, above note 128.

193 *Ibid* at para 21.

concurrent operation of both laws … Conflict must be defined narrowly, so that each level of government may act as freely as possible within its respective sphere of authority.[194]

To sum up, there is a very limited potential of the paramountcy doctrine applying to provincial GHG emissions reduction regulations. The doctrine might operate, for instance, if a federal law creates an entitlement for unpriced GHG emissions or a GHG emissions price ceiling at a level lower than what a province requires, and a court finds that the higher provincial price frustrates the purpose of the federal entitlement. Even there, however, it is not clear that a federal law creating an entitlement for unpriced GHG emissions would be *intra vires* of Parliament, since *GGPPA References* created jurisdiction over *minimum* (not maximum) standards of price stringency. Also, as discussed in Chapter 4, it is possible that *Charter* protections and/or unwritten constitutional principles will eventually require a baseline level of GHG reductions by all orders of government, which could limit efforts to enact policies encouraging increases in GHG emissions.

Occupying a space to not regulate

The conferral of law-making authority to provinces is confined to jurisdiction "in the province," as per the opening language of section 92.[195] This is uncontroverted. The Supreme Court of Canada in the *Interprovincial Cooperatives* decision confirmed that "a province has no authority to legislate in relation to acts done outside the province, even if those acts cause damaging pollution to enter the province."[196] While none of

---

194　*Moloney*, above note 126 at para 27.

195　*Continental Shelf*, above note 162 at 128.

196　*GGPPA References*, above note 1 at para 99, referring to the *Interprovincial Co-operatives* decision. In *Interprovincial Co-operatives Ltd v R*, 1975 CanLII 212 (Supreme Court of Canada) [*IPCO*] a three-judge majority penned by Pigeon J found that the authority to legislate in relation to interprovincial pollution could not be provincial, which meant it was federal by default. Ritchie J concurred in the result but drew upon the fisheries power to justify the reasons and found that the impugned provincial legislation was valid but inapplicable to the polluting facilities outside the province. See also Andrew Leach, "I Can't Believe It's Not Butter: Supply Management and the Constitutionality of Carbon Budget Legislation in Canada" (2023) 56:2 UBC L Rev 493 at 521–22 [Leach, "I Can't Believe It's Not Butter"].

the judges in the *Interprovincial Cooperatives* case expressly cited POGG, all judges endorsed the view that the federal government has jurisdiction over interprovincial rivers and courts and scholars have since noted that POGG best explains the result.[197] It is clear that provinces cannot regulate activity in other provinces, even when such activity causes harm (i.e., through pollution into waterways or airways that cross provincial boundaries).[198] Assuming constitutional authority is fully allocated so that there are no gaps, Parliament should logically have jurisdiction to regulate interprovincial pollution.

A different but related question arises as to whether authority to regulate within the province equates to exclusive occupation of the relevant jurisdictional space. It is logical to assume that a province may choose whether to legislate on a matter within its jurisdiction, and if so, select the means. The more interesting question is whether a province, in occupying that jurisdictional space, can preclude Parliament from regulating in relation to that matter. This question harkens back to a watertight interpretation of powers that results in a zero-sum allocation whereby authority to legislate can only exist in one place.[199] In his dissent in *GGPPA References*, Brown J stated that provinces can decide whether "to act – or not act – as they see fit" within areas of provincial jurisdiction.[200] He went further in stating that the exclusive allocation of powers in the Constitution means provinces are sovereign within their sphere of jurisdiction, meaning they can occupy that space, even if only to preclude federal regulation.

The only way a province could occupy the space would be to convince a court that the doctrine of interjurisdictional immunity applies to protect the core of that matter against any incursion, even incidental, by Parliament. In light of the fact that the doctrine of interjurisdictional immunity has not been applied to amorphous areas of jurisdiction such as health or environment, that there are few/no precedents of it being applied to

---

197 *IPCO*, above note 196 at 514, cited in the majority opinion in *GGPPA References*, above note 1 at para 99.

198 Leach, "I Can't Believe It's Not Butter," above note 196 at 521.

199 Chalifour, Oliver & Wormington, above note 94 at 181.

200 *GGPPA References*, above note 1 at para 394.

provincial jurisdiction, and its overall decrease in importance, it would not likely offer any support to this argument.[201]

The majority provided a different answer to the question of whether a province's choice not to price GHG emissions could preclude Parliament from doing so, based on the interplay of cooperative federalism and the double aspect doctrine. While provinces can choose not to legislate on a matter within their jurisdiction, they cannot preclude Parliament from regulating an aspect of that matter that lies within Parliament's jurisdiction. The majority's clarification of the provincial inability test under POGG (discussed later) as applying when a province *choosees* to do nothing (rather than lacking the ability to do something), and this choice has extra-provincial effects, was an important clarification that addresses this point. A province's choice not to legislate does not preclude Parliament from legislating in relation to a matter within the latter's authority, even if it touches on the same activity.

To conclude, provinces have wide-ranging jurisdiction over activities within the province that generate GHG emissions. Their jurisdiction is limited when it comes to regulating emissions that come from outside the province.

3.3.2 Federal Jurisdiction

The main powers used to justify federal environmental legislation include fisheries, navigable waters, criminal law, POGG national concern, and to some extent jurisdiction over Indigenous issues.[202] Other powers that may be relevant but have not been used overtly used to justify environmental legislation (or have had only tangential relevance) include trade and commerce, international treaties, aviation and shipping, and the emergency branch of POGG.[203] The main sources of jurisdiction

---

201 See *Canadian Western Bank*, above note 120 at para 44.

202 See, e.g., *Fisheries Act*, RSC 1985, c F-14; *Canadian Navigable Waters Act*, RSC 1985, c N-22; *Oceans Act*, SC 1996, c 31;*CEPA*, above note 29 *Yahey v British Columbia*, 2021 BCSC 1287. See generally Benidickson, above note 26.

203 See, e.g., *GGPPA References*, above note 1 at paras 399–403 (dissenting Brown J discussing the emergency branch); *Crown Zellerbach*, above note 3 at paras 33–34; *Hydro-Quebec* above note 3 at para 64 (dismissing use of the emergency branch).

for federal climate-related laws to date have been criminal law and the national concern doctrine of POGG. This section discusses both sources of jurisdiction and also explores the emergency branch, trade and commerce, taxation, and declaratory power. It begins with a discussion of the authority to spend.

3.3.2.1 Spending

There is no explicit power to spend within the Constitution. However, the courts have confirmed that federal and provincial governments are free to spend the money they raise (whether through taxation or their property) on matters even if they fall outside of their legislative authority, provided they do not, in so doing, seek to regulate matters outside their jurisdiction.[204] The federal government uses its spending power extensively.

As explained by the Supreme Court of Canada in *YMHA Jewish Community Centre of Winnipeg Inc v Brown*, the federal spending power "must be inferred from the power to levy taxes (s. 91(3)), to legislate in relation to public property (s. 91(1A)), and to appropriate federal funds (s. 106)."[205] While the act of spending itself falls within the jurisdiction of the executive branch, such spending must always be approved by Parliament (as per the notion of responsible government). In the case of federal transfers of funds to provinces, Parliament can either approve the amount and give the federal government full discretion to impose any conditions to such transfers or it can approve the amount and legislate certain conditions other parties must fulfill in order to receive such transfers.[206]

The federal government currently has four major federal transfer programs that provide financial support to provinces and territories, including

---

204 See, e.g., *Winterhaven Stables Ltd v Canada (Attorney General)*, 1988 ABCA 334 [*Winterhaven*], 21262, leave to appeal to Supreme Court of Canada refused, [1989] 1 SCR xvi; *Confédération des syndicats nationaux c Canada (Procureur général)*, 2006 QCCA 1454, revd in part in *Confédération des syndicats nationaux v Canada (Attorney General)*, 2008 Supreme Court of Canada 68. See also *Reference re Canada Assistance Plan (BC)*, 1991 CanLII 74 at 567 (Supreme Court of Canada) [*Canada Assistance Plan Reference*]; Hogg & Wright, above note 152 at 6.8–6.9.

205 Hogg & Wright, above note 152 at 6.8.

206 *Ibid*.

the Canada Health Transfer, the Canada Social Transfer, Equalization, and the Territorial Formula Financing.[207] The latter two programs provide unconditional transfers, whereas the former two may have conditions attached.[208] The federal government also transfers all funds raised through the carbon pricing system either directly to individuals through a carbon rebate (known as the Climate Action Incentive Payment) or through programs funding Indigenous governments and small- and medium-sized businesses.[209]

The federal government has long funded health and social programs in the provinces, even though many aspects of these subjects are outside their competence to legislate. The funding has also sometimes been associated with conditions, meaning that the federal government has influence over provincial policy by, for instance, dictating the conditions under which funding will be released. For example, approximately half of the recent health care transfers to provinces have been conditioned upon provincial commitments to improve collection, sharing, and use of health data.[210] These transfers are the basis for Canada's entire system of socialized health care.

Several cases have considered whether spending directed at matters of provincial jurisdiction, especially when associated with conditions, is an intrusion on provincial jurisdiction. In *Winterhaven Stables v Canada*,[211] the Court acknowledged that the spending power gives the federal government a great deal of influence over matters of provincial concern, particularly when the transfers are tied to federal conditions. However, the

---

207 Government of Canada, "Federal Transfers to Provinces and Territories" (last modified 13 November 2024), online: canada.ca [perma.cc/ES2Z-FUXN].

208 For an overview and critique of federal transfers, see Jonathan Deslauriers et al, *Federal Transfers to the Provinces: Setting the Record Straight* (Montréal: Centre for Productivity and Prosperity, 2021), online: cpp.hec.ca [perma.cc/4H42-H569].

209 The *Greenhouse Gas Pollution Pricing Act* mandates that all revenue generated from the fuel charge be returned to the province of origin (see *Greenhouse Gas Pollution Pricing Act*, SC 2018, c 12, ss 186 and 188(1) [*GGPPA*]). See also Government of Canada, "Government of Canada Returning $872.4 million in 2024-25 Fuel Charge Proceeds to Indigenous Governments and Small- and Medium-Sized Businesses" (last modified 16 February 2024), online: canada.ca [perma.cc/U5FX-RFMA].

210 See Health Canada, "Working Together to Improve Health Care for Canadians" (last modified 15 February 2023), online: canada.ca [perma.cc/S4SG-Y8RP].

211 *Winterhaven*, above note 204 at para 20.

Court found that as long as any conditions relate to how the government spends its money and is not legislation that, in pith and substance, seeks to regulate an area of exclusive provincial jurisdiction, it will be *intra vires* of Parliament. In addition, the courts noted that they will not evaluate the economic influence or effectiveness of valid conditional spending legislation.[212] In *Reference Re Canada Assistance Plan (BC)*, the Court rejected the argument that the withdrawal of federal funding to a province amounted to regulation.[213]

There has been a practice since the 1970s to establish cost-shared programs where the federal government dictates criteria for the program but generally only after establishment of a broad national consensus around the key elements of such program. In addition, it was always understood that a province could opt out of the program should it choose to.[214] As a result, the tendency over the last few decades has been for federal funding to allow as much flexibility as possible to the provinces and focus on requiring public reporting and accountability to voters. However, there are also examples of conditional spending where legislation authorizes the withholding of transfers based on established criteria. One such example comes from the health care sector. To preserve "the principle of free and universal access to health services throughout the country,"[215] Parliament unanimously adopted the *Canada Health Act*[216] in 1984. This Act introduced conditions to federal health grants in order to implement national norms. The legislation was generally opposed by most provinces as they feared increased public costs due to these new conditions. Nonetheless, following the enactment of the *Canada Health Act*, provinces amended noncompliant legislation to continue receiving funding.[217]

---

212 *Canada Assistance Plan Reference*, above note 204 at 566.

213 *Ibid* at 567.

214 See Hogg & Wright, above note 152 at 6.7.

215 Odette Madore, Parliamentary Information and Research Service, *The Canada Health Act: Overview and Options* (Ottawa: Library of Parliament, Parliamentary Research Branch, 2005) at 5.

216 *Canada Health Act*, RSC 1985, c C-6, s 7.

217 Morris Manning, "Canada Health Act: Unpalatable Carrot, Unconstitutional Stick" (1986) 134:10 CMAJ 1166 at 1166.

Under the *Canada Health Act*, federal funding to the provinces is contingent upon several criteria, including public administration, universality, and accessibility.[218] Each of the criteria is elaborated in some detail.[219] If a provincial government defaults on one of the five conditions, the federal Minister of Health is required to refer the matter to the Governor in Council, at which point a consultation process with the province is engaged.[220] Ultimately, the Governor in Council has the discretion to withhold funding.[221] In practice, the federal government has never used its discretionary power to sanction a provincial government in relation to a default on one of the conditions in spite of evidence to suggest the conditions have been violated.[222]

The federal government has committed over \$100 billion to climate-related initiatives between 2016 and 2027, with more than half of this spending still to come at the time of writing.[223] The funding ranges from investments in clean energy, transportation and public transit to adaptation and international climate finance.[224] Some of the expendi-

---

218 *Canada Health Act*, above note 216, s 7.

219 For instance, section 11 elaborates on the portability criteria by stating that "the health care insurance plan of a province; (a) must not impose any minimum period of residence in the province, or waiting period, in excess of three months before residents of the province are eligible for or entitled to insured health services" (see *Canada Health Act*, above note 216, s 11).

220 *Canada Health Act*, above note 216, s 14.

221 Madore, above note 215 at 11.

222 Natalie Mehra, *Eroding Public Medicare: Lessons and Consequences of For-Profit Health Care Across Canada* (Ottawa: Canadian Health Coalition, 2008) at 51–88. It has withheld funding from provinces when they used "extra billing" or "user charges" in accordance with the legislation (see, e.g., Health and Welfare Canada, *Canada Health Act, Annual Report*, 1993-95; Health and Welfare Canada, *Canada Health Act, Annual Report*, 1995-96; Health and Welfare Canada, *Canada Health Act, Annual Report*, 1998).

223 House of Commons, Standing Committee on Environment and Sustainable Development, *Evidence*, 44-1, No 9 (24 March 2022) at 1105 (Hon Steven Guilbeault). See also Marc Lee, Caroline Brouillette & Hadrian Mertins-Kirkwood, *Spending What it Takes: Transformational Climate Investments for Long-Term Prosperity in Canada* (Canadian Centre for Policy Alternatives & Climate Action Network, 2023), online: policyalternatives.ca [perma.cc/YW95-5DBH].

224 Lee, Brouillette & Mertins-Kirkwood, above note 223 at 8.

tures are sent directly to provinces and territories through initiatives such as the Green Infrastructure Fund and the Low Carbon Economy Fund.[225]

The climate funding that the federal government has spent or committed to date has not been conditioned upon a certain level of GHG emissions reductions. There have been recommendations to use federal funding to leverage initiatives that support GHG emissions reductions in the provinces.[226] Constitutionally, the federal government is relatively unconstrained to use climate financing in such a way that supports GHG reductions. The only legal caveat is that the spending must not be a disguised attempt to regulate in areas not within federal competence.[227] The jurisprudence clearly shows that this is a high threshold, which means that climate spending made conditional upon performance standards, for instance (such as actual emissions reductions or cost-effectiveness of those reductions) could withstand challenge, depending upon its design and purpose.

3.3.2.2 Criminal Law Power

The criminal law power has been a key source of jurisdictional authority for federal legislation relating to the environment for decades.[228] It has been a key source of authority for federal regulations relating to GHG

---

225 *Ibid* at 9. The International Institute for Sustainable Development recommended that funding in support of recovery after the COVID-19 pandemic be tied to seven environmental principles (see Vanessa Corkal, Philip Gass & Aaron Cosbey, *Green Strings: Principles and Conditions for a Green Recovery from Covid-19 in Canada* (Winnipeg: International Institute for Sustainable Development, 2020), online: iisd.org [perma.cc/TL9E-Z9CK].

226 Corkal, Gass & Cosbey, above note 225.

227 This would be considered colourable, which means that in form the spending measure would appear to be federal, but in substance it relates to a matter within provincial jurisdiction. See *Quebec (Attorney General) v Canada (Attorney General)*, 2015 Supreme Court of Canada 14 at para 31. See also footnotes 233–34 and accompanying discussion, below.

228 See, e.g., *Hydro-Quebec*, above note 3, upholding provisions of the *CEPA*, above note 29, under criminal law. See also *Groupe Maison Candiac Inc v Procureur general du Canada*, 2020 FCA 88, leave to appeal to Supreme Court of Canada refused, 2020 CanLII 97859, upholding the constitutionality of provisions of the *Species at Risk Act* under the criminal law power.

emissions since 2005, when GHGs were added to the list of toxic substances under CEPA. Since then, CEPA has been used to regulate a variety of GHGs. For instance, there are regulations requiring a minimum content of renewable fuels in gas and diesel,[229] reductions for the release of methane in the oil and gas sector,[230] reductions in GHG emissions from automobiles and light trucks,[231] reductions in GHG emissions from heavy-duty vehicles and engines,[232] reductions of CO_2 emissions from natural gas-fired electricity generation[233] and from coal-fired electricity generation,[234] and clean fuel regulations that require gasoline and diesel suppliers to reduce the carbon intensity of fuels.[235]

Jurisdiction under the criminal law power is broad. Aside from the *Charter*, the only qualification that applies to Parliament's plenary power over criminal law is that — once a measure satisfies the test for being a criminal provision — it cannot be used colourably.[236] The Federal Court

---

229 *Renewable Fuels Regulations*, SOR/2010-189 [*Renewable Fuel Regulations*].

230 *Regulations Respecting Reduction in the Release of Methane and Certain Volatile Organic Compounds (Upstream Oil and Gas Sector)*, SOR/2018-66. Note the equivalency agreements with BC, Saskatchewan, and Alberta (see *Order Declaring that the Provisions of the Regulations Respecting Reduction in the Release of Methane and Certain Volatile Organic Compounds (Upstream Oil and Gas Sector) Do Not Apply in British Columbia*, SOR/2020-60; *Order Declaring that the Provisions of the Regulations Respecting Reduction in the Release of Methane and Certain Volatile Organic Compounds (Upstream Oil and Gas Sector) Do Not Apply in Saskatchewan*, SOR/2020-234; *Order Declaring that the Provisions of the Regulations Respecting Reduction in the Release of Methane and Certain Volatile Organic Compounds (Upstream Oil and Gas Sector) Do Not Apply in Alberta*, SOR/2020-233).

231 *Passenger Automobile and Light Truck Greenhouse Gas Emission Regulations*, SOR/2010-201.

232 *Heavy-duty Vehicle and Engine Greenhouse Gas Emission Regulations*, SOR/2013-24.

233 *Regulations Limiting Carbon Dioxide Emissions from Natural Gas-fired Generation of Electricity*, SOR/2018-261.

234 *Reduction of Carbon Dioxide Emissions from Coal-fired Generation of Electricity Regulations*, SOR/2012-167. Note that this regulation does not apply in Nova Scotia or Saskatchewan due to equivalency agreements with those provinces (see *Order Declaring that the Reduction of Carbon Dioxide Emissions from Coal-fired Generation of Electricity Regulations Do Not Apply in Nova Scotia*, SOR 2014-265; *Order Declaring that the Reduction of Carbon Dioxide Emissions from Coal-fired Generation of Electricity Regulations Do Not Apply in Saskatchewan*, SOR/2019-167).

235 *Clean Fuel Regulations*, SOR/2022-140.

236 *Hydro-Quebec*, above note 3 at para 121.

of Appeal addressed the colourability argument in the *Syncrude* decision. The company argued that the renewable fuel regulations at issue were an incursion into the province's authority over non-renewable resources. The Court rejected this argument, pointing to the fact that the regulations applied to all consumers of diesel fuel and gasoline, not only those in the natural resources sector. Syncrude also argued that the regulations were a colourable attempt to regulate provincially controlled natural resources. In refuting this argument, the Court reiterated that "colourability is not lightly inferred, nor is it a backdoor to a reconsideration of the wisdom or efficacy of the law." Colourability is a high threshold, in that it requires Parliament's declared purpose to be mere pretense for intruding into provincial jurisdiction. It did not find that to be the case, since the regulations were clearly enacted pursuant to the broader goal of addressing climate change.[237]

While the criminal law power is sure to remain an important basis for federal climate legislation, there are some points of tension and uncertainty, especially after a decision of the Federal Court ruling that an order adding plastic manufactured items to Schedule 1 of *CEPA* was *ultra vires*. While that decision was under appeal at the time of writing, it sheds light on the debate. The tension is underpinned by the potential breadth of the power since a broadly interpreted criminal law power widens the reach of federal jurisdiction over matters that are also regulated provincially. The criminal law power has been described as plenary, in the sense that it is broad and interpreted flexibly so it can respond to new and emerging matters.[238] Like with POGG, the tension is mediated by the extent to which courts tolerate overlap and concurrency through doctrines such as the double aspect, pith and substance, and limited application of paramountcy. More tolerance protects provincial autonomy by allowing provincial laws on provincial aspects to co-exist, subject to the risks of paramountcy trumping provincial laws in the case of a direct conflict or frustration. However, cooperation and overlap are not a full answer for

---

237 *Syncrude* FCA, above note 30 at paras 87–93.

238 *GGPPA References*, above note 1 at para 121. See also *RJR-MacDonald*, above note 71 at paras 28 and 32; *Hydro-Quebec*, above note 3 at para 213; *Genetic Non-Discrimination*, above note 67 at para 69.

some provinces concerned about federal overreach through the criminal law power — some provinces seek exclusive authority to determine the extent of requirements to reduce GHG emissions emanating from within their borders. As Parliament continues to roll out GHG-related regulations, the scope of the criminal law power will remain highly relevant and will likely be subject to challenges. This section will consider its past use and the prospect of it justifying a variety of federal GHG regulations.

3.3.2.2.1 Criminal Law Power and Environmental Regulation

The first major Supreme Court of Canada decision interpreting the criminal law power in the environmental context was the 1997 *Hydro Quebec* decision. In that 5-4 decision, the majority upheld provisions of the *Canadian Environmental Protect Act (CEPA)* regulating toxic substances under section 91(27).[239] The Supreme Court of Canada articulated and applied a three-part test to determine whether a law or provision can be constitutionally justified as criminal law. The test is now well-established and was recently applied in the *Reference re Genetic Non-Discrimination Act*.[240] For a provision to be justified under Parliament's criminal law power, it must be: (1) a prohibition; (2) accompanied by a penalty; and (3) backed by a criminal law purpose.[241] Some justices have grouped the first two criteria together and characterized them as formal requirements, with the criminal law purpose representing a more substantive element.[242]

In *Hydro Quebec*, the Supreme Court of Canada stated that "[t]he purpose of the criminal law is to underline and protect our fundamental values ... [T]he stewardship of the environment is a fundamental value of our society and ... Parliament may use its criminal law power to underline

---

239 *Hydro-Quebec*, above note 3.

240 *Genetic Non-Discrimination*, above note 67.

241 *Ibid* at para 67. See also *Assisted Human Reproduction*, above note 71 at paras 35–36; *Hydro-Quebec*, above note 3 at paras 34–36; *Firearms Reference*, above note 67 at para 27.

242 *Assisted Human Reproduction*, above note 71 at para 233, minority decision. See also *Firearms Reference*, above note 67 at para 27; *Genetic Non-Discrimination*, above note 67 at para 229 (noting that the tripartite test "ensures Parliament cannot use its authority improperly to invade upon provinces' areas of competence, thus ensuring the balance of federalism is respected").

that value. The criminal law must be able to keep pace with and protect our emerging values."[243] This section examines both parts of the test and considers their application in the context of climate laws.

3.3.2.2.2 A Criminal Law Purpose

There is no debate in the courts that environmental protection is a valid criminal law purpose. While they disagreed on other matters, all nine Supreme Court Justices in *R v Hydro Quebec* agreed that environmental protection is a valid criminal law purpose.[244] Justice La Forest for the majority characterized environmental protection as a "public purpose of superordinate importance,"[245] stating that "pollution is an 'evil' that Parliament can legitimately seek to suppress." He added that "[i]t would be surprising indeed if Parliament could not exercise its plenary power over criminal law to protect this interest and to suppress the evils associated with it by appropriate penal prohibitions."[246]

The courts have been consistent in holding that the environment is sufficient as its own valid criminal law purpose and does not need to be affiliated with one of the other recognized purposes, such as health or security.[247] When articulating the list of harms to the public interest which could justify the use of the criminal law power in its most recent decisions since *Hydro Quebec*, the courts often reference the environment.[248]

The d/evil is in the details

Although it has not (to date) proven to be difficult to meet this part of the test in environmental cases, establishing that legislation has a criminal law purpose has been controversial in other cases. The debate has

---

243 *Hydro-Quebec*, above note 3 at para 127.

244 *Ibid* at paras 43 and123.

245 *Ibid* at para 123.

246 *Ibid*.

247 *Ibid* at para 132; *Syncrude* FCA, above note 30 at para 49. See also *Responsible Plastic Use Coalition*, above note 51 at para 139.

248 See, e.g., *Genetic Non-Discrimination*, above note 67 at para 68. A law is said to be backed by a criminal law purpose if it is intended to respond to a threat of harm to a public interest, such as peace, order, security, health, morality, and the environment (see *Genetic Non-Discrimination*, above note 67).

centred on the extent to which legislation must be aimed at suppressing an "evil" or serious harm to qualify as federal criminal law. The rationale for resorting to the concept of evil is to focus the power on more traditional forms of criminal law rather than ordinary (sometimes beneficial) activities that would otherwise fall within provincial jurisdiction.[249] This tension animated the debate in the *Assisted Human Reproduction* reference and ultimately divided the court. Chief Justice McLachlin's reasons focused on identifying a valid criminal law purpose, holding that there is no required threshold of harm to trigger criminal law jurisdiction, as long as there is some reasonable apprehension of harm.[250] She justified her approach based, in part, on a need for flexibility, offering pollution as an example, noting that "the list of toxic substances capable of harming the populace is ever-changing. It is unrealistic to expect Parliament to enact new laws every time a change occurs, and the criminal law power does not require it to do so."[251] In contrast, Lebel and Deschamps JJ held that a criminal law provision must be directed at a "real evil" where there is a "real apprehension of harm."[252] The provisions of the legislation in the *Assisted Human Reproduction case* were ultimately upheld as valid criminal law by a tie-breaking set of reasons that did not take a position on the necessity of evil.[253]

---

249 Eric M Adams, "Touch of Evil: Disagreements at the Heart of the Criminal Law Power" (2022) 104 SCLR 67 [Adams, "Touch of Evil"]. The broader approach, which stems from Rand J's jurisprudence, does not require the purpose to target evil but rather a criminal law purpose that engages the public nature of its concern (e.g., peace, order, health, morality, environment) and the public nature of the harm it seeks to limit or suppress or the public interest it aims to safeguard. In other words, the focus is on public harms rather than evil (see *ibid*).

250 *Syncrude* FCA, above note 30 at para 75. See also *Assisted Human Reproduction*, above note 71 at paras 55–56.

251 *Assisted Human Reproduction*, above note 71 at para 36.

252 *Ibid* at paras 39–48 and 240.

253 *Ibid* at paras 282–94 (reasons by Cromwell J). While Cromwell J did not dwell on the necessity of evil, he noted that the law criminalized "negative practices" that "fall within the traditional ambit of the federal criminal law power" (see *ibid* at para 291). See also Joshua Ginsberg, *Biodiversity and the Canadian Constitution: Opportunities for Broader Federal Action to Protect and Restore Nature* (LLM Research Paper, University of Ottawa, 2024).

The debate continued in the *Genetic Non-Discrimination* case which divided the court at both the characterization and classification stages of the analysis. With respect to the criminal law purpose, Moldaver J (with Côté J) summarized the difference of opinion in that case between the majority (Karakatsanis J, with Abella and Martin JJ) and Kasirer J (with Wagner, Rowe, and Brown JJ) as follows:

> Justice Karakatsanis would hold that "[a]s long as Parliament is addressing a reasoned apprehension of harm to one or more of [the public interests protected by the criminal law], no degree of seriousness of harm need be proved before it can make criminal law" (para. 79). By contrast, Justice Kasirer would require something more — he would hold that Parliament must be responding to a "threat [that is] 'real', in the sense that Parliament had a concrete basis <u>and</u> a reasoned apprehension of harm" (para. 234 (emphasis added)).[254]

Justice Moldaver declined to weigh in on this question, concluding that the legislation satisfied the requirement under either approach.[255] The legislation was ultimately upheld as valid criminal law with the purpose of protecting autonomy, privacy, equality, and public health.[256]

The threshold that divides the two camps relates to risks of harm. What animates the narrower camp's desire for a more limited criminal law purpose is the need to find some way to limit the scope and breadth of the criminal law power. Justice Kasirer explained in the *Genetic Discrimination* case that the word "evil" helps distinguish between merely undesirable effects and those that are aimed at suppressing some real threat.[257] According to this perspective, if the criminal law power extends to regulation of merely undesirable effects, it would be too broad.

Justice Kasirer drew upon reasons in the *Margarine Reference* and LeBel and Deschamps JJ's reasons in the *AHR Reference* suggesting that

---

254 *Genetic Non-Discrimination*, above note 67 at para 138.
255 *Ibid* at para 138.
256 *Genetic Non-Discrimination*, above note 67 at paras 4 and 80.
257 *Ibid* at para 233.

three questions must be considered in determining whether a law rests on a valid criminal law purpose: (1) does the provision relate to a "public purpose," such as health, security, morality or public peace; (2) did Parliament identify a well-defined threat the provision would be aimed at preventing; and, (3) did Parliament have a concrete basis and reasonable apprehension of harm when enacting the provision or legislation (in other words, is the threat "real")?[258] While Kasirer J agreed that health relates to a public purpose, satisfying the first prong, he held that the promotion of health is not sufficiently precise to meet the second prong of the test, since this did not identify a well-defined threat to be suppressed.[259] He offered tobacco,[260] dangerous food products,[261] illicit drugs,[262] firearms,[263] and toxic substances[264] as examples of threats that are sufficiently well-defined to meet the test.

One of the key factors that troubled Kasirer J (and the minority in the *AHR Reference*) was the positive aspect of the legislation. The law aimed (in Kasirer J's view) at promoting health, and some of what the *AHRA* regulated were beneficial practices. According to this perspective, regulating beneficial and health-promoting practices does not constitute a valid criminal law purpose. Rather, the criminal law power should justify only legislation that punishes or provides the stick that motivates behaviour changes, rather than the carrot. The implications for environmental regulation are important. Under this narrower view, a public criminal purpose aimed at protecting environmental health generally could be considered insufficiently related to a well-defined threat. It is likely that regulations aimed at reducing GHG emissions to a level consistent with the science would meet even this stricter test, since the harm from climate change is existential and largely irreversible on a human timescale. However, it is widely recognized that a precautionary approach that prevents

---

258 *Ibid* at para 236.
259 *Ibid* at para 242.
260 *RJR-MacDonald*, above note 71.
261 *R v Wetmore*, 1983 CanLII 29 (Supreme Court of Canada) [*Wetmore*].
262 *R v Malmo-Levine*, 2003 Supreme Court of Canada 74.
263 *Firearms Reference*, above note 67.
264 *Hydro-Quebec*, above note 3.

environmental harm is better than trying to remediate damage once it has occurred.[265] As such, this narrower approach may be limiting for environmental regulations aimed at preventing harm.

A recent decision of the Federal Court suggests the more restrictive interpretation of a criminal law purpose may be gaining favour. In the *Responsible Plastic Use* case, the Federal Court held that an order adding plastic manufactured items (PMIs) to Schedule 1 of *CEPA* exceeded the jurisdictional bounds of the criminal law power since not all PMIs are toxic. The Court pointed to examples of products that the government itself had not identified as environmentally problematic, such as some disposable personal care items, and determined that the inclusion of all PMIs on the list of toxic substances in an unqualified manner — regardless of the potential for those items to cause harm — exceeded Parliament's authority.[266] The Federal Court clarified that the criminal law purpose of environmental protection must be calibrated to harm, interpreting the *Hydro-Quebec* decision as finding environmental protection to be a criminal law purpose only insofar as it protected the environment through the prohibition against toxic substances. It then held that PMIs did not pose environmental harm as a broad group. The Federal Court held that to employ criminal law, what is being restricted has to actually be dangerous, signalling an affinity for the more narrow approach to the criminal law power.[267] The Court distinguished carbon dioxide as an example of a substance that while not inherently toxic may have aspects or uses that are toxic.[268] In contrast, the broad category of PMIs included in the Order under review included items that had, as per the Court, no reasonable apprehension of environmental harm.[269] It remains to be seen whether the finding that not all PMIs have the potential to cause environmental harm will be maintained on appeal.

---

265 Jose Felix Pinto-Bazurco, "The Precautionary Principle: Still Only One Earth – Lessons from 50 years of UN Sustainable Development Policy" (23 October 2020), online: iisd. org [perma.cc/L42H-WHJ4].

266 *Responsible Plastic Use Coalition*, above note 51 at para 119.

267 *Ibid* at para 179.

268 *Ibid* at para 184.

269 *Ibid*.

Reducing GHG emissions a valid criminal law purpose

To date, the courts have been consistent in holding that laws aimed at reducing air pollution satisfy even the narrower approach to the criminal law power. For instance, in *Hydro-Quebec*, the majority ensured there was no ambiguity about the criminal law purpose of legislating for a clean environment, characterizing pollution as an "evil" that Parliament can seek to suppress through its criminal law power.[270] In the *Assisted Human Reproduction* case — although the Supreme Court of Canada was split over the determination of whether the legislation in that case had a valid criminal law purpose — both the majority and the minority (which subscribed to the more narrow view) pointed to laws aims at reducing pollution as an example of laws that would have a valid criminal purpose, since polluting has the potential to harm the environment.[271]

In the *Syncrude* case, the Federal Court of Appeal confirmed that addressing climate change is a valid criminal law purpose. In that case, Syncrude had challenged the constitutionality of section 5(2) of the *Renewable Fuel Regulations* under the *Canadian Environmental Protection Act (CEPA)*, which require a minimum content of renewable fuel in gasoline and diesel.[272] Interestingly, Syncrude conceded from the beginning that the *Regulations* would address a valid criminal purpose if their dominant purpose was to combat climate change. However, Syncrude argued that their dominant purpose was to create demand for biofuels in the Canadian marketplace (a matter within provincial jurisdiction).

The Federal Court of Appeal found the pith and substance of the regulations to be directed at "maintaining the health and safety of Canadians, as well as the natural environment upon which life depends."[273] Recall that GHGs had been added to schedule 1 of CEPA in 2005, after a determination that GHGs posed a risk to human health and the environment.[274] In

---

270 *Hydro-Quebec*, above note 3 at para 123.

271 *Assisted Human Reproduction*, above note 71 at paras 36 and 235.

272 *Renewable Fuel Regulations*, above note 229.

273 *Syncrude* FCA, above note 30 at para 41.

274 Adding a substance to Schedule 1 of CEPA enables federal regulations under the Act to apply to that substance;, Section 64 of CEPA defines toxic substances (under Schedule 1) as a substance that is or "may enter the environment in a quantity or concentration or under conditions that (a) have or may have an immediate or long-term

drawing a link between the regulations and the criminal law purpose of protecting health and the environment, Rennie J for the Federal Court of Appeal stated that "it is uncontroverted that *GHGs* are harmful to both health and the environment and as such, *constitute an evil* that justifies the exercise of the criminal law power" (emphasis added).[275] While the regulations may be aimed at creating demand for renewable fuels, the driving purpose for doing so is to lower GHG emissions.

The Alberta Court of Appeal briefly considered the criminal law power in the *Reference re the Impact Assessment Act*. While it found that the impugned section 7(1) satisfied the criteria of constituting a prohibition coupled with a penalty, the Court held that it did not contain a criminal law purpose.[276] The Court identified several concerns. First, the provision did not exclusively prohibit harmful effects but rather any change to components of the environment (thus being vague and overbroad). Second, the provision only applies to projects on the designated list, which creates a checkerboard of criminal liability, something the Court held is not contemplated by the criminal law power. Finally, the Court also held that the prohibition and accompanying penalties are ancillary to what it concluded is a regulatory (versus a criminal) regime in the *IAA*.[277] While the Supreme Court of Canada agreed that core provisions of the *IAA* were *ultra vires*, it did not address the criminal law power in any depth.

As such, the preponderance of jurisprudence to date suggests that safeguarding the climate by reducing GHG emissions is a valid criminal law purpose even under the narrower test, given clear findings about the grave impacts of GHG emissions. Legislation aimed at reducing GHG emissions

---

harmful effect on the environmental or biological diversity; (b) constitute or may constitute a danger to the environment on which life depends, or (c) constitute or may constitute a danger in Canada to human life or health" (see *CEPA*, above note 29). The Government of Canada's justification for adding GHGs to Schedule 1 was explained in the report, Environment Canada, *The Kyoto Protocol Greenhouse Gases (GHGs) and the Canadian Environmental Protection Act: A Synthesis of Relevant Science from the IPCC Third Assessment Report in the Context of CEPA Section 64* (Gatineau: Environment Canada, 2005), online: publications.gc.ca [perma.cc/JL4X-R82E].

275 *Syncrude* FCA, above note 30 at para 62.
276 *IAA Reference*, above note 5 at paras 403–7.
277 *IAA Reference*, above note 5 at paras 405–7.

to further this purpose is thus likely to satisfy the first step in the analysis. However, there is ongoing uncertainty about using criminal law to justify regulations that promote environmentally responsible behaviour — especially when broadly worded legislation gives the executive branch wide discretion, as illustrated by the ongoing plastics litigation.

3.3.2.2.3 Prohibition Coupled with a Penalty

To be justified as a matter of criminal law, a provision must also satisfy the requirement that it constitutes a prohibition coupled with a penalty. There is an important distinction to make between the criminal law purpose and the means used to achieve that purpose. The means relate to this other part of the test requiring a prohibition and penalty. The means used to achieve the criminal law purpose have been the source of discussion within the climate jurisdiction cases. In the *Syncrude* case, the company argued that the regulations at issue were not valid criminal law because they were not prohibitions, as they simply required a minimum content (2 percent or 5 percent) of renewable fuel and even allowed for some exceptions. The Federal Court rejected this argument. First, the Court emphasized that a prohibition need not be complete — it can admit exceptions.[278] Second, the court clarified that criminal law is meant to modify behaviour. Syncrude had argued that the use of the regulations to create demand for renewable fuels was too indirect a way to achieve the GHG reduction goal. The Court disposed of this argument by noting that increased market demand for renewable fuel due to the regulation reinforced its dominant purpose. The federal government regularly uses its criminal law power to regulate activities tangential to behaviours it seeks to discourage, such as smoking and prostitution. Instead of outright banning smoking or prostitution, Parliament influences behaviour by criminalizing related activities, such as advertising. Similarly, it is impossible to completely prohibit all

---

278 The Court offered examples of regulations that set limits or concentrations of listed substances in not only the environmental context but also in the food industry (see *Syncrude* FCA, above note 30 at para 73).

GHG emissions, so Parliament regulates them in a way that reduces their public harm, sometimes using indirect means (such as pricing) to do so.[279]

In the *Genetics Discrimination* case, Kasirer J agreed that Parliament can exercise its criminal law power using indirect means to achieve its purpose, such as deterring the use of tobacco through a variety of measures aimed at marketing and sale of the product. In contrast, Brown J in the *GGPPA References* characterized the carbon pricing system as one that discourages, rather than prohibits, activities, which he held removes it from the prospect of justification as criminal law.[280] The majority in *GGPPA References* did not engage in a criminal law analysis.

Tolerance for overlap

It is not uncommon for jurisdiction over criminal law to grant to Parliament authority to regulate in areas that are also within provincial jurisdiction. For example, public health was recognized as a valid criminal purpose in *Morgentaler v The Queen*,[281] even though provinces are considered to have general jurisdiction over public health as matters of a purely local nature.

In *Hydro-Quebec*, the Court noted that the criminal law's plenary power affords Parliament broad discretion to determine what harms it wishes to suppress and what interests it wishes to safeguard.[282] It also noted, however, that this broad power does not interfere with provincial powers, since provinces are not precluded from exercising their extensive powers to regulate the environment under section 92.[283] The Court offered an analogy to the protection of health where Parliament has regulated food and drugs safety through the criminal law power, which has not prevented the provinces from extensively regulating and prohibiting many activities relating to health.[284]

---

279 *Syncrude* FCA, above note 30 at paras 69, 84, and 91.
280 *GGPPA References*, above note 1 at para 404.
281 1975 CanLII 8 at 625 (Supreme Court of Canada)[*Morgentaler*].
282 *Hydro-Quebec*, above note 3 at para 119.
283 *Ibid* at para 131.
284 *Ibid* at para 131.

The *Clean Fuel Regulations* (CFR), which came into force in 2023 are aimed at decreasing pollution by reducing the carbon intensity of fuels produced and sold in Canada.[285] They are similar to the *Renewable Fuel Regulations* (RFR) that were upheld by the Federal Court of Appeal in *Syncrude* but introduce additional elements, such as a wider range of compliance options. The RFR will eventually be phased out as the CFR supersede them. Given their similarity, there is a good chance the CFR would be upheld if challenged, unless there is a significant shift in approach to interpreting the criminal law power. Similarly, federal regulations pertaining to methane emissions under CEPA would likely withstand challenge, assuming the regulations are aimed at addressing the harms of climate change.

The *Clean Electricity Regulations (CER)* aim to regulate the emissions of CO_2 from electricity generation by establishing an emissions performance standard related to GHG intensity, with prohibitions associated with emissions that exceed the standards.[286] Established under CEPA, the regulations aim to achieve net-zero emissions from electricity grids by 2050 (a date pushed back from 2035 in response to opposition from provinces for which the transition would be especially difficult). Despite rolling back the deadline for decarbonizing the electricity grid, Alberta and Saskatchewan have each promised to challenge the constitutionality of the regulations.[287] Saskatchewan used its *Saskatchewan First Act* to draw its own conclusion about the encroachment of the federal regulations on provincial jurisdiction over electricity generation, declaring it would refuse to comply with the regulations.[288]

In November 2024, the federal government released a draft of the *Oil and Gas Sector Greenhouse Gas Emissions Cap Regulations* under CEPA. The regulations would cap GHG emissions from certain activities in the oil and

---

285 Environment and Climate Change Canada, "What are the Clean Fuel Regulations?" (last modified 7 July 2022), online: canada.ca [perma.cc/AE9X-CHCV].

286 Environment and Climate Change Canada, "Proposed Frame for the Clean Electricity Regulations" (last modified 26 July 2022), online: canada.ca [perma.cc/86QF-TLMV].

287 See, e.g., Matthew Black, "Ottawa Pushes Back Net-Zero Electricity Target to 2050 as Alberta Readies Court Challenge" *Edmonton Journal* (17 December 2024), online: edmontonjournal.com [perma.cc/2RFW-ZJ8B].

288 Government of Saskatchewan, "Saskatchewan Rejects Federal Clean Electricity Regulations" (18 December 2024), online: saskatchewan.ca [perma.cc/QMW9-YWG3].

gas sector through the distribution of emissions allowances to regulated entities for free starting in 2029, based on historical emissions. By limiting the total number of allowances to the emissions cap, the regulation would limit GHG emissions to the level of the cap. Regulated entities would be allowed to use certain flexible compliance units, such as those earned through eligible offset credits or contributions to a decarbonization fund.[289]

The province of Alberta has already declared its intention to challenge the constitutionality of the regulations, arguing that it is a measure aimed at capping oil and gas production that encroaches upon the province's jurisdiction over natural resources under section 92A of the *Constitution Act*.[290] The province will likely emphasize not only its jurisdiction over natural resources but also the fact that the regulations have the effect of targeting an industry of central importance to their economy. If the regulations were found to be intended to regulate an industry, this would strengthen Alberta's argument. However, the federal government will argue that the measure is aimed not at regulating production of oil and gas but rather GHG emissions to address climate change from an industry that is the largest source of Canada's GHG emissions.[291] The federal government will likely aim to justify the measure constitutionally under the criminal law power, though it could possibly rely upon other sources of authority, such as POGG, discussed next. Unless current interpretations of the criminal law power take a significant turn or the courts interpret the pith and substance of the cap as something other than regulations aimed at reducing GHG emissions to address the existential harms of climate change, the regulations would likely be upheld in a federal court challenge.

---

289 Proposed Regulation (Oil and Gas Sector Greenhouse Gas Emissions Cap Regulations), (2024) C Gaz I, 3264 (established under the *Canadian Environmental Protection Act, 1999*), online: canadagazette.gc.ca [perma.cc/6XBJ-JVJM].

290 Chris Varcoe, "Varcoe: 'They're Going to Come After the Industry' – Oilpatch and Province Push Back Against Federal Emissions Cap," *Calgary Herald* (4 November 2024), online: calgaryherald.com [perma.cc/2EZH-5YGS].

291 Environment and Climate Change Canada, *Greenhouse Gas Emissions: Canadian Environmental Sustainability Indicators* (Gatineau: Environment and Climate Change Canada, 2022), online: canada.ca [perma.cc/7THB-M5H6].

To sum up, the jurisprudence to date suggests that the criminal law power will continue to offer a solid basis upon which to regulate GHG emissions in Canada. While some uncertainty remains in the wake of the plastics decision and ongoing debates about what level of harm is needed to justify criminal provisions, courts have consistently held that climate change presents a grave threat and that it is essential to reduce GHG emissions to mitigate the threat. As such, criminal law justification for regulatory measures related to GHG emissions remains solid. The test for this power — and for the resilience of our Constitution — will involve regulations such as the oil and gas cap which could have significant economic implications for provinces like Alberta and Saskatchewan.

3.3.2.3   POGG – National Concern Branch

A second source of jurisdiction to enact laws relating to GHG emissions reductions is the National Concern branch of POGG. This test for the national concern branch was first elaborated by the Supreme Court of Canada in the *Crown Zellerbach* case and updated in *GGPPA References*. This section reviews the test and discusses its potential to justify future climate regulations.

3.3.2.3.1   Step One – Is the Matter Inherently of National Concern?

The first part of the test poses the preliminary threshold question of whether a matter can be characterized as inherently of national concern. The Court in *GGPPA References* clarified that recognition of a matter of national concern must be based on evidence and refuted the idea that a matter must be new.[292] The Court held that historical "newness" is irrelevant to the analysis. The key requirement is that the matter, by its nature, transcends the provinces and is, thereby, inherently national in character.[293] As examples, the Supreme Court of Canada cited its prior holding that marine pollution is a matter of national concern since it is predominantly extra-provincial and international in character, as well

---

292 *GGPPA References*, above note 1 at para 133.
293 *Ibid* at paras 134, 136, and 141.

as the development of the national capital region since it is (as per the Supreme Court of Canada) of concern to Canada as a whole.[294]

3.3.2.3.2 Step Two – Does the Matter Have the Requisite Singleness, Distinctiveness, and Indivisibility?

The second part of the test involves assessing the infamous trilogy of singleness, distinctiveness, and indivisibility.[295] Two principles underpin this part of the analysis. First, there must be a *specific and readily identifiable matter* that is qualitatively different from matters of provincial concern. The matter may be predominantly extra-provincial and international in character, given its inherent nature and its effects.[296] The content of international agreements relating to the matter may be helpful in identifying matters with a national and international component. Also, identifying a distinct role for federal legislation that is not simply duplicative of the provinces is helpful in identifying national matters.[297]

The second principle underpinning the singleness, distinctiveness, and indivisibility analysis is *provincial inability to deal with the matter*. This means that the legislation is such that provinces would be constitutionally incapable of enacting it, and that the failure to include one or more provinces in the legislative scheme would jeopardize the success of the legislation in other parts of the country or at a national level.[298] In *GGPPA References*, the Supreme Court of Canada added a third element to this part of the test, noting that a "province's failure to deal with the matter must have grave extra-provincial consequences."[299] The Court also clari-

---

294 *Ibid* at para 140.

295 *IAA Reference*, above note 5 at para 145.

296 *GGPPA References*, above note 1 at para 148.

297 *Ibid* at para 151. For a thoughtful argument about inverted subsidiarity (where there is a unique role for Parliament in matters best dealt with at a national level), see Jean Leclair, "'Tis a Rock – a Crag – a Cape? A Cape? Say Rather a Peninsula!' The Supreme Court of Canada's Revisitation of the National Concern Doctrine" (2023) 108 SCLR (3d) 3.

298 *GGPPA References*, above note 1 at para 152.

299 *Ibid* at para 153.

fied that establishing provincial inability is an essential part of the test for national concern and is not optional.[300]

For the grave extra-provincial consequences, the Supreme Court of Canada cited the range of environmental, economic, and human harm caused by climate change and noted that these harms would have profound implications for Indigenous Peoples, the Canadian Arctic, and coastal regions.[301] The Court explained that these consequences may be demonstrated by proof of actual harm or a serious risk of harm in the future to human life or health or the environment.[302] Adding a threshold of seriousness was intended to set a high bar for a finding of provincial inability.[303]

3.3.2.3.3 Step Three – Scale of Impact on Provincial Jurisdiction

The third step in the national concern analysis involves determining whether a finding of national concern would have a scale of impact on provincial jurisdiction that is reconcilable with the fundamental distribution of legislative power in the Constitution.[304] The goal of this step is to prevent federal overreach and protect against unjustified intrusions on provincial autonomy. In *GGPPA References*, the Supreme Court of Canada clarified this is ultimately a balancing test that requires weighing the impacts on provincial interests against the impacts if Parliament were constitutionally unable to legislate the matter.[305] Pointing to the interests that could be harmed if Parliament were unable to address the matter at a national level and finding the impact on provincial jurisdiction to be modest, the Court held this part of the test to be satisfied. The Court pointed to the profound, irreversible consequences for vulnerable communities and regions, including Indigenous Peoples and the Canadian Arctic, to support this conclusion.[306]

---

300 *Ibid* at para 156.
301 *Ibid* at para 187.
302 *Ibid* at para 155.
303 *Ibid* at para 155.
304 *Ibid* at para 160.
305 *GGPPA References*, above note 1 at para 161.
306 *Ibid* at para 206.

3.3.2.3.4 Applying the National Concern test to Legislation Reducing GHG Emissions

In *GGPPA References* case, several factors influenced the Court's finding that the carbon pricing backstop satisfied the national concern test. Application of the double aspect doctrine plus the design of the measure as a backstop limited its potential to intrude into provincial jurisdiction. Additionally, the gravity of the matter in question justified some incursion, as did the national aspect of the issue where the actions of one province (including the decision not to act) could have detrimental impacts on another province or the national interest. As an example, the Court noted that the decreases in GHG emissions in one province were mostly offset by increases in two other provinces.[307]

Would other legislation aimed at reducing national GHG emissions meet the updated national concern test? It is impossible to conduct a thorough division of powers analysis in the absence of specific legislation. However, it is possible to discuss factors that might influence the analysis one way or the other.

Step one of the test would likely be easy to meet for legislation aimed at addressing the threat of climate change. In *GGPPA References*, the Supreme Court of Canada characterized climate change as an existential threat that poses a grave threat to human life in Canada and around the world and held that it is essential for Canada to reduce its GHG emissions.[308] As the Court noted, "the undisputed existence of a threat to the future of humanity cannot be ignored."[309] Based on the pith and substance analysis, the Court found that the matter of the legislation was "establishing minimum national standards of GHG price stringency to reduce GHG emissions" and held that this readily passed the threshold test as being of national concern.[310] It is reasonable to think that most legislation clearly aimed at reducing national GHG emissions to address climate change would be of national concern, especially given

---

307 *Ibid* at para 24.
308 *Ibid* at paras 2, 167, and 171.
309 *Ibid* at para 167.
310 *Ibid* at para 171.

international commitments such as the *Paris Agreement*. GHGs are predominantly extra-provincial and international in character and certainly in implications, since they accumulate in the atmosphere to cause climate change. In *GGPPA References*, the Court held that "GHG emissions represent a truly global pollution problem that demands a coordinated international response"[311] and determined that reducing GHG emissions through minimum national standards concerned Canada as a whole and readily passed the threshold test.[312] Catastrophic risk seems to be a relevant factor. In *Ontario Hydro v Ontario (Labour Relations Board)*, the Supreme Court of Canada held that atomic energy is a matter of national concern within federal jurisdiction given the potential for catastrophic interprovincial and international harm.[313] Legislation aimed at addressing an issue that is inherently national and global in nature and has the potential to cause catastrophic harm globally should pass the threshold test in step one of the analysis without difficulty.

What about the singleness, distinctiveness, and indivisibility portion of the test?[314] One of the features of GHG emissions is that they are identifiable, measurable, and quantifiable. Similarly, the extent to which a given level of GHG emissions contributes to a national target is also measurable. In *GGPPA References*, the Supreme Court of Canada pointed to its consistent holding that "interprovincial pollution is constitutionally different from local pollution and that it may fall within federal jurisdiction on the basis of the national concern doctrine," citing *Interprovincial Cooperatives, Crown Zellerbach, Hydro Quebec* and *Morguard Investments*.[315] In *Interprovincial Co-operatives*, a case concerning one province's emission of pollutants into an interprovincial river, Pigeon J observed that the Court was "faced with a pollution problem that is not really local in scope but truly interprovincial."[316] GHG emissions are

---

311 *GGPPA References*, above note 1 at para 194.

312 *Ibid* at para 171.

313 *Ontario Hydro*, above note 162 at 340, 352, 379–80 (Supreme Court of Canada), cited in *GGPPA References*, above note 1 at para 107.

314 *IAA Reference*, above note 5 at para 145.

315 *GGPPA References*, above note 1 at para 195; *Hydro-Quebec*, above note 3 at para 76.

316 *IPCO*, above note 197 at 514.

a quintessential global, transboundary pollutant. They are much more limited in scope (and more easily quantified) than "toxic substances," the broad category that the Supreme Court of Canada in *Hydro-Quebec* determined to be insufficiently specific to satisfy this part of the POGG test.[317] GHG emissions are a specific set of pollutants that can be named and counted. Indeed, there are robust accounting standards aimed specifically at reporting on GHG emissions.[318]

National GHG reduction goals are, by definition, extra-provincial and contribute to an international agenda. The targets are linked to the *Paris Agreement*, illustrating their international character. There is a distinct role for federal legislation here since no single province can regulate emissions at a national level, nor impose GHG reduction requirements on other provinces. If one province fails to cooperate, the burden of GHG reduction is pushed onto other jurisdictions who must make up the shortfall, or the national interest of reducing emissions to net-zero is jeopardized.[319]

What about provincial inability? It may be possible to satisfy the provincial inability test by noting that the provinces are constitutionally incapable, acting together or alone, of establishing and enforcing national GHG emissions levels. Like in the *2018 Securities Reference* (where the Supreme Court of Canada upheld national securities regulation that addressed systemic risk)[320] and *GGPPA References*, the federal government has established

---

317 *Hydro-Quebec*, above note 3 at para 69.

318 See, e.g., Intergovernmental Panel on Climate Change, *2006 IPCC Guidelines for National Greenhouse Gas Inventories*, Volume I: General Guidance and Reporting, Simon Eggleston et al, eds (Hayama: Institute for Global Environmental Strategies, 2006).

319 The Supreme Court of Canada illustrated this point in the context of carbon pricing in *GGPPA References*, above note 1 at para 184, noting that the decreases in emissions in some provinces were offset by increases in other provinces.

320 The 2018 case followed a 2011 reference where the Supreme Court of Canada held that Parliament did not have authority to establish a national securities regulator with broad powers that would encroach upon provincial jurisdiction over securities but could establish a national law aimed at addressing systemic risk and collecting national data under the trade and commerce power (see *Reference re Securities Act*, 2011 Supreme Court of Canada 66 [*Securities Act Reference*]; *Pan-Canadian Securities*, above note 67).

national goals (in this case, GHG reduction targets) to address a systemic risk (climate change), and there is no guarantee or even likelihood that all provinces will enact legislation to ensure emissions comply with the national target. Provinces and territories are constitutionally incapable of establishing binding GHG emissions reduction targets that apply in all provinces and territories at all times.[321] It would also be possible to show that a province's failure to deal with the matter within its own borders would have grave extra-provincial consequences for other provinces and the national interest. Projections of GHG emissions from provinces and their trajectories are available and show that emissions from some provinces jeopardize meeting the national targets.[322]

Finally, a court would need to consider whether upholding jurisdiction would have a scale of impact on provincial jurisdiction that is reconcilable with the fundamental distribution of legislative power in the Constitution. In *GGPPA References*, the Supreme Court of Canada explained that this is a balancing exercise where the impact on provincial jurisdiction must be weighed against the interests that would be harmed if Parliament were unable to regulate GHG emissions. Considering the catastrophic impacts of climate change and its collective action nature (which requires every country to do its part in reducing emissions), the interests harmed if Parliament cannot effectively regulate to ensure the country meets its GHG reduction targets are significant.

Unlike *GGPPA References*, where the carbon pricing system was implemented as a backstop, regulation of GHG emissions in other contexts might have a bigger impact on provincial jurisdiction. This may have been a concern in the *IAA Reference* case which struck down provisions of the *IAA* which established a procedure for evaluating the impact of projects, including the extent to which a project might jeopardize Canada's ability to meet its GHG emissions commitments. While the Supreme Court of

---

321   See *Pan-Canadian*, above note 67 at para 113. See also *GGPPA References*, above note 1 at para 182.

322   See, e.g., the evidence summarized by the Supreme Court of Canada in *GGPPA References* showing how growth of emissions in Alberta and Saskatchewan offset decreases in emissions in other provinces (see *GGPPA References*, above note 1 at para 184).

Canada held that it was within Parliament's authority to consider a wide range of impacts in assessments (including GHG implications), the federal government was much more limited at other stages of the process including decision-making.[323] While the federal government did not advance a POGG argument to justify this aspect of the law, it is interesting to consider what might have happened if it had done so.

Consideration of GHG emissions in the *IAA* (as it then was) could, conceivably, result in Parliament imposing stringent conditions or even refusing to approve a project based on its projected GHG emissions if those exceed certain thresholds. As such, establishing a justifiable scale of impact was a higher threshold in the *IAA Reference* compared to *GGPPA References*. However, it is important to bear in mind that the goal of what the court called "the interprovincial effects clause" in the *IAA Reference* was to reduce transboundary impacts (i.e., interprovincial air pollution, including GHG emissions) and not to limit production. As such, the impact on provincial jurisdiction would likely be modest for all but a small number of the very largest emitting projects. Provinces could still regulate the oil and gas industry as they wish, as long as the GHG emissions conform with conditions imposed under the *IAA*.

In a rapidly decarbonizing world, there is less scope for projects that lock in high levels of GHG emissions. Courts around the world — including many at the highest level — are increasingly holding governments to account for mitigating climate change.[324] Potential impacts on provincial jurisdiction would need to be weighed against the importance of Canada

---

323   *IAA Reference*, above note 5.

324   See, e.g., Hoge Raad [Supreme Court of the Netherlands], The Hague, 20 December 2019, *Netherlands v Urgenda* (13 January 2020), ECLI:NL:HR:2019:2007 (Netherlands), the first case where the human rights dimensions of climate change were recognized by a court; Tribunal Administratif de Paris, Paris, 14 January 2021, *Notre Affaire à Tous and Others v France* (3 February 2021),No 1904967, 1904968, 1904972, 1904976/4-1 (France); Bundesverfassungsgericht [Federal Constitutional Court], Karlsruhe, 6 February 2020, *Neubauer v Germany*, 00362/19/R/H (Germany) (constitutional complaint) (holding the German government responsible for mitigating climate change). For a review of climate litigation, see United Nations Environment Program & Sabin Center for Climate Change at Columbia Law School, *Global Climate Litigation Report: 2023 Status Review* (Nairobi: United Nations Environment Programme, 2023), online: wedocs.unep.org [perma.cc/8MTP-GHZJ].

being able to meet its national targets in the face of the worsening climate crisis and increasing accountability for reductions. The interests of provinces that would need to make up the shortfall if emissions from other provinces exceeded their share of the national emissions budget are also relevant.

Finally, as the devastation from climate change worsens, the scale and magnitude of grave impacts increases, weighing the scales more heavily in favour of federal jurisdiction, even if there are impacts on provincial jurisdiction. Offering an analogy between *Crown Zellerbach* and the case before it, the Supreme Court of Canada in *GGPPA References* stated that the "case for finding that the matter is of national concern is even stronger here than in *Crown Zellerbach*" given the existential nature and severity of the impacts of climate change and the "uncontested evidence of grave extra-provincial harm" from provincial inability.[325] A similar argument could be made in the context of other GHG legislation.

Courts have been careful to interpret the national concern branch of POGG in a way that avoids tipping the scales too far in favour of federal jurisdiction, and it is clear that Parliament does not have unfettered jurisdiction over GHG emissions. That said, as long as the double aspect doctrine applies, given the global nature of GHGs and the threat of excessive emissions to the future of humanity, it is reasonable to think that Parliament could rely upon this source of authority to justify a number of laws clearly aimed at the specific goal of reducing national GHG levels. The national concern branch is especially helpful in the context of uneven efforts by provinces to reduce emissions, where a province's choice not to act has extra-provincial impacts and jeopardizes Canada's ability to meet its GHG targets.

As long as the criminal law power continues to authorize federal GHG emissions regulations, reliance on the national concern branch of POGG will likely remain modest. However, should the criminal law power be interpreted in a more restrictive way, Parliament may turn to POGG to justify more of its GHG emissions measures.

---

325 *GGPPA References*, above note 1 at para 195.

3.3.2.4 POGG Emergency Branch

POGG's emergency branch may have once seemed like a last resort as a source of jurisdictional authority for federal GHG regulations, but the rapidly evolving state of climate change makes it an increasingly credible possibility.[326] Several interveners in *GGPPA References* argued the legislation could be justified under the emergency branch.[327] While counsel for Canada did not make arguments to this effect, their factum closed with the statement that the Act could be justified, in the alternative, under the emergency branch of POGG.[328] In his dissenting reasons in *GGPPA References*, Brown J expressed surprise that the majority did not pursue this source of jurisdiction, given that — in his view — the temporary requirement of emergency jurisdiction would "do less damage to Canadian federalism."[329]

The emergency branch of POGG grants the federal government broad legislative powers to address situations of emergency on a temporary basis.[330] It is distinct from the national concern branch in that it justifies legislation that intrudes on matters of provincial jurisdiction if needed to deal with the emergency. In other words, there is no requirement for the impact on provinces to be reconcilable, given that the federal jurisdiction is meant to address an emergency with temporary measures.[331] Unlike the national concern branch of POGG which conveys jurisdiction over a "subject matter," the emergency power only authorizes the specific legislation

---

326 See Nathalie J Chalifour, "Canadian Climate Federalism: Parliament's Ample Constitutional Authority to Legislate GHG Emissions through Regulations, a National Cap and Trade Program, or a National Carbon Tax" (2016) 36 NJCL 331 at 355–61; Nathalie J Chalifour, "Jurisdictional Wrangling over Climate Policy in the Canadian Federation: Key Issues in the Provincial Constitutional Challenges to Parliament's *Greenhouse Gas Pollution Pricing Act*" (2019) 50:2 Ottawa L Rev 197.

327 These included the David Suzuki Foundation, Canadian Labour Congress, Intergenerational Climate Coalition, Athabasca Chipewyan First Nation, and the National Association of Women and the Law and Friends of the Earth Canada (see *GGPPA References*, above note 1 at para 399).

328 Counsel for Canada stated at the end of her oral arguments that Canada would not object to a finding of jurisdiction under this branch.

329 *GGPPA References*, above note 1 at para 400.

330 See *Re: Anti-Inflation Act*, 1976 CanLII 16 at 461 (Supreme Court of Canada) [*Anti-Inflation Reference*].

331 *Ibid* at 461.

in question. No enduring jurisdiction is created. Two requirements must be met to justify legislation under this branch. First, there must be an emergency. Second, the legislation responding to that emergency must be temporary.

The emergency power has been interpreted cautiously in the past[332] but broadly enough to justify a variety of legislative measures in times of war, apprehended insurrection, and to deal with inflation.[333] For instance, the *War Measures Act* has been justified under this power and used to authorize federal price controls and rent control after the first and second world wars.[334] This Act was also used to justify policing powers in Quebec's October crisis of 1970.[335] Parliament also relied on the emergency branch of POGG to justify enacting a suite of economic measures to address months of double-digit inflation in the 1970s and high levels of unemployment.[336] The purpose of the *Anti-Inflation Act* was to comprehensively contain and reduce inflation.[337] In the *Anti-Inflation Reference*, the Supreme Court of Canada affirmed the measures as valid under the emergency branch, holding that they were "temporarily necessary to meet a situation of economic crisis imperiling the well-being of Canada as a whole and requiring Parliament's stern intervention in the interest of the country as a whole."[338]

---

332 For instance, most of the federal measures under the New Deal legislation enacted in the 1930s to address the Depression were held to be unconstitutional on the basis that the Depression did not constitute an emergency (in the Privy Council's view). Hogg suggests that the court may also have been influenced by the permanent character of much of the legislation (see Hogg & Wright, above note 152 at 17.7, 17.11).

333 See, e.g., *Fort Frances Pulp and Paper Co v Manitoba Free Press Co*, 1923 CanLII 429 (UK JCPC) [*Fort Frances Pulp*]; *Reference re Wartime Leasehold Regulations*, 1950 CanLII 27 (Supreme Court of Canada); *Reference to the Validity of Orders in Council in relation to Persons of Japanese Race*, 1946 CanLII 46 (Supreme Court of Canada).

334 See *Fort Frances Pulp*, above note 333; *Reference re Wartime Leasehold Regulations*, 1950 CanLII 27 (Supreme Court of Canada).

335 See CBC Learning, "The October Crisis" (last visited 14 January 2024), online: www.cbc.ca [perma.cc/UTB2-M37H].

336 *Anti-Inflation Reference*, above note 330.

337 *Anti-Inflation Reference*, above note 330.

338 *Ibid* at 425.

3.3.2.4.1 A Climate Emergency

Could climate change be considered an emergency that could justify federal GHG emissions regulation? Given that the Supreme Court of Canada has characterized climate change as an existential crisis that poses a grave threat to humanity, it should not be difficult to meet this part of the test. At the planetary level, we are now experiencing profound impacts from the current level of warming, as discussed in Chapter 1.[339] There is ample evidence to characterize the issue as a national emergency. Surface temperatures in Canada are warming at twice the global average, with warming in the north even more pronounced.[340] The extensive impacts are outlined in Chapter 1 and are captured by several of the youth climate lawsuits discussed in Chapter 4.

Climate change can also be characterized as an economic crisis. The head of the International Monetary Fund called climate change "the greatest economic challenge of the 21st century" more than a decade ago. Emerging research shows that the macroeconomic damages from climate change are six times worse than initially thought, with warming of one degree Celsius reducing world GDP by 12 percent.[341] Climate change is projected to cause some $38 trillion of damage every year by mid-century, with average incomes falling almost a fifth relative to what they would have been without the climate crisis.[342] In Canada, climate change is estimated to cause annual losses of $35 billion to GDP.[343] The Fort McMurray wildfire of 2016 was the most expensive disaster in Canadian history,

---

339 See also UN Environment Programme, "The Climate Emergency" (last visited 20 December 2024), online: unep.org [perma.cc/VY5Y-2W2T]; United Nations, "The Climate Crisis – A Race We Can Win" (last visited 20 December 2024), online: un.org [perma.cc/4GKN-2DHS].

340 See *GGPPA References*, above note 1 at paras 10–11.

341 Adrien Bilal & Diego R Känzig, "The Macroeconomic Impact of Climate Change: Global vs Local Temperature" (2024) National Bureau of Economic Research, Working Paper No 32450, online: nber.org [perma.cc/AMM5-3NWL].

342 Maximilian Kotz, Anders Levermann & Leonie Wenz, "The Economic Commitment of Climate Change" (2024) 628 Nature 551.

343 Dave Sawyer et al, *Damage Control: Reducing the Costs of Climate Impacts in Canada* (Canadian Climate Institute, 2022), online: climateinstitute.ca [perma.cc/JQ9N -U7EK].

with insured damages estimated at $5.96 billion.[344] Insured damage from severe weather and natural catastrophes exceeded $3 billion in both 2022 and 2023.[345]

As of September 2024, 2,364 jurisdictions in forty countries had declared a climate emergency.[346] This includes many Canadian jurisdictions. In 2019, Parliament held an emergency debate on the matter and declared that Canada is in a national climate emergency.[347] As of 2022, 650 Canadian municipalities had declared a climate emergency.[348] The climate crisis is perhaps most evident for Indigenous communities, who have close ties with lands and waters and are disproportionately impacted by climate change. The Assembly of First Nations declared a climate emergency in July 2019.[349]

The requirement for an emergency in this first part of the test is not a high threshold. The court need only find a "rational basis" for the existence of such an emergency.[350] There is, for instance, no requirement for the word "emergency" to appear in the impugned legislation. In the *Anti-Inflation Reference*, high rates of inflation combined with relatively high levels of unemployment satisfied the Court that there was an emergency. As noted earlier, Brown J in his dissent in *GGPPA References* said it was curious that the majority did not consider the emergency branch as a potential source of jurisdiction, given its characterization of climate change as an existential threat to Canada and the world.[351] Justice Brown observed that Parliament would not have lacked a rational basis upon

---

344   Insurance Bureau of Canada, "Severe Weather in 2023 Caused Over $3.1 Billion in Insured Damage" (8 January 2024), online: ibc.ca [perma.cc/56Q6-E7T9].

345   *Ibid.*

346   Climate Emergency Declaration, "Climate Emergency Declarations in 2,364 jurisdictions and Local Governments Cover 1 Billion Citizens" (22 September 2024), online: climateemergencydeclaration.org [perma.cc/3MD7-MFFL].

347   House of Commons, *Votes*, 42-1, No 1366 (17 June 2019).

348   Partners for Climate Protection, "The State of Climate Action in Canadian Municipalities" (Last visited 6 April 2025), online: pcp-ppc.ca [perma.cc/HVM9-RC54].

349   Assembly of First Nations, "Assembly of First Nations (AFN) Advances First Nations Climate Strategy at 3rd National Climate Gathering" (8 October 2024), online: afn.ca [perma.cc/636Q-ACLZ].

350   *Anti-Inflation Reference*, above note 330 at 375, 423.

351   See *GGPPA References*, above note 1 at para 399.

which to enact the *GGPPA* to respond to the climate emergency, but that the Attorney General had not done the work to show Parliament was justified in relying on the emergency power to justify the legislation.[352]

3.3.2.4.2 Temporary Measures

The second requirement to establish jurisdiction under the emergency branch of POGG is that the measure in question be temporary. There is little guidance from the courts to help define what a temporary measure is, though the nature of the emergency will be relevant. According to the Supreme Court of Canada, under the emergency branch Parliament's jurisdiction "knows no limit other than those which are dictated by the nature of the crisis," though "one of those limits is the temporary nature of the crisis."[353] Interpretation of the criterion of temporariness is, therefore, of great importance.

The very nature of an emergency or crisis is that it is temporary. While it may be difficult or impossible to predict the precise moment when an emergency may end, the nature of an emergency is that it will, at some point, end.[354] In fact, as Brown J explained in the *GGPPA References* dissent, the point of acting under the emergency branch "is presumably to do what is necessary to ensure that the emergency *will* end."[355]

One of the easiest ways to satisfy the criterion for temporariness may be to enact legislation that is explicitly temporary. This could be achieved by establishing a "best before" date or expressly specifying that provisions of legislation are only operable when an emergency is declared or certain conditions are met.[356] However, it is also possible to satisfy this criterion in the absence of express provisions for temporary operation.[357] As Brown J notes in the context of the *GGPPA*, legislation that contemplates an end by

---

352 *GGPPA References*, above note 1 at para 401.

353 *Anti-Inflation Reference*, above note 330 at 461.

354 *GGPPA References*, above note 1 at para 400.

355 *Ibid* at para 400.

356 For instance, the *Anti-Inflation Act* specified that it ended on a specified date or an earlier date fixed by proclamation (see *Anti-Inflation Reference*, above note 330 at 383–84).

357 *GGPPA References*, above note 1 at para 400. See also *Anti-Inflation Reference*, above note 330 at 427.

the character of its provisions can satisfy this requirement.[358] The *GGPPA*, for instance, aims to change behaviour in a way that supports decarbonization of the economy. Once decarbonized and GHG mitigation targets are met, the mitigation emergency is addressed.

Justice Brown noted that the lack of any temporal constraint in the *GGPPA* was problematic but not fatal to its potential justification under the emergency branch.[359] While there is no "best before" date in the legislation, the *Act* does contemplate an end-point since the behavioural changes required by the *Act* will eventually not be needed anymore (once these changes are ingrained in the economy, presumably).[360] Justice Brown acknowledged that not knowing the exact end-point of an emergency, which can be difficult to identify with any precision, does not preclude justifying the legislation under the emergency branch.[361] The Supreme Court of Canada's words about the anti-inflation measures being "temporarily necessary to meet a situation of economic crisis imperiling the well-being of Canada as a whole and requiring Parliament's stern intervention in the interest of the country as a whole" could be just as appropriate to the climate context: national measures are urgently needed to set the Canadian economy on a path of decarbonization in order to address the climate crisis imperiling the well-being of Canadians and indeed the world, and Parliament's stern intervention in the interest of the country as a whole is required.

However, the fact that the Attorney General of Canada did not present arguments on this point, along with the fact that Parliament does not appear to have passed the *Act* relying upon this legislative authority, were determinative for Brown J.[362] He concluded by stating that "the role of this Court … is not to root around the Constitution or constitutional doctrine to scrounge up some basis, *any* basis, to rescue federal legislation," especially when the exceptional authority of POGG is contemplated as the basis for jurisdiction.[363]

---

358 *Ibid* at para 400.
359 *GGPPA References*, above note 1 at para 400.
360 *Ibid*.
361 *Ibid*.
362 *Ibid* at para 403.
363 *Ibid*.

The unique nature of climate change raises interesting questions in terms of what time limits might be considered temporary. The climate crisis is a long-lasting emergency, since it will only lose its emergency characteristic once the world is on a solid path to decarbonization (achieving net zero emissions) and the climate begins to stabilize. The time-frame for addressing the climate emergency could therefore be longer than that in other emergencies.[364] It would be logical that the context of a given emergency be considered in assessing what is temporary. Climate change is a planetary level process with built-in inertia due to the interactions of complex atmospheric systems. As such, addressing it will take longer than tackling factors that may be more easily manipulated. That said, reducing emissions to net-zero by 2050 — a goal articulated in the *Canadian Net-Zero Accountability Act* — has quantifiable timelines.

Could some federal climate laws meet the temporary criteria? There are several plausible arguments that could be proffered to qualify certain GHG mitigation laws as temporary. One way would be to tie the measures to the desired outcome, which is to achieve net-zero emissions by 2050. As such, legislation would be required for approximately twenty-five years. Another credible option would be to set the timeframe as five years (to 2030). This may be more palatable than the longer time frame. It is also defensible, since it corresponds with national targets established under the *Paris Agreement* and is the timeframe identified by the IPCC as the crucial one within which rapid decarbonization must take place.[365]

Another way to frame GHG emissions reduction regulations as temporary would be to enact the given legislation with an explicit sunset clause and/or interim points for re-evaluation. An interim point could mark a moment in time at which time Parliament must re-evaluate whether an emergency still exists to justify the legislation.

In the absence of explicit time-frames built into legislation, could a law's temporary character be deduced from the nature of the legislation

---

364 For instance, the *Anti-Inflation Act* was in place for four years.

365 Intergovernmental Panel on Climate Change, *Climate Change 2023: Synthesis Report*, Contribution of Working Groups I, II and III to the Sixth Assessment Report of the Intergovernmental Panel on Climate Change, Core Writing Team et al, eds (Geneva: Intergovernmental Panel on Climate Change, 2023) [IPCC, *Climate Change 2023*].

and/or the emergency? In *GGPPA References*, Brown J addressed submissions made by an intervenor suggesting the legislation's temporary character could be implied through the preambular references to Canada's 2030 commitment under the *Paris Agreement*.[366] While Brown J described this as an "intriguing proposition'" he found that "considering time-delimited jurisdiction in the emergency doctrine analysis would require a departure from this Court's jurisprudence."[367] That said, there is also jurisprudential support for the idea that the timeframe is determined based on circumstances of the emergency. As Lord Haldane noted in *The Fort Frances Pulp and Paper Company, Ltd v The Manitoba Free Press Company Ltd and Others*, "it may be that it has become clear that the crisis which arose is wholly at an end and that there is no justification for the continued exercise of an exceptional interference which becomes *ultra vires* when it is no longer called for."[368]

The strongest basis for establishing jurisdiction under the emergency branch would be for Parliament to clearly state its intention to rely upon this power when enacting legislation and specify an explicit temporary time-frame for the law. While there is no requirement in the jurisprudence that the time-frame be short (only that it be temporary), chances are that courts will be more willing to support a finding of emergency jurisdiction for a shorter window of time. Of course, temporary measures can always be continued, while permanent measures can be repealed.[369] The more meaningful safeguards to protect against abuse of the emergency power are the need for a rational basis, the requirement of evidence

---

366 See *GGPPA References*, above note 1 at para 402. Other intervenors that raised the emergency branch include Canadian Labour Congress, the Intergenerational Climate Coalition, the Athabasca Chipewyan First Nation, and the National Association of Woman and the Law and Friends of the Earth (see *GGPPA References*, above note 1 para 399).

367 *GGPPA References*, above note 1 at para 402.

368 *Fort Frances Pulp*, above note 333.

369 See Hogg & Wright, above note 152 at 17.11. See also, *References re Greenhouse Gas Pollution Pricing Act*, 2021 Supreme Court of Canada 11 (Factum, Intervenor, David Suzuki Foundation) [David Suzuki Factum].

of a continuing emergency, and proportionality between the severity of the emergency and the level of intervention in provincial jurisdiction.[370]

In sum, there is a reasonably strong argument to be made that the emergency branch could justify federal GHG regulations if crafted appropriately. The use of the emergency branch will presumably not be welcomed by provinces, given it authorizes Parliament to legislate in the manner it chooses to address the emergency, regardless of whether the subject matter of the legislation is normally within provincial jurisdiction. It is the temporal limit that is meant to soften the blow of this intrusion on provincial jurisdiction. However, there are two approaches that courts could take to improve the emergency branch's fit into our federalist structure. First, if the courts accept a lengthy time frame for the legislation, the threshold for determining that the legislation responds to an emergency could be correspondingly stricter (limiting the scope for actual conflict or frustration of the federal measures and the paramountcy doctrine). Second, there is no reason why the double aspect doctrine should not also continue to operate in the case of the emergency branch. This would allow provincial GHG-related laws to exist alongside federal law as long as the provincial law was validly enacted under a provincial power and did not conflict with, or frustrate the purpose of, the federal emergency law. While the use of the emergency branch is relatively rare, the stakes of climate change are high enough that Parliament may opt to use it.

The emergency branch offers certain advantages for Parliament. The first is its breadth. The emergency power authorizes Parliament to intrude in areas of provincial jurisdiction for the duration of the emergency. This means that Parliament would not have to be concerned (at least legally and within the confines of the emergency) with designing the legislation to avoid subjects within provincial control. The power would authorize any regulation aimed at addressing the emergency.

A second advantage relates to the burden of proof. Like for all *vires* cases, the burden of proving that a given provision is unconstitutional lies with the challenger.[371] When a court must choose between competing,

---

370 David Suzuki Factum, above note 369.

371 *Re The Farm Products Marketing Act*, 1957 CanLII 1 at 255 (Supreme Court of Canada).

plausible interpretations of a law, it should opt for the characterization that supports the law's validity.[372] Combined with the fact that a court need only find a "rational basis" for a finding of fact that there is an emergency, the threshold is relatively low.[373] If federal emergency legislation to address GHG emissions was challenged, the challenging party would bear the burden of proving that there is no rational basis for the federal law. Given the robust scientific evidence of catastrophe without rapid decarbonization, it would be difficult for challenging parties to prove that there was no rational basis. As already noted, the debate would likely center on the temporariness of the measures. The disadvantage is that it could be perceived as a threat to federalism. Cooperation and collaboration, when possible, are always preferred.

3.3.2.5 Trade and Commerce Power

Section 91(2) grants Parliament the power to make laws in relation to "the regulation of trade and commerce." This authority has been interpreted to confer two branches of authority: (1) interprovincial or international trade and commerce and (2) the general regulation of trade and commerce.[374] The Supreme Court of Canada in *GGPPA References* observed that the national concern doctrine and trade and commerce power both pose similar challenges to federalism, given their potential scope and breadth which overlaps with the provincial power over property and civil rights.[375] Both powers have been, therefore, interpreted narrowly to avoid undermining provincial autonomy.[376] In fact, the Supreme Court of Canada drew upon the jurisprudence on the general branch of the trade and commerce power

---

372 *Siemens v Manitoba (Attorney General)*, 2003 Supreme Court of Canada 3 at para 33.

373 *Anti-Inflation Reference*, above note 330 at 375. Recall that where a law is subject to both a wide and narrow interpretation, where the wider interpretation would intrude on the other jurisdiction's powers, the presumption of constitutionality requires the Court to choose the more narrow interpretation, or "read down" that law (see, e.g., *Oldman River*, above note 22 (where a federal environmental assessment directive was read down to limit the process to areas within federal jurisdiction)).

374 *Citizens' and the Queen Insurance Cos v Parsons*, [1881] UKPC 49.

375 *GGPPA References*, above note 1 at para 113. See also Mathen & Macklem, above note 57 at 334; *References re Securities Act*, 2011 SCC 66 at para 70.

376 *Ibid* at para 113.

(namely the two Securities References)[377] in its analysis of the national concern doctrine in *GGPPA References*.[378]

3.3.2.5.1 Interprovincial or International Trade and Commerce

The first branch of the trade and commerce power has been less controversial, since it is unsurprising that Parliament has jurisdiction over international and interprovincial trade and commerce. The court's interpretation of this branch has been mainly concerned with the extent to which regulation of interprovincial or international trade extends (and potentially intrudes) into intra-provincial matters. The court has upheld a number of initiatives under this branch despite their incursion into provincial matters, as long as the incursion is incidental to an extra-provincial issue.[379] The courts have upheld federal agricultural product marketing quotas allocated to provinces, for instance, for products such as eggs and chickens.[380] In *Reference re Agricultural Products Marketing*, even though 90 percent of all eggs were consumed within provinces and not traded interprovincially, the court upheld the national scheme in part because it was built upon federal-provincial collaboration.[381] However, the courts have consistently declined to uphold laws under this power that regulate a single trade or industry, even if the industry is dominated by a few firms that market on a nation-wide basis.[382]

The extent to which this power could be relied upon to justify hypothetical national carbon budget legislation whereby emissions quotas are allocated among the provinces was explored in detail by Professor Leach. He concluded that the supply management cases do not provide a satisfactory analogue for carbon budget allocations since emissions are not traded in the same way tangible goods such as are eggs or grain.[383]

---

377 *Pan-Canadian Securities*, above note 67; *Securities Act Reference*, above note 320.

378 *GGPPA References*, above note 1 at para 113.

379 See, e.g., *Reference re Agricultural Products Marketing*, 1978 CanLII 10 (Supreme Court of Canada) [*Agricultural Products Reference*].

380 *Ibid*; *Fédération des producteurs de volailles du Quebec v Pelland*, 2005 Supreme Court of Canada 20. See also Mathen & Macklem, above note 57 at 379–82.

381 *Agricultural Products Reference*, above note 379.

382 See, e.g., *Reference re National Securities Act*, 2011 Supreme Court of Canada 66.

383 See Leach, "I Can't Believe It's Not Butter," above note 196 at 497.

Whereas federal regulation under the trade and commerce power could restrict extra-provincial trade of goods such as milk or butter (goods can be stopped at the border), Leach argues that Parliament could not justify controlling emissions in the same way under this power, since emissions cannot similarly be stopped at the border; to regulate them would require regulating production or trade within the province.[384] While this argument is compelling, it may be possible to conceive of emissions following a path from source to atmosphere where — at some point on their journey — they necessarily leave provincial jurisdiction. An analogy could be made to jurisprudence holding that aircraft flying over provinces are not "within the province," suggesting that there is a point at which emissions exit a province. Federal jurisdiction over the extra-provincial movement of those emissions may be plausible under this interpretation.[385] While emissions leaving a province cannot be stopped in the same way a traded good at a border crossing can, this may have more to do with its intangible quality and the vertical, versus horizontal, trajectory of the crossing than the fact that it is not extra-provincial. In any event, as Professor Leach notes, there are other (possibly better) sources of jurisdiction for federal carbon budget legislation, such as POGG and criminal law.[386]

3.3.2.5.2 The General Regulation of Trade and Commerce

The second branch of the trade and commerce power authorizes Parliament to legislate aspects of intra-provincial matters, since interprovincial and international matters are covered under the first branch. Like the national concern branch of POGG, it has been carefully and narrowly interpreted given its potential to overtake provincial jurisdiction over property and civil rights. It has been rejected as a general authority for regulating the economy, though it was used to uphold the *Competition Act*

---

384 *Ibid* at 498.
385 See, e.g., *The Queen (Man) v Air Canada*, 1980 CanLII 16 (Supreme Court of Canada).
386 See, Leach, "I Can't Believe It's Not Butter," above note 196 at 498.

in the 1989 *General Motors of Canada Ltd v City National Leasing* case[387] and the federal trademark law in *Kirkbi AG v Ritvik Holdings Inc.*[388]

The Court in the *General Motors* case held that to fall under the second branch of section 91(2), the legislation in question must engage the national interest in a way that is different from provincial concerns. Building on previous cases, the Court created a non-exhaustive and non-inclusive list of factors that should be considered when the federal government invokes section 91(2) to justify legislation: (1) the law must be part of a general regulatory scheme; (2) the scheme should be under the oversight of a regulatory agency; (3) the legislation should be concerned with trade as a whole and not a particular industry; (4) the legislation should be of a nature that the provinces, jointly or severally, would be constitutionally incapable of enacting; and (5) failure to include one or more provinces in the regulatory scheme would jeopardize its successful operation.[389]

This five-part test was applied again by the Supreme Court of Canada in a more recent case in 2005[390] and again in the two National Securities references.[391] In the first *National Securities Reference*, the Court was asked to determine whether a national system of securities regulation was within the general trade and commerce power (no other power was assessed). The proposed Act allowed provinces to opt into the federal scheme, thereby permitting them to maintain jurisdiction over securities provincially if so desired. The Court rejected the scheme, holding that the general trade and commerce power cannot be used in a way that denies provinces the power to regulate local matters and industries within their boundaries.[392] In applying the GM criteria, the Court held that the proposed national scheme met the first two criteria, and that the fourth was partially met. But the court held that the legislation was not concerned with trade as a whole (criteria 3) and emphasized the

---

387 1989 CanLII 133 (Supreme Court of Canada) [*General Motors*].

388 2005 Supreme Court of Canada 65 [*Kirkbi AG*].

389 *General Motors*, above note 387 at 643.

390 *Kirkbi AG*, above note 388 at paras 16–17.

391 *Pan-Canadian Securities*, above note 67; *Securities Act Reference*, above note 320 at paras 108 and 110–23.

392 *Securities Act Reference*, above note 320 at para 89.

long-time involvement of the provinces in securities regulation. The thrust of the decision was that the federal legislation went too far. While the economic importance and character of the securities market may justify federal intervention that is distinct from what the provinces can do, it does not warrant a "takeover" of the entire area of securities regulation. To be valid, Parliament would have needed to develop a scheme that dealt with genuine national concerns (such as the management of systemic risk and national data collection) while leaving the provinces to deal with the local securities regulation. As designed at that time, the main thrust of the legislation was not focused on matters of national importance and trade as a whole in a way that could be distinguished from provincial concerns. The Court also emphasized that the area of securities has not been so transformed that it now needed to be regulated nationally.[393]

In the second *Securities Reference*, the Court examined revised draft legislation aimed at controlling systemic risks that posed a threat to the stability of the country's overall financial system. Instead of displaying provincial securities laws, the federal act aimed to complement provincial legislation by addressing national objectives.[394] As such, it was upheld.

It is possible that a national cap or limit on GHG emissions could be justified under POGG or the criminal law power, as discussed earlier.[395] If entities subject to the cap were allowed to buy and sell allowances with each other, it is possible that these trading provisions could also be justified under POGG and/or criminal law as part of the means of achieving the measure.[396] Could the general branch of the trade and commerce power offer an additional source of authority? It is not uncommon for federal legislation to be justified under more than one head of power. Indeed, portions of federal legislation like the *Food and Drug Act* have been

---

393 *Securities Act Reference*, above note 320 at para 116.

394 *Ibid* at paras 92 and 96.

395 *Syncrude* FCA, above note 30, affg *Syncrude* FC, above note 30. See also Leach, "I Can't Believe It's Not Butter," above note 196.

396 See Leach, "I Can't Believe It's Not Butter," above note 196 (noting that a federal tradeable quota system for renewable fuels was upheld as *intra vires* the criminal law power in the criminal law power in Syncrude).

justified under the criminal law power, while its marketing provisions have been justified by the trade and commerce power.[397] Environmental impact assessment offers another example of legislation justified under multiple heads of power. Both branches of the trade and commerce power may be relevant to a national emissions trading program. If the bulk of the trading was expected to be interprovincial (and have an international component, if linked with systems in other countries), then the first branch of section 91(2) may justify the trading portions of the legislation. Courts have accepted that a predominantly interprovincial trading system may still be justified under this branch if it has some intraprovincial components.[398] However, if more than an incidental amount of trading were intra-provincial, could the general branch apply?

Answering this question requires considering the criteria established in the GM case. The first two criteria of the second branch would easily be met, since the rules governing trading would be part of a general regulatory scheme for the cap and trade program and under the oversight of an agency to manage the program.[399] The third criterion requiring that the legislation be concerned with trade as a whole, versus a particular industry, would likely be satisfied if the cap and trade system was designed to apply across sectors, versus applying only to particular industries. This criterion may not be met if the government takes a piecemeal, industry by industry approach to the trading system. However, a cap on GHG emissions could designed to apply across industrial sectors.[400] Whether the trading component of a cap and trade system would satisfy this criterion is the subject of some debate. As noted earlier, Professor Leach distinguishes the tradable form of emissions quotas or credits from the emissions themselves, characterizing the quotas as a single commodity. As such, he argues the third criterion would not be satisfied for the trading component of the

---

397 *Wetmore*, above note 261 at 288.
398 See Andrew Leach, "Environmental Policy is Economic Policy: Climate Change Policy and the General Trade and Commerce Power" (2022) 52:2 Ottawa L Rev 295 at 323 [Leach, "Trade and Commerce"].
399 *Ibid.*
400 Professor Leach concludes that "regulatory charges applied broadly to GHGs across the economy would satisfy the third *General Motors* criterion of a truly national policy" (see Leach, "Trade and Commerce," above note 398 at 330).

system since it would be regulating the trade of a particular commodity (emissions quotas) not emissions more generally.[401]

The fourth criterion requires that the provinces be unable to effectively enact a similar regime, while the fifth criterion considers whether the omission of any given province from the system would jeopardize its success. While it is certainly possible for the jurisdictions to cooperate, it is impossible for any given province to bind other provinces to their system. No province can legislate activities outside its borders nor can they regulate extra-provincial trade (aside from what falls within section 92A). Only the federal government can do this. As such, refusal to cooperate with the national scheme jeopardizes its success. The distinction between provincial inability to legislate and provincial unwillingness to participate in a national scheme is important and has been the subject of debate in the *Securities* and *GGPPA References* cases. In *GGPPA References*, the Supreme Court of Canada clarified that provincial inability includes a province's decision not to cooperate in a national scheme when that choice has extra-provincial consequences.[402] Assuming this reasoning is applied in the trade and commerce analysis, the fourth criterion would likely be met.

There is ample evidence from economics to show that a national system would be more cost-effective and avoid leakage from one jurisdiction to another, in comparison to a series of sub-national systems.[403] Leach offers compelling economic analyses of carbon pricing and emissions trading in several of his publications.[404] The court in the *Securities References* emphasized the importance of fair, efficient, and competitive markets in a national context, factors that are relevant to an emissions trading system.[405] The fact that some jurisdictions are the source of disproportionately high emissions would mean that their omission would have a correspondingly disproportionate impact on the success of the system.

---

401 *Ibid* at 334.

402 *GGPPA References*, above note 1.

403 See generally Centre for Climate and Energy Solutions, "Cap and Trade Basics" (last modified May 2024), online: c2es.org [perma.cc/S73F-BLJB].

404 See, e.g., Leach, "I Can't Believe It's Not Butter," above note 196; Leach, "Trade and Commerce," above note 398.

405 *Securities Act Reference*, above note 320 at para 123; *Pan-Canadian Securities*, above note 67 at para 115.

In sum, it is possible that both the cap and trading portions of a national cap and trade program could be justified by the trade and commerce power, though some uncertainty remains. This power has not been used before to authorize federal environmental legislation and much would depend on the specific design elements of a given regulation.[406]

3.3.2.6 Taxation Power

Parliament has broad powers of taxation under section 91(3) of the *Constitution Act, 1867*, which authorizes Parliament to raise money by any mode or system.[407] These powers authorize Parliament to impose direct and indirect taxes, raising revenue for a variety of purposes. There is no requirement for revenue raised through taxation to be earmarked for particular purposes. In fact, one of the distinguishing characteristics of a tax is that it raises revenue for general purposes.

Those unfamiliar with the jurisprudence interpreting the taxation power might have expected Parliament to justify its national carbon price (often referred to by its opponents as a "carbon tax") in the *GGPPA* under the taxation power.[408] Saskatchewan even conceded in its factum at the Court of Appeal in its challenge to the law that Parliament has authority to impose a federal carbon tax, stating that it "would have no constitutional objection if the federal government adopted a national carbon tax that

---

406 Professor Andrew Leach examines whether this power could be used to justify a federal allocation of emissions quotas by province and concluded the parallel to supply management was unsatisfying. Not only are emissions distinct from agricultural produce (in that the latter can be physically stopped from crossing borders, whereas the former cannot), but GHG emissions lack the shared political goals that characterized agricultural supply management schemes (see Leach, "I Can't Believe It's Not Butter," above note 196 at 498).

407 See, e.g., *Winterhaven*, above note 204. The distinction between direct and indirect taxes is important in that the provincial government's main power of taxation is limited to direct taxation. Provincial power to tax under section 92(2) is limited to direct taxes within the province. Section 92A grants broad taxation powers (direct and indirect) over natural resources in the province.

408 Opponents of the federal carbon pricing system have consistently called the levy established in the *Greenhouse Gas Pollution Pricing Act* as a carbon tax (see, e.g., Sara Jabakhanji, "Federal Carbon Tax 'Has to Go', says Ontario Premier," *CBC News* (2 April 2024), online: cbc.ca [perma.cc/5N4T-P8TQ].

applied uniformly all across the country ... [or that] provided for variations based on objective criteria."[409]

However, Canada did not attempt to justify the *GGPPA* under the taxation power. Why not, given broad interpretations of this federal power and the province of Saskatchewan's concession to this effect?[410] The main reason is that the pith and substance of a tax is to raise revenue. The purpose of the carbon price is to reduce GHG emissions by changing behaviour, not to raise revenue. In fact, the carbon pricing system generates no net revenue by design, since the legislation obligates Parliament to return all revenues raised under the Act back to the provinces.[411] As such, it is not a tax but rather a regulatory charge or levy that must be justified under another head of power.

Characterizing the carbon price as a constitutional tax would also open the door to the application of section 125 of the *Constitution Act, 1867*, which would exempt provincially owned property from a federal carbon tax.[412] This could result in exemptions from the tax for GHG emissions generated by provincially owned utilities, such as SaskPower. Such exemptions would reduce the effectiveness of the tax as a means of modifying behaviour and could lead to uneven application across provinces, depending on the relative proportion of Crown-owned GHG generating utilities. Since section 125 only applies to taxes, characterizing the carbon price as a regulatory charge avoids this exemption. Similarly, when characterized as a regulatory charge, the arguments about section 53 (which

---

409 Saskatchewan argued that the national carbon price as enacted in the GGPPA is indeed a tax but an unconstitutional one that infringes section 53 of the *Constitution Act, 1867*. In contrast, Ontario did not concede Parliament's authority to implement a carbon tax and argued that the GGPPA is an attempt to implement a tax that violates section 53. Both provinces also argued that if the GGPPA is found to create a regulatory charge, it does not have the required nexus to the regulatory scheme in question and is thus invalid.

410 For an extensive discussion of jurisdictional authority to impose carbon pricing, see Chalifour, "Making Federalism Work," above note 180; Nathalie J Chalifour, "The Constitutional Authority to Levy Carbon Taxes" in Thomas J Courchene & John R Allan, eds, *Canada: The State of the Federation 2009 – Carbon Pricing and Environmental Federalism* (McGill-Queen's University Press, 2010) 177.

411 *GGPPA*, above note 209, s 188(1).

412 This provision exempts provincial Crown resources from federal taxation and vice versa. Hogg & Wright, above note 152 at 31.26.

requires that taxes originate from the House of Commons, rather than by executive order) also become irrelevant, since that section only applies to taxation.[413]

In 2021, the Supreme Court of Canada confirmed that the levies imposed in Parts 1 and 2 of the *GGPPA* are, in pith and substance, aimed at changing behaviour (not raising revenue). The Court observed that the legislation "could fully accomplish its objectives ... without raising a cent."[414] To arrive at this conclusion, the Court relied upon extensive jurisprudence distinguishing between a tax and other measures, such as regulatory charges. This finding also meant that the regulatory charge had to be justified under some other power, which was the National Concern doctrine of POGG in the end.[415]

3.3.2.6.1 Distinguishing Between Taxes and Regulatory Charges

The leading case distinguishing between a tax and a regulatory charge is *Westbank First Nation v British Columbia Hydro and Power Authority*.[416] Because many levies have elements of taxation and regulation, the courts must determine the primary purpose of the measure. As a starting point, the Supreme Court of Canada in *Westbank* provided a list of three possibilities for what might be the dominant purpose of a given levy: (1) to tax, i.e., to raise revenue for general purposes; (2) to finance or constitute a regulatory scheme, i.e., to be a regulatory charge or to be ancillary or adhesive to a regulatory scheme; or (3) to charge for services directly rendered, i.e., to be a user fee.[417]

Interestingly, this list of categories does not capture the full range of purposes for which levies may be implemented. In particular, the list fails to provide a clear space for measures whose primary purpose is to create a price signal through an economic instrument intended to change behaviour. Economic instruments are increasingly used by governments

---

413   See Peter W Hogg, "Can the Taxing Power be Delegated?" (2002) 16:10 SCLR, online: digitalcommons.osgoode.yorku.ca [perma.cc/NAR8-SG5X].

414   *GGPPA References*, above note 1 at para 219, citing *GGPPA References* SKCA, above note 32 at para 87.

415   See discussion of POGG National Concern in section 3.2.3 of this Chapter.

416   1999 CanLII 655 (Supreme Court of Canada) [*Westbank*].

417   *Ibid* at para 30.

to influence behaviour. Courts should add a fourth element, namely "to impose a price signal intended to change economic behaviour, i.e., to internalize an environmental externality" to the *Westbank* list of possibilities for the purpose of a levy.

Since the carbon price is clearly not a charge for services rendered (eliminating category three above), the Supreme Court of Canada had to determine whether it was a tax or regulatory charge. In *Westbank*, the Court drew upon a long line of cases to define a tax as meeting five criteria: (1) compulsory and enforceable by law; (2) imposed under the authority of the legislature; (3) levied by a public body; (4) intended for a public purpose; and (5) unconnected to any form of a regulatory scheme.[418] The Court noted that many regulatory charges and taxes share the first four characteristics. As a result, the main feature that distinguishes a tax from a regulatory charge is whether the levy is connected to a regulatory scheme (the last criterion).

3.3.2.6.2 Connection to a Regulatory Scheme

To determine whether a levy is connected to a regulatory scheme, the courts must first identify a relevant regulatory scheme (step 1) and then determine whether there is a nexus between the regulatory scheme and the revenues generated from it (step 2). To identify a relevant regulatory scheme, the court suggests looking for the following criteria, though it emphasizes that the list is meant as a guide, not a rigid or exhaustive list: (1) a complete, complex, and detailed code of regulation; (2) a regulatory purpose which seeks to affect some behaviour; (3) the presence of actual or properly estimated costs of the regulation; (4) a relationship between the person being regulated and the regulation, where the

---

418 *Ibid.* The first four characteristics of a tax come from the Supreme Court of Canada's 1931 decision in *Lawson v Interior Tree Fruit and Vegetable Committee of Direction*, 1930 CanLII 91 (Supreme Court of Canada) [*Lawson*]. The fifth characteristic was added by the Court in *Eurig Estate (Re)*, 1998 CanLII 801 (Supreme Court of Canada). This definition was recently confirmed by the Supreme Court of Canada in *620 Connaught Ltd v Canada (Attorney General)*, 2008 Supreme Court of Canada 7 at paras 19 and 22 [*620 Connaught*]. In this case, the Court considered whether the portion of a business license fee based on a percentage of annual purchases of alcohol paid by hotel, restaurant and bar owners in Jasper National Park was a tax or a license fee.

person being regulated either benefits from, or causes the need for, the regulation.[419]

In *GGPPA References*, the Supreme Court of Canada accepted that the *GGPPA* constitutes a complete, complex, and detailed code of regulation for the charge.[420] The explicit regulatory purpose of the *GGPPA* is to affect behaviour, namely to encourage choices that lead to reduced GHG emissions. The costs of the measure will be estimated by the Department of Finance Canada. The entities being regulated cause the need for the regulation. This latter point could be characterized in one of two ways: one, provinces to which the backstop will apply created the need for the backstop by not opting to impose their own carbon price (though note that the charge applies to fuel distributors, not the province itself); or two, GHG emitting entities cause the need for the regulation by contributing to rising GHG emissions.

The second step requires determining whether there is a relationship or nexus between the regulatory charge and the regulatory scheme. This is the part of the test that the challenging provinces argued is not met.[421] The Supreme Court of Canada, however, held that this nexus existed.

This connection may be established in one of two ways. First, the connection may be shown to exist by demonstrating a link between the revenue generated by the charge and the costs of the regulatory framework. Alternatively, the connection may be established by showing that the charge has a regulatory purpose — that the charge itself constitutes the regulatory purpose. The challenging provinces argued that the regulatory charge in the *GGPPA* fails to meet the first test. They pointed to the legislated requirement that revenue generated by the charge be returned to the province of origin, rather than being earmarked for a regulatory purpose, such as GHG mitigation. Since the *GGPPA* requires revenue generated by the charge to be returned to the relevant province, the revenue would not be used to defray the costs of implementing the law nor to advance its

---

419 *620 Connaught*, above note 418 at para 25.
420 *GGPPA References*, above note 1 at paras 212–15.
421 *Ibid* at para 214.

regulatory purpose. As such, the provinces argued, it would not meet this first branch of the test.

Canada argued that the regulatory charge meets the second branch of the test because the charge itself constitutes the regulatory purpose. The very purpose of the charge is to send a price signal intended to reduce GHG emissions. Courts have recognized the role that charges can play in influencing behaviour, such as the levy on landfill deposits to discourage waste or a deposit refund charge to encourage recycling.[422] They have also accepted that a nexus can be established not just where the charge is used to defray the costs of the regulatory scheme but also "where the purpose of the regulatory charge is to proscribe, prohibit or lend preference to certain conduct."[423] As established in *620 Connaught*, regulatory charges may be aimed at altering individual behaviour, which means the fee can be set at the level at which it can effect that influence.[424] However, the Court in *620 Connaught Ltd v Canada (AG)* left open the question of whether the costs of a regulatory scheme are a limit on the fee revenue generated, when the purpose of the regulatory charge is to change behaviour.[425] In *GGPPA References*, the Supreme Court of Canada answered this question in the negative, confirming that the amount of a behaviour-modifying regulatory charge is not tied to the recovery costs of a scheme and that the revenues collected need not be used to further the purposes of the scheme.[426] The key is whether the charges have a regulatory purpose.[427]

While the taxation power grants broad authority to Parliament, it is necessary to carefully consider the pith and substance of levies, charges, and taxes that might not qualify as taxes in the constitutional sense. As illustrated by *GGPPA References*, the federal taxation power was not a

---

422 See *620 Connaught*, above note 418 at para 20, citing *Westbank*, above note 416 at para 29, referring to *Re Ottawa-Carleton (Regional Municipality) By-law 234-1992*, [1996] OMBD No 553 and *Cape Breton Beverages Ltd v Nova Scotia (Attorney General)*, 1997 CanLII 9915 (NS SC).

423 *620 Connaught*, above note 418 at para 48.

424 *GGPPA References*, above note 1 at para 215, citing *620 Connaught*, above note 418 at para 20.

425 *620 Connaught*, above note 418 at para 48.

426 *GGPPA References*, above note 1 at para 216.

427 *Ibid* at para 216, citing *620 Connaught*, above note 418 at para 44.

source of constitutional authority for the federal carbon pricing law. That said, it is certainly possible to use revenue-generating taxes in a way that would have the effect of changing behaviour in a way that is more aligned with environmental goals. The question of whether taxes implemented with a dual purpose of revenue raising and environmental goals on equal footing could be justified under the taxation power is an open one that is worth further exploration.[428]

3.3.2.7 Declaratory Power Under Section 92(10)(c)

Section 92(10)(c) of the *Constitution Act, 1867* grants the provinces legislative jurisdiction over local works and undertakings, except for specific situations, including:

> [s]uch Works as, although wholly situate within the Province, are before or after their Execution declared by the Parliament of Canada to be for the general Advantage of Canada or for the Advantage of Two or more of the Provinces.[429]

The resulting federal declaratory power effectively allows Parliament to assume jurisdiction over a qualifying project that would normally be under provincial responsibility. The federal declaratory power was used extensively during the early days of Confederation to support the country's nation-building efforts. The power was mainly used in areas of transportation, telecommunications, and labour disputes of national importance, with over half of the declarations relating to railways.[430] The federal government used the declaratory power to claim jurisdiction over nearly the entire supply chain for the grain trade beyond the farm gate, including elevators and processing plants. It also declared all works and undertakings constructed for the production and use of nuclear energy

---

428 See Chalifour, "Making Federalism Work," above note 180 at 151–54 (for a discussion of the different ways taxes can be characterized, including Pigouvian style taxes aimed at correcting environmental externalities).

429 *Constitution Act, 1867* (UK), 30 & 31 Vict, c 3, s 92(10)(c) reprinted in RSC 1985, Appendix II, No 5.

430 The power has also been used to support federal authority over railways, canals, bridges, dams, tunnels, harbors, telephones, mines, mills, grain elevators, hotels, restaurants, and factories, among others (see Hogg & Wright, above note 152, at 22.1).

to be for the general advantage of Canada, bringing nuclear facilities —
even if wholly within a province — within federal jurisdiction.[431] It has
occasionally been used in the modern context, such as the Samuel Champlain Bridge that connects the island of Montreal to the city of Brossard.[432]
There is a private members' bill proposing using the declatory power on
the Chignecto Isthmus Dykeland System that sits on the land between
Nova Scotia and New Brunswick.[433] While legally possible, it is unlikely
that Parliament would use its declaratory power to justify legislation over
works and undertakings within a province or territory without the latter's
consent, given the political ramifications of doing so. However, it is a constitutional power that entails explanation.

The power is significant in several ways. For example, as long as a
declaration meets the criteria of relating to a work and that such work is
for the general advantage of Canada, it is dispositive.[434] In other words, it
is a final decision by Parliament that is not subject to judicial review.[435] The
Supreme Court of Canada has stated that "Parliament is not limited either
as to time or as to occasion in resorting to s. 92(10)(c)"[436] and that Canadian
courts "have never shown any disposition to limit its operation."[437] The
power thus essentially allows Parliament, by a unilateral (declaratory) act,
to increase its legislative power over the provinces.[438] Additionally, any
work (or class of works) that falls under the federal declaratory power
creates exclusive legislative powers for Parliament in relation to that work,

---

431 Bora Laskin, "Tests for the Validity of Legislation: What's the 'Matter'?" (1955) 11:1
      UTLJ 114 at 120. See *Nuclear Energy Act*, R.S.C. 1985, c. A-16, s 18(a).
432 See s. 3 *New Bridge for the St. Lawrence Act*, S.C. 214, c. 20, s. 375.
433 See Bill S-273, *The Chignecto Isthmus Dykeland System Act*.
434 See Kenneth Hanssen, "The Federal Declaratory Power under the British North
      America Act" (1968) 3:1 Man LJ 87. See also Irvin Studin, "Constitution and Strategy:
      Understanding Canadian Power in the World," (2009) 5:1 Comp Research in L & Political Econ Research Paper Series 46.
435 Hanssen, above note 434 at 102–5.
436 *Jorgenson v Attorney General of Canada*, 1971 CanLII 136 at 729 (Supreme Court of
      Canada) [*Jorgenson*].
437 *Ontario Hydro*, above note 161 at 370.
438 Hogg & Wright, above note 152 at 22.1; Hanssen, above note 434 at 92.

ousting provincial jurisdiction over that work.[439] As Professor Andrew Leach notes, the declaratory power is largely unfettered.[440]

What can Parliament declare under this power? While section 92(10)(a) confers jurisdiction on Parliament for any works and undertakings that connect provinces or extend beyond their limits, section 92(10)(c) allows Parliament to declare works wholly situate within a province to be for the general advantage of Canada or two or more provinces, thereby bringing such works within Parliament's jurisdiction. While subsections (a) and (b) of section 92(10) refer to both "works and undertakings," subsection (c) refers only to "works." This suggests that section 92(10)(c) may not apply to "undertakings." However, the courts have not interpreted the section with this limitation. The courts have offered guidance on the difference between works and undertakings, characterizing the former as a tangible thing and the latter as an intangible "arrangement," "organization," or "enterprise."[441] The majority in *Ontario Hydro* defined a work as "a going concern or functioning unit."[442] The courts have also suggested that the distinction is not meaningful in interpreting the declaratory power. Indeed, no declaration under section 92(10)(c) has ever been invalidated because it dealt with an undertaking rather than a work. In one case, the Supreme Court of Canada held that the fixing of tolls was an undertaking and yet entirely valid within Parliament's declaratory power.[443] More recently, the Supreme Court of Canada in *Ontario Hydro* acknowledged that while the absence of only the word undertakings in section 92(10)(c) would appear to limit the declaratory power to "works," the Court concluded that the distinction was immaterial and that "Parliament may validly declare an undertaking to be a work for the general advantage of

---

439 *Ontario Hydro*, above note 161 at para 52.

440 See Leach, "I Can't Believe It's Not Butter," above note 1963 at 514. See also Jean Leclair, "The Supreme Court of Canada's Understanding of Federalism: Efficiency at the Expense of Diversity" (2002) 28 Queen's LJ 411 at 443.

441 *Ibid*. See also *City of Montreal v Montreal Street Railway Company (The Attorney-General For the Dominion of Canada and the Attorney-General For the Province of Quebec Intervening)*, 1912 CanLII 352 (UK JCPC).

442 *Ontario Hydro*, above note 161 at 329.

443 *Town of Beauport v Quebec Railway, Light & Power Co/Quebec Railway, Light & Power Co v Town of Beauport*, 1944 CanLII 49 (SCC).

Canada."[444] The limiting factor is that the undertaking needs to be linked to a work.

Because of this expansive interpretation, the declaratory power has been held to enable Parliament to regulate activities relating to a given work. For instance, the Supreme Court of Canada held that a federal declaration over nuclear power plants in Ontario conferred jurisdiction on Parliament over the labour relations related to the declared work, since the declaration gave Parliament "control over operation and management" of the work.[445] In support of its expansive interpretation, the Court held that a declaration "must surely be to bring within federal authority not only the physical shell of the activity but also the integrated activity carried on therein."[446] This decision is in keeping with earlier decisions which found that Parliament has the authority to regulate activities relating to a given work subject to a valid declaration. For instance, in the 1940s, Parliament declared all grain elevators in Canada to be works for the general advantage of Canada in the *Canada Grain Act* and all flour mills and warehouses to be such in the *Canadian Wheat Board Act*. In separate cases, the Supreme Court of Canada concluded that these declarations were valid and had the effect of authorizing Parliament to regulate activities relating to the grain trade, such as the delivery, receipt, storage, and processing of grains.[447] The Supreme Court of Canada has also held that declarations referring to a class of works, and/or future works, were valid. In other words, a work does not need to be specifically noted by name — a class of works will suffice.[448] As well, Parliament is not limited to existing works, but the power extends to works that will be constructed in future.[449] As

---

444 *Ontario Hydro*, above note 161 at 73–74.

445 *Ibid*. Of Ontario Hydro's operating plants, five were nuclear powered and were therefore declared to be for the benefit of Canada under the *Atomic Energy Act* (see *ibid* at paras 329 and 341).

446 See Bora Laskin & Neil Finklestein, *Laskin's Canadian Constitutional Law*, 5th ed (Toronto: Carswell, 1986) Vol 1 at p 629; Hogg & Wright, above note 152 at 22.17. See also Hanssen, above note 428 at 93–95.

447 *Jorgenson*, above note 436 at 729; *R v Chamney*, 1973 CanLII 197 (Supreme Court of Canada).

448 *Jorgenson*, above note 436 at 729.

449 *Ibid* at 733–34.

per the Court, "[i]t matters not that new members of the described class may come into existence after the declaration is made, for the declaration can be made before or after the execution of the work."[450]

The Supreme Court of Canada has not defined or contextualized the words "for the general advantage of Canada."[451] It seems that to be valid, a declaration by Parliament must simply state explicitly that the work or undertaking is for the general advantage of Canada (in other words, no work will be justified by implication).[452] Presumably there is some implicit normative content inherent in the concept of "general advantage," but this has not been elaborated upon in the jurisprudence.

Could the declaratory power be used by the federal government to justify climate legislation? Leaving aside the obvious and serious political challenges of using this power, the power may be available to justify certain types of legislation aimed at reducing GHG emissions. Much would turn on the type of legislation under consideration and the extent to which courts were willing to apply an expansive interpretation of "works." It seems clear that Parliament would be within its authority under section 92(10)(c) to declare power-generating works and infrastructure to be for the general advantage of Canada. Given the significant economic, social, and environmental implications of GHG emissions associated with power-generation and the international obligation to decarbonize energy production, Parliament would have a solid argument for declaring such works to be for the general advantage of Canada. Were Parliament to use this power, it would undoubtedly wish to carefully circumscribe the declaration to limit intrusion into provincial matters. However, the declaratory power does not require this, as it is not subject to a colourability analysis. As such, it would be within Parliament's authority — following such a declaration — to impose caps and or efficiency standards on GHG emissions from power generating plants across the country.

---

450 *Jorgenson*, above note 436 at 734.

451 Hogg & Wright, above note 152.

452 *Ibid*. See also *YMHA Jewish Community Centre of Winnipeg Inc v Brown*, 1989 CanLII 53 at 1552 (Supreme Court of Canada).

Given the broad definition of "works," the range of projects that Parliament could declare to be works for the general advantage of Canada is extensive. Consider resource extraction projects, electricity grids, and refineries, for instance. The economic, social, and environmental implications of the GHG emissions relating to these activities are such that Parliament could conceivably declare them to be for the general advantage of Canada and justify imposing GHG reduction targets on these entities. Since classes of works and future works are within the scope of this power, Parliament might even conceivably declare jurisdiction over all works and undertakings that produce GHG emissions over a certain significant threshold.

There are many uncertainties regarding the use of this power given the limited judicial interpretation in the last several decades. Additionally, Parliament is not likely to use the power given the political repercussions of doing so. However, the risks posed by climate change are increasingly significant and the context in which we live is vastly different than it was just a decade ago. Provinces with abundant fossil fuel reserves are understandably reluctant to commit to significant GHG emissions reductions, but that does not erase the imperative to do so. The declaratory power was essential for building the country in the days of Confederation. Perhaps it will be needed to deal with the global climate threat that risks being the instrument of "unbuilding" the country. Considering the political context, the strength and relevance of the declaratory power may lie in its negotiating power. The possibility of Parliament using the declaratory power to assume control of oil and natural gas during the 1970s OPEC oil embargo is said to have been an instrumental bargaining chip, leading to negotiations with the western provinces (and the eventual enactment of section 92A of the Constitution).[453] Perhaps it will serve a similar function in the context of ongoing resistance to nation-wide GHG emissions reduction

---

453 See Peter J Meekison & Roy J Romanow, "Western Advocacy and Section 92A of the Constitution" in Peter J Meekison et al, eds, *Origins and Meaning of Section 92A: The 1982 Constitutional Amendment on Resources* (Montreal: Institute for Research on Public Policy, 1985) at 18. See also Studin, above note 434 at 64.

policies (assuming political will to reduce GHG emissions continues at the national level).

3.3.2.8 Conclusion

The above discussion does not purport to comprehensively address all potential sources of jurisdictional authority to enact laws aimed at reducing GHG emissions. As Brown J pointed out in his dissent in *GGPPA References*, aspects of GHG reduction laws could come within other federal powers including navigation and shipping.[454] Federal authority over fisheries could have an important influence over GHG emissions through regulations on effluent to water from industrial activities, including oil and gas production. Similarly, federal jurisdiction over bankruptcy and insolvency could have significant influence over large industrial actors, including fossil fuel producers, responsible for creating liabilities through abandoned wells and more.[455]

3.4 Looking Ahead – POGG 2.0

Leading up to *GGPPA References*, there were suggestions in scholarship and in the arguments presented in the case that the POGG test established in *Crown Zellerbach* needed to be updated.[456] In *GGPPA References*, both Ontario and Canada explicitly recognized the need to modernize the POGG test in their submissions, proposing two different approaches that each drew on the Court's jurisprudence on the general trade and commerce power. Ontario submitted that the Court should determine whether the POGG matter is qualitatively different from provincial matters and relegate the provincial inability test to jurisdictional inability.[457] Canada suggested that the test boils down to three elements: (1) whether there is a new matter or transformation that makes a local matter

---

454 *GGPPA References*, above note 1 at para 379.

455 Martin Olszynski, "Inextricably Linked: Climate Policy and the Oil and Gas Sector's Closure Liabilities" (21 February 2024), online: ablawg.ca [perma.cc/CUG8-3WJ4].

456 See, e.g., Chalifour, Oliver & Wormington, above note 93 at 181.

457 *References re Greenhouse Gas Pollution Pricing Act*, 2021 Supreme Court of Canada 11 (Factum, Appellant at para 44).

national; (2) whether the matter is distinctly national; and (3) whether the impact of recognizing federal jurisdiction is reconcilable with the distribution of powers.[458] In addition, some interveners, such as the Athabasca Chipewyan First Nation (ACFN) and the National Association of Women and the Law (NAWL) and Friends of the Earth Canada (FOE), proposed changes to the test. The ACFN proposed a fluid approach to interpreting the branches of POGG, rejecting the trinity of national concern, national emergency, and new matters. NAWL and FOE proposed that the POGG analysis be more flexible and purposive, allowing a court to draw upon the insights and qualities of the emergency and national concern doctrines in a proportional way to align the division of powers with broader constitutional values, such as substantive equality.[459] The Supreme Court of Canada did clarify and update the test in a modest way in *GGPPA References*. This section considers other ways in which the test could be modified in light of the evolving context.

Legal academics have suggested that the traditional dimensions of POGG — gap, emergency, and national concern — are incomplete. Patrick J. Monahan, Byron Shaw, and Padraic Ryan proposed a fourth branch of POGG that "involves the regulation of matters which have interprovincial impact or effects and which cannot be regulated on the basis of any of the enumerated powers in s. 91 of the *Constitution Act, 1867*." Monahan and his co-authors point to judicial commentary by La Forest J in *Morguard Investments Ltd v De Savoye* noting support for the idea that authority to recognize and enforce judgments throughout Canada be read into the Constitution as part of POGG, almost like a fourth branch. The language supporting the idea of a fourth branch of POGG is further developed in *Hunt*, also penned by La Forest J. An application of *Morguard* in *Hunt* led La Forest J to "effectively suggest that Parliament may, pursuant to the POGG power, enact legislation providing for the free movement of persons or for the enhancement of the economic union." For Monahan and

---

458   *References re Greenhouse Gas Pollution Pricing Act*, 2021 Supreme Court of Canada 11 (Factum, Respondent at para 69).

459   *References re Greenhouse Gas Pollution Pricing Act*, 2021 Supreme Court of Canada 11 (Factum, Interveners, National Association of Women and the Law & Friends of the Earth at para 3).

others, this represents a departure from the national concern doctrine that leads them to conclude "there is a fourth, independent source of federal authority under POGG ... [which] permit[s] Parliament to legislate in matters of interprovincial concern, impact, or significance."[460] The *Interprovincial Co-Operatives* decision lends support to this argument, since it struck down a provincial law on the basis that it was aimed at an interprovincial matter.

Given that GHG emissions impact the atmosphere at a global level and are thereby not confined to the borders of a specific province, they are necessarily interprovincial (and international). As such, a fourth branch of POGG focused on matters with interprovincial impacts could be an added source of justification for authority under POGG. It would be necessary to impart some kind of threshold (i.e., significance) to such a fourth branch, to avoid suggesting federal jurisdiction is triggered for trivial effects. The need for a fourth branch may also no longer be necessary given the clarified approach to provincial inability in *GGPPA References* which explicitly includes provincial inaction or non-cooperation that has extra-provincial impacts.

Alternatively, it may be that segmenting authority under POGG into discrete "branches" may not be the optimal way to define its parameters. Instead of discrete branches, another option would be to consider jurisdiction under POGG as a spectrum and adopt a proportionality assessment that would evaluate the degree of intrusion (scale of impact) that is tolerable in relation to the magnitude of the national concern at hand, especially its implications for foundational constitutional values. The Supreme Court of Canada did this to some extent in *GGPPA References* when it incorporated a counterbalance to the scale of impact on provincial jurisdiction — namely the gravity of *not* recognizing federal jurisdiction.[461] The next section explores approaching POGG as a spectrum versus discrete branches.

---

460 Patrick J Monahan, Byron Shaw & Padraic Ryan, *Constitutional Law*, 5th ed (Toronto: Irwin Law, 2017) at 284.
461 *GGPPA References*, above note 1 at para 206.

3.4.1 POGG As a Spectrum Versus Discrete Branches

Instead of operating with distinct branches of jurisdiction under POGG, it is interesting to consider the possibility of approaching POGG as conferring jurisdiction on a spectrum within definable parameters.

As noted above, POGG has traditionally been defined as having three branches: gap, national concern (or national dimensions), and emergency. The gap branch of POGG, also known as the residual branch, is engaged when a specific matter can neither be tied to an enumerated section 92 power nor a federal power under section 91. Matters that are not listed within the *Constitution Act, 1867, Constitution Act, 1982* and that are not local and private thus become a federal power as per the gap branch of POGG. Examples of the gap branch of POGG in action include recognizing federal jurisdiction over the continental shelf of Newfoundland in *Reference re Seabed & Subsoil of Continental Shelf Offshore Newfoundland* and upholding the *Official Languages Act* of the federal government in *Jones v AG New Brunswick*.[462] Since courts interpret provincial property and civil rights and local matters under sections 92(13) and (16) of the *Constitution Act, 1867* broadly, the gap branch of POGG is seldom applied and has faded into obscurity.

The courts developed the national concern and emergency branches in order to distinguish between matters that qualify as federal either because they transcend provincial jurisdiction (i.e., they "concern the Dominion as a whole," address matters of public order and safety, or extra-provincial impacts) or because they amount to an emergency. They have treated these branches as related yet distinct sources of jurisdiction, employing different strategies to address concerns about their potential to upset the balance of powers in the federation.

Within the national concern doctrine, the courts have addressed the risk of overreach by requiring the matter to be clearly and narrowly circumscribed ("single, distinct, and indivisible") and considering whether the scale of impact of the ensuing jurisdiction would be reconcilable with the distribution of powers in the federation. In the case of an emergency, the Supreme Court of Canada has deferred to

---

462 1975 Supreme Court of Canada 30.

Parliament's judgment in finding a rational basis for the existence of the emergency, imposing only the requirement that the federal legislation be of a temporary nature. In other words, subject to the temporal limit, Parliament is empowered under emergency jurisdiction to intrude into matters of provincial jurisdiction as much as needed to deal with the emergency. The tolerance for intrusion is justified because of what is at stake in emergencies, such as death, injury, poverty, economic instability, or national security. The temporary requirement mitigates the intrusion by ensuring it lasts only as long as needed to address the emergency.

While these branches may be helpful ways to delineate jurisdiction in some cases, they are ill-suited to long-lasting emergencies. Because there is no middle ground between the poles of these branches, there is a risk of creating a void for matters that fall in between. Deconstructing the existing analytical framework illustrates this point.

The national concern and emergency branches are both defined by the same two factors: the *duration* and *scope* of jurisdiction. The courts have treated the relationship between these two factors as inversely related: emergency powers are sweeping but temporary, whereas national concern powers are narrow but permanent. There is no middle ground to recognize jurisdiction that is somewhere in between. The treatment of these doctrines as mutually exclusive, inversely related on the polarities of duration and scope, creates a hole in POGG for the middle ground. The absence of a middle ground opens the risk of certain subjects falling into a jurisdictional void.

Another concern with the existing analytical framework for POGG is how it deals with the need to balance powers in the federation. There is an all-or-nothing flavour to the analysis that allows sweeping, unrestricted powers under the emergency branch (no need for any balancing), compared to very careful balancing in the national concern branch. The Supreme Court of Canada in *GGPPA References* adjusted the scale of impact test in the national concern branch analysis to include consideration of the implications of denying federal jurisdiction on the interests at stake. As a result, there is more of a middle ground now for national concern, though not for the emergency power.

The current approach has hallmarks of formalism and rigidity, perhaps explained by the era from which they emerged. This could be addressed by introducing a degree of flexibility and a normative element that would fold the POGG analysis into its broader constitutional framework. What if courts drew upon the virtues of both doctrines, viewing the factors of *duration* and *scope* independently on a spectrum in light of the particular circumstances of each case? Such an adaptation would anchor POGG more clearly in the modern paradigm of cooperative federalism that embraces flexibility.

3.4.2 Updated POGG Test

The modification to the POGG analysis proposed here requires three changes, which each introduce a degree of flexibility and proportionality to the analysis.[463] First, the adaptation requires tempering the *duration* factor so that the length of time for which jurisdiction is authorized is commensurate with the significance and urgency of the issue. The more consequential and urgent the issue, the more tolerance for jurisdiction that is long-lasting. Long-lasting jurisdiction need not be permanent in the sense of "adding a new sub-head of power" to the federal jurisdictional column, but it could be something more than the temporary jurisdiction envisioned in the emergency context. Essentially, jurisdiction would last as long as needed to address the problem.

While this might raise concerns about the introduction of uncertainty and/or potential of requiring more court oversight, it is already the case from a practical standpoint. Additionally, our judicial system is accustomed to dealing with this kind of constitutional ebb and flow without requiring judicial intervention at every step. While still deferring to Parliament's judgment about the existence of an emergency, a court would recognize that jurisdiction lasts as long as needed to address the emergency, even if that approximates a duration that is somewhere between what we might think of as temporary and permanent. This would be especially helpful in the case of a longer-lasting emergency like climate change.

---

463 See David Suzuki Factum, above note 367 (arguing for proportionality in interpretating the emergency branch).

Second, the adaptation would require tempering the *scope* factor. While efforts should always be made to clearly identify the contours of jurisdiction (i.e., single, distinct, and indivisible), the scope should be defined in accordance with the features of the national concern (i.e., its interprovincial, national, or international aspects) rather than driven by the need to clearly distinguish it from matters of provincial concern. Here again a proportionality test can be helpful: the more significant and urgent the national concern, the greater the tolerance for jurisdiction that may overlap with provincial powers. The breadth of jurisdiction would rarely need to reach the end of the spectrum, where there are no limits to Parliament's jurisdiction to legislate on subject matters and where it is even possible to reach those that are entirely intra-provincial and that fall within enumerated provincial powers (as is currently possible under the emergency branch). But a more flexible approach to *scope* could reduce some of the difficulty courts face in trying to assign black and white contours to a subject matter that is comprised of grey hues, as is the case with GHG emissions.

Third, the adapted POGG test would consider whether the scale of impact on provincial jurisdiction is reconcilable not only with the fundamental distribution of legislative powers (federalism) but with other important constitutional values such as Indigenous rights and *Charter* values. This would once again allow the court to apply a proportionality analysis that would balance the potential degree of intrusion into provincial jurisdiction with not only the values of federalism but also the extent of extra-provincial impacts. This would, for example, allow the court to consider the potential infringement of Indigenous or equality rights that might ensue if jurisdiction is *not* recognized. The Supreme Court of Canada effectively incorporated this modification at the scale of impact stage of the national concern analysis in *GGPPA References*.[464]

Additionally, the modified test would be explicit about the fact that courts should, at all times, give effect to the double aspect doctrine to recognize the potential for concurrent operation of legislation directed to the same objective, whether a national concern or emergency, and apply the

---

464 *GGPPA References*, above note 1 at para 206.

paramountcy doctrine in the strictest sense (i.e., the most restrained) to maximize the constitutional space for legislation by both provincial legislatures and Parliament. For instance, the potential impacts on federalism from a broader *scope* of jurisdiction could be addressed through explicit recognition that the double aspect doctrine applies to the subject matter, especially when more broadly described.

This updated approach to POGG would allow the court to engage in a flexible analysis that accounts for the unique circumstances of climate change. Those unique circumstances (outlined in Chapter 1) include the need to act quickly to avoid irreversible tipping points, the magnitude of the extra-provincial impacts on other jurisdictions (provincial and federal), and the related potential for climate change to lead to serious infringements of Indigenous rights and human rights, including the equality rights of vulnerable groups. A more flexible approach that engages proportionality would supplant the rigid, dualistic approach that risks creating a void for issues that fall in the middle of the spectrum.

In the context of GHG emissions reductions laws, such an adaptation to POGG would do three things. First, it would allow the court to consider the fact that climate change is a long-lasting emergency without this leading to the inevitable conclusion that emergency branch jurisdiction is unavailable (because climate-related laws are not seen to qualify as temporary legislation). Second, it would allow the court to factor in the particular nature of the climate change problem and of GHG emissions (i.e., the fact that they are generated by activities on the ground but create global harm when they reach the atmosphere) in determining the scope of jurisdiction, without having to be overly focused on the need to find a firm dividing line between GHG emissions that are provincial versus federal.

Third, by broadening the scale of impact analysis to consider the impact not only on provincial jurisdiction but also on Indigenous and equality rights, as the Supreme Court of Canada recently did in *GGPPA References*, the modified test would allow the court to balance concerns about federalism with concerns about fundamental rights and our relationship with Indigenous Peoples. Additionally, the modified test would ensure that the scale of impact on provincial jurisdiction is evaluated in a way that accounts for the application of the double aspect doctrine.

By explicitly recognizing the applicability of the double aspect doctrine, the modified approach would acknowledge that there are different facets to GHG emissions that both levels of government can respectively (and, *de facto*, concurrently) address. This would address the primary concern raised by provinces that upholding GHG regulations would eviscerate provincial jurisdiction over the same: when it is clear that the double aspect doctrine applies, any provincial law affecting GHGs that is anchored within an enumerated provincial power will apply at the same time as federal regulations impacting GHGs subject only to the paramountcy doctrine (which would be applied with the utmost restraint). The modified test would not only help resolve most of the tensions that surfaced in the carbon pricing references but would also bring POGG into the era of cooperative federalism.

Further analysis of this adaptation to POGG is needed, especially in terms of ensuring certainty and predictability with respect to jurisdiction. However, the flexibility afforded by such an approach is warranted especially in the context of major public policy challenges that require regulatory responses from all orders of government. Canada needs a Constitution that is nimble enough to support laws intended to address those challenges.

3.5 MOVING FORWARD — THE CLIMATE-READY CONSTITUTION

A foundational premise of this book is that climate change should not be a cause for the country's dissolution or failure. As such, it is a further premise that the division of power should be interpreted in such a way that each level of government is equipped and empowered to enact climate legislation within its respective spheres of jurisdictional authority. A climate-ready Constitution will enable all levels of government to enact responsive policies within their respective spheres of jurisdiction and reject, as the Supreme Court of Canada has already done, the "either-or," water-tight compartment version of constitutional interpretation advanced by some provinces in the *GGPPA References* that would constrain responses to climate change and create potential gaps in jurisdiction.

3.6 CONCLUSION

The way in which law-making authority was allocated in the Canadian Constitution over 150 years ago is ill-suited to the complexities of today's society. However, the judiciary has navigated this challenge by developing numerous doctrines and principles to support an interpretation of the division of powers that aims to strike a balance between provincial autonomy and national unity. While it is a challenging journey with imperfect results, the horse is still on the track.

The last fifty years of experience regulating environmental issues in the federation in the context of this ambiguous, often contentious, allocation has shown that it can be done. The Supreme Court of Canada's decisions clarifying the contours of authority to enact environmental regulations combined with the ongoing evolution of several interpretative principles, such as cooperative federalism, double aspect, and the pith and substance doctrines, have shaped the jurisdictional framework as it relates to environmental laws and increasingly climate legislation.

Jurisdiction over climate policy will continue to generate debate and controversy as courts address an ongoing wave of challenges to legislation aimed at reducing GHG emissions and other climate goals. The highly charged political context of conflicting views over the future of fossil fuels in a federation with diverse regional economies makes it especially difficult for courts. However, the existential and urgent nature of the climate crisis requires courts to interpret the Constitution in a way that enables the federation to address this grave threat. The imperative to rapidly reduce GHG emissions is not debatable, legally or morally. The division of powers must be interpreted in a way that allows each jurisdiction to enact climate-related legislation aimed at mitigating GHG emissions, as well as adapt to the impacts of climate change and deal with the inevitable loss and damage. The Supreme Court of Canada has consistently recognized that all levels of government must take action in addressing the environmental problems that plague our country.[465] It must also do so for climate

---

465 See *Hydro-Quebec*, above note 3 at para 116, where the Court noted that "[t] he general thrust of Friends of the *Oldman River Society v Canada (Minister of Transport)*

policy. Interpretations that hamstring jurisdictions from taking steps to address this existential threat have no place in our federation, lest they be the instrument of our demise.

The analysis in this chapter suggests an enabling interpretation is entirely possible. The constitutional division of powers can be interpreted in a manner that deals with what may be the greatest public policy challenge of our time. In fact, climate change is perhaps the quintessential issue for engaging the tools of cooperative federalism and a progressive interpretation of our Constitution. All orders of government must be allowed to enact laws to reduce GHG emissions. This is possible if the division of powers is interpreted in a way that is enabling, expansive, and evolving, versus an inhibiting, restrictive, and stalled way.

As governments and industries fight jurisdictional battles in this country, they do so in the shadow of a ticking existential clock. The devastating wildfires, heat waves, floods, and storms we see every year offer a hint of what is to come, but are just the tip of the proverbial (and rapidly melting) iceberg given the time lag between emissions and their impacts.[466] Our actions are jeopardizing humanity's life support systems. We must not keep our heads in the sand about this looming crisis when faced with the challenge of decarbonizing in a country with regionally significant fossil fuel economies. Canada's ability to meet its national climate targets — something it has consistently failed to do — is tied to its jurisdictional capacity to regulate significant GHG emissions. The jurisdictional authority to address the climate crisis in Canada exists. Whether we succeed will be a matter of judicial commitment to progressive interpretations of the Constitution. We must also interpret our *Charter of Rights and Freedoms* in a similarly progressive way — the subject of the next chapter.

(1992) 1 SCR 3 ... is that the Constitution should be so interpreted a to afford both levels ample means to protect the environment while maintaining the general structure of the Constitution."

466 IPCC, *Climate Change 2023*, above note 365.

Additional Resources

Eric M Adams, "Canadian Constitutional Identities" (2015) 38:2 Dal LJ 311.

Eric M Adams, "Touch of Evil: Disagreements at the Heart of the Criminal Law Power" (2022) 104 SCLR (2d) 67–89.

Eric M Adams & Erin RJ Bower, "Notwithstanding History: The Rights-Protecting Purposes of Section 33 of the Charter" (2022) 26:2 Rev of Const Studies 121.

Nigel Bankes & Andrew Leach, "Preparing for a Mid-Life Crisis: Section 92A at 40" 60:4 (2023) Alta L Rev 853.

Philip Barton, "Economic Instruments and the Kyoto Protocol: Can Parliament Implement Emissions Trading without Provincial Co-operation?" (2002) Alta Law Rev 417 at 441.

Joseph E Castrilli, "Legal Authority for Emissions Trading in Canada" in Elizabeth Atkinson, ed, *The Legislative Authority to Implement a Domestic Emissions Trading System* (Ottawa: National Roundtable on the Environment and the Economy, 1998).

Nathalie J Chalifour, "Making Federalism Work for Climate Change: Canada's Division of Powers over Carbon Taxes" (2008) 22:2 NJCL 119.

Nathalie J Chalifour, "Canadian Climate Federalism: Parliament's Ample Constitutional Authority to Legislate GHG Emissions through Regulations, a National Cap and Trade Program, or a National Carbon Tax" (2016) 36 NJCL 331.

Nathalie J Chalifour, "Jurisdictional Wrangling over Climate Policy in the Canadian Federation: Key Issues in the Provincial Constitutional Challenges to Parliament's Greenhouse Gas Pollution Pricing Act" (2019), 50:2 Ottawa L Rev 197.

Elisabeth DeMarco, Robert Routliffe & Heather Landymore, "Canadian Challenges in Implementing the Kyoto Protocol: A Cause for Harmonization" (2004) Alta Law Rev 209.

Stewart Elgie, "Kyoto, the Constitution, and Carbon Trading: Waking a Sleeping BNA Bear (or Two)" (2007) 13:1 Rev Const Stud 67.

Kathyrn Harrison, "Climate Governance and Federalism in Canada" in Alan Fenna, Sébastien Jodoin & Joana Setzer, eds, *Climate Governance and Federalism: A Forum of Federations Comparative Policy Analysis* (Cambridge University Press, 2023) ch 4.

Peter W Hogg, "Constitutional Authority over Greenhouse Gas Emissions" (2008) 46 (2) Alberta Law Review 507–20.

Peter Hogg & Wade Wright, *Constitutional Law of Canada*, 5th ed (Toronto: Carswell, 2019) (loose-leaf updated 2024, release 1).

Shi-Ling Hsu & Robin Elliot, "Regulating Greenhouse Gases in Canada: Constitutional and Policy Dimensions" (2009) 54:3 McGill LJ 463.

Anna Johnston, "Federal Jurisdiction and the Impact Assessment Act: Trojan Horse or Rational Ecological Accounting?" in Meinhard Doelle & A John

Sinclair, eds, *The Next Generation of Impact Assessment: A Critical Review of the Canadian Impact Assessment Act* (Toronto: Irwin Law, 2021) 97.

Anna Johnston, "Two Wins, a Loss, and a Question Mark: What the Impact Assessment Act Reference Case Means for the Environment" (October 18, 2023), West Coast Environmental Law.

Andrew Leach, *Between Doom and Denial: Facing Facts about Climate Change* (2023) McGill Max Bell Lectures.

Andrew Leach, "I Can't Believe It's Not Butter: Supply Management and the Constitutionality of Carbon Budget Legislation in Canada" (2023) 56:2 UBC L Rev Art 4.

Andrew Leach & Eric M Adams, "Seeing Double: Peace, Order, and Good Government, and the Impact of Federal Greenhouse Emissions Legislation on Provincial Jurisdiction" (2020) 29:1 Const Forum Const 1.

Jean Leclair, "Federalism as Rejection of Nationalist Monisms" in Dimitrios Karmis & François Rocher, eds, *The Trust/Distrust Dynamic in Multinational Democracies: Canada in Comparative Perspective* (Montreal and Kingston: McGill-Queen's University Press, 2018) pp. 210–47.

Jean Leclair, "Tis a Rock – a Crag – a Cape? A Cape? Say Rather a Peninsula!" The Supreme Court of Canada's Revisitation of the National Concern Doctrine (2023) 108 SCLR (2d) 3.

Alistair R Lucas & Cheryl Sharvit, "Constitutional Powers" in Alastair R Lucas and Roger Cotton, eds, *Canadian Environmental Law*, 3rd ed (Toronto: LexisNexis, 2023) (looseleaf updated October 2023, release 196) 31.

Alastair R Lucas & Jenette Yearsley, "The Constitutionality of Federal Climate Change Legislation" (2011) 4 The School of Public Policy Publications, online: https://dx.doi.org/10.2139/ssrn.1970007.

Nathan Murray & Martin Olszynski, "Locating the Constitutional Guardrails on Federal Environmental Decision Making after Reference re: Impact Assessment Act" (30 January 2024), online: http://ablawg.ca/wp-content/uploads/2024/01/Blog_NM_MO_Constitutional_Guardrails.pdf.

Peter C Oliver, "The Busy Harbours of Canada Federalism: The Division of Powers and its Doctrines in the McLachlin Court" in David A Wright & Adam M Dodek, eds, *Public Law and the McLachlin Court: The First Decade* (Toronto: Irwin Law, 2011)167.

Martin Olszynski, "General Rules and Principles of Federal Legislative Authority in Relation to Oil and Gas Development in Canada" (IISD, 2025) (in press).

Chris Rolfe, *Turning Down the Heat: Emissions Trading and Canadian Implementation of the Kyoto Protocol* (Vancouver, Canada: West Coast Environmental Law Research Foundation, 1998).

Dayna Nadine Scott, 'The Environment, Federalism, and the Charter' in Peter Oliver, Patrick Macklem & Nathalie Des Rosiers (eds), *The Oxford Handbook of the Canadian Constitution* (OUP 2017) 493.

Jocelyn Stacey, *The Constitution of the Environmental Emergency* (Hart Publishing, 2018).

David V Wright, "Constitutional Caution, Correction, and Abdication: The Proposed Amendments to the Impact Assessment Act" (10 May 2024), online: http://ablawg.ca/wp-content/uploads/2024/05/Blog_DW_IAA _Amendments.pdf.

Wade K Wright, "Canadian Federalism's Underlying Question: What It Is and Why It Matters" (2020) 53:2 UBC L Rev 531 at 556.

chapter 4

CLIMATE CHARTER LITIGATION[*]

> "The Canadian judiciary, in tandem with the other branches
> of government, has an important role to play in protecting
> the 'right to a safe environment.'"[1]
> *Chief Justice Richard Wagner*

4.1 INTRODUCTION

The existential threat posed by climate change combined with its effect
of worsening inequalities makes it one of the most important human
rights issues of our time. A landmark decision in 2015 by a Dutch court

---

[*] The author wishes to thank the many people who have had a positive impact on this
chapter, from those with whom I have engaged in discussions about the topics and
those with whom I have co-authored previous pieces of related research, to those
who reviewed and commented upon the chapter (or sections of it), and the excellent
research assistants who helped with editing and footnotes. These include Professor
Martha Jackman, Professor Lynda Collins, Professor Chris Tollefson, Professor Sheila
McIntyre, Professor Sanda Rodgers, Larissa Parker, Erin Dobbelsteyn, Laura Macintyre, Jessica Earle, Gérick Girard (JD & LLL Candidate, uOttawa), and Alexandria
Peacock (JD Candidate, uOttawa).

[1] *Reference re Impact Assessment Act*, 2023 SCC 23 [*IAA Reference*], at para 1, citing
Ontario v Canadian Pacific Ltd, [1995] 2 SCR 1031 at para 55.

was the first case holding a government accountable for failing to reduce greenhouse gas (GHG) emissions in line with what the science dictated was required to avert dangerous levels of warming.[2] In 2019, the Supreme Court of the Netherlands affirmed the decision. Since then, a growing number of courts around the world — including many Supreme and Apex courts — have joined the Netherlands in holding governments accountable to do their part in responding to climate change.[3] In recent years, cases against governments for climate inaction have been won in France, Germany, Belgium, Colombia, Brazil, Pakistan, and South Korea, as well as in international venues such as the European Court of Human Rights and the International Tribunal on the Law of the Sea, to name a few.[4]

The successful decisions have been hard fought, often landing at the highest level of court after multiple appeals. This is unsurprising, as cases about climate change raise a variety of challenging questions, from whether the claims they raise are justiciable to how courts should evaluate the adequacy of climate policies against human rights and what remedies are appropriate. Climate rights cases, in particular, require courts to forge new judicial ground as they engage in the adjudication of one of the most important and consequential issues of our time.

While the *Canadian Charter of Rights and Freedoms*[5] includes a range of fundamental rights and freedoms, including a guarantee of life, liberty and security of the person and a right to equality, climate litigants have thus far been unsuccessful in their attempts to hold Canadian governments to account for climate-related decisions they argue threaten their constitutional rights, joining environmental litigants who have similarly been

---

2   Rechtbank Den Haag [The Hague District Court], The Hague, 24 June 2015, *Urgenda Foundation v The State of the Netherlands (Ministry of Infrastructure and the Environment)*, ECLI:NL:RBDHA:2015:7196, C/09/456689 / HA ZA 13-1396 (Netherlands), online (unofficial English translation): uitspraken.rechtspraak.nl [perma.cc/LU7H-XU7X] [*Urgenda* District Court].

3   Joanna Setzer & Catherine Higham, *Global Trends in Climate Change Litigation: 2024 Snapshot* (London, UK: Grantham Research Institute on Climate Change and the Environment and Centre for Climate Change Economics and Policy, 2024) at 1.

4   *Ibid*. See also Larissa Parker, "Climate Litigation and Emerging Environmental Dimensions of Human Rights: An Opportunity in Canada" (2021) 5 PKI Global Justice J 42.

5   *Canadian Charter of Rights and Freedoms*, Part I of the Constitution Act, 1982, being Schedule B to the *Canada Act 1982* (UK), 1982, c 11 [*Charter*].

unsuccessful for decades in their *Charter* claims. However, this is very much a live issue. Only one climate *Charter* case has been fully decided at the time of writing (*Environnement Jeunesse*) and it was not decided on the merits of *Charter* rights but rather on whether a class proceeding should be certified.[6] The four other climate *Charter* cases filed as of the time of writing are at various stages of litigation.[7] While none have succeeded thus far, the most recent decisions have been increasingly favourable, suggesting a win for youth claimants may be imminent. It is almost certain that more cases invoking *Charter* rights in the context of climate change will be filed in the coming years, and that the ones progressing through the courts will be appealed, likely to the Supreme Court of Canada.[8]

Just like climate change is a destabilizing force for the planet, it is legally disruptive for the judiciary. Courts have for decades recognized that the *Charter* must be construed flexibly and adapted to reflect changing realities as per the famous living tree analogy.[9] Courts have had to confront troubling precedents that no longer align with evolving social values, adapting their interpretation of *Charter* rights and even sometimes reversing themselves on matters such as access to abortion, availability of supervised drug injection sites, the recognition of rights for same-sex partners, and medically-assisted dying, as societal views shift toward inclusivity and tolerance.[10]

---

6   *Environnement Jeunesse c Canada (PG)*, 2021 QCCA 1871 [*Enjeu* CA]. An unofficial English translation of the decision is available at Climate Change Litigation Databases, "ENVironnement JEUnesse v. Procureur General du Canada" (13 December 2021): climatecasechart.com [perma.cc/NMR7-RPFQ].

7   See, e.g., *La Rose v Canada*, 2023 FCA 241 [*La Rose* FCA]; *Mathur v Ontario*, 2024 ONCA 762 [*Mathur* CA]. See also Table 1 (in Section 4.1.1.2, below in this chapter for a full list of cases).

8   The Ontario government filed a leave to appeal to the Supreme Court of Canada in the *Mathur v Ontario* case in December 2024 (see *Ontario v Mathur* (16 December 2024), Ottawa (SCC) (Application for leave to appeal, Ontario) [*Mathur*, Leave to appeal]). The Supreme Court of Canada denied the request for leave to appeal (see *Ontario v Mathur* (1 May 2025), Ottawa (SCC) [*Mathur*, Leave to appeal denied].

9   See, e.g., *Reference re Same-Sex Marriage*, 2004 SCC 79 at paras 22–30 [*Marriage Reference*].

10   See, e.g., *R v Morgentaler*, [1988] 1 SCR 30 (SCC) [*Morgentaler*]; *Canada (AG) v PHS Community Services Society*, 2011 SCC 44 [*PHS*]; *Marriage Reference*, above note 10; *Carter v Canada (AG)*, 2015 SCC 5 [*Carter*].

Climate change is poised to challenge the courts in their evaluation of *Charter* rights perhaps like never before. From justiciability to positive obligations, climate *Charter* cases push courts into unfamiliar territory. Canadian courts are used to adjudicating human rights cases involving complex public policy issues in areas relating to labour rights, health care, housing, and more. However, climate change presents a set of unique circumstances, including that it involves a human-caused global phenomenon that poses a largely irreversible, existential threat for which the window of time to act is rapidly closing, and the solutions involve fundamentally restructuring the industrial carbon-based economy. The complexity and uniqueness of climate change will require courts to confront novel and often difficult questions. The decisions made by the judiciary over the next few years will mark the course of history for the future of modern human society.[11]

This chapter considers how Canadian courts have evaluated *Charter* rights in climate cases to date and what questions the judiciary will need to grapple with when adjudicating these cases in the future. The chapter does not purport to be comprehensive, as the intersection of climate change and human rights presents a nearly limitless set of questions. Rather, it considers the emerging jurisprudence related to the "systemic framework cases" (defined below), identifying points of convergence and divergence in the Canadian climate *Charter* cases to date and discussing the implications of different judicial interpretations.

The chapter is structured as follows. Section 4.1 offers a brief introduction to the *Charter* and rights-based framing in climate cases by courts around the world, followed by an overview of the climate-related cases brought under the Canadian *Charter*. Section 4.2 discusses the issue of justiciability of the Canadian climate cases, unpacking the way in which the courts have analyzed justiciability and identifying some concerns arising from the cases, including the key issue of how the cases are framed. Section 4.3 examines how the section 7 *Charter* claims have been evaluated, delving into question of causation, the positive/negative rights dichotomy, and principles of fundamental justice. Section 4.4 assesses how section 15(1)

---

11 *References re Greenhouse Gas Pollution Pricing Act*, 2021 SCC 11 at para 2 [*GGPPA References*].

Charter claims have been addressed, focusing on causation, positive rights, intersectionality, and the ground of age. Section 4.5 briefly addresses the issue of section 1 of the *Charter*, and Section 4.6 considers remedies.

Throughout this chapter, I identify pivot points where the judiciary's approach to the interpretation of *Charter* rights will be highly influential in determining the outcome of the cases and the interpretation of the *Charter* in the context of environmental and climate-related rights.

4.1.1 Constitutional Rights and Climate Change

4.1.1.1 An Overview of the *Charter*

The *Charter* safeguards basic rights and freedoms that are core to a free and democratic society. Its main function is to ensure that Canadian governments (and their agents) do not interfere with these rights and freedoms.[12] *Charter* rights are not absolute: in accordance with section 1, freedoms may be limited by law for a reason that is justifiable in a free and democratic society, where the limitation is reasonable and proportionate.[13] For example, we accept that it is reasonable to limit someone's right to liberty if they have committed a crime, allowing jail time commensurate with the nature of the crime and the circumstances.

The *Charter* protects seven categories of rights: fundamental freedoms, such as freedom of expression and religion; democratic rights, such as the right to vote; mobility rights that allow freedom of movement into, through, and out of the country; legal rights that ensure the justice system is fair; equality rights that protect against discrimination; language rights that create equality between French and English; and minority language education rights.

There is no provision in the *Charter* guaranteeing a right to a healthy environment or a stable climate. While there were multiple proposals to

---

12   See generally Carissima Mathen & Patrick Mecklem, eds, *Canadian Constitutional Law*, 6th ed (Toronto: Emond Publishing, 2022) at ch 16; Robert J Sharpe & Kent Roach, *The Charter of Rights and Freedoms*, 7th ed (Toronto: Irwin Law, 2021) at Introduction and ch 1; Peter Hogg & Wade Wright, *Constitutional Law of Canada*, 5th ed (Toronto: Carswell, 2019) (loose-leaf updated 2024, release 1) at ch 36.

13   Section 1 of the *Charter* states that the rights and freedoms that it sets out are subject only to such "reasonable limits prescribed by law as can be demonstrably justified in a free and democratic society" (see *Charter*, above note 5, s 1).

include such a right during the constitutional reform era of the 1970s and 1980s, such efforts failed. Perhaps this is a sign that Canadians take ecological stability and environmental health for granted. However, as Professor Lynda Collins has eloquently noted, "[w]ithout an ecological consciousness, the constitution is a paper temple — an aspirational blueprint for political community with no real guarantee of its survival over time."[14] In other words, without the stability and predictability of the Earth's functional ecological systems, including a stable climate, Canadians may not be in a position to enjoy the other rights. The *Charter's* rights and fundamental freedoms are hollow without a liveable planet. As such, the right to a healthy environment should be part of our Constitution.

As pressure on ecological and planetary systems mount, more than 100 countries have responded by explicitly protecting some version of a right to a healthy, safe, and clean environment in their constitutions.[15] In 2022, the United Nations General Assembly adopted a resolution recognizing the right to a clean, healthy, and sustainable environment.[16] A small number of countries have even amended their constitutions to include provisions specific to the climate.[17] Dozens of nations argued before the

---

14   Lynda Collins, *The Ecological Constitution: Reframing Environmental Law* (Abingdon: Routledge, 2021) [Collins, *The Ecological Constitution*].

15   David Boyd, *The Environmental Rights Revolution: A Global Study of Constitutions, Human Rights and the Environment* (Vancouver: UBC Press, 2011); *Report of the Special Rapporteur on the Human Right to a Clean, Healthy and Sustainable Environment: Overview of the Implementation of the Human Right to a Clean, Healthy and Sustainable Environment*, UNGA, 79th Sess, UN Doc A/79/270 (2024).

16   *The Human Right to a Clean, Healthy and Sustainable Environment,* UNGA, 76th Sess, UN Doc A/RES/76/300 (2022) GA Res 76/300.

17   The Ecuadorian constitution, for example, was amended in 2009 to include a provision stating that "[t]he State shall adopt adequate and cross-cutting measures for the mitigation of climate change, by limiting greenhouse gas emissions, deforestation, and air pollution ... it shall protect the population at risk" (see "Republic of Ecuador Constitution of 2008" (last modified 31 January 2011), art 414, online: pdba.georgetown.edu [perma.cc/A8LC-SHTR]). Zambia's constitution states that the country "shall ...establish and implement mechanisms that address climate change" (see "Constitution of Zambia (amendment)" (last visited 17 December 2024), art 257, online: parliament.gov.zm [perma.cc/JN58-RXY8]). While these examples are avant garde, they are largely untested and remain uncommon (see Collins, *The Ecological Constitution*, above note 14 at 107).

International Court of Justice in a pending case that the right to a healthy environment is part of customary international law.[18]

Canada recently passed federal legislation that incorporates the right to a healthy environment.[19] However, the right to a healthy environment remains unrecognized by the Canadian Constitution, including the *Charter*. While polling data suggests Canadians would support constitutionalizing the right, constitutional amendments are no simple matter in Canadian politics and no amendments are forthcoming.[20] A central question for *Charter* climate litigation then is whether the right to a stable climate is inherent in the other rights and freedoms of the *Charter* or in unwritten principles underpinning the Constitution. An answer to this question is emerging as Canadian courts address the claims by youth and others in the climate *Charter* cases.

Professor Lynda Collins has convincingly argued that environmental rights are implicit within many existing *Charter* rights: "[t]he protections provided within [the *Charter*] are broad enough, and our judicial traditions of interpretation robust enough, to allow these crucial interests to benefit from the protection of the supreme law of the land."[21] Courts in other jurisdictions without an explicit right to a healthy environment have also recognized that such a right is inherent within other existing rights. For instance, in the first climate case to crystalize the human rights dimensions of climate change, a Dutch court interpreted the right to life and

---

18   See IISD, "Healthy Environment: A Human Right and Customary International Law, January 29, 2025, online: https://sdg.iisd.org/commentary/guest-articles/healthy-environment-a-human-right-and-customary-international-law/.

19   *Canadian Environmental Protection Act*, SC 1999, c 33. Some form of a right to a healthy environment is also recognized in Ontario, Quebec and the three territories (see *Environmental Bill of Rights, 1993*, SO 1993, c 28; Quebec *Environmental Quality Act*, CQLR c Q-2; Nunavut, *Consolidation of Environmental Rights*, RSNWT 1988, c 83 (Supp); *Yukon Environmental Act*, RSY 2002, c 76; Northwest Territories, *Environmental Rights Act*, SNWT 2019, c 19).

20   See David Suzuki Foundation, "Blue Dot Movement" (last visited 17 December 2024) online: davidsuzuki.org [perma.cc/GK2K-W7ZF]; Jason Maclean, "You Say You Want An Environmental Rights Revolution: Try Changing Canadians' Minds Instead (of the Charter)" (2019) 49:1 Ottawa L Rev 183.

21   Lynda Collins, "An Ecologically Literate Reading of the Canadian Charter of Rights and Freedoms" (2009) 26 Windsor Rev Legal Soc Issues 7 at 48.

the right to private and family life — two rights protected by the European Convention on Human Rights — to effectively provide a right to a safe environment.[22] Upholding the decision, the Dutch Appellate and Supreme Courts relied upon Articles 2 (right to life) and 8 (right to private and family life) of the European Convention on Human Rights, reasoning that climate change creates risks jeopardizing the lives and well-being of Dutch citizens.[23]

In a similarly framed case, a Belgian court found the government breached its duty of care for failing to enact good climate governance (also citing Articles 2 and 8 of the European Convention on Human Rights). While the lower court declined to order an injunction to set specific targets to avoid treading on the powers of the executive, the Belgian Court of Appeal overturned this finding and granted an injunction ordering a specific level of GHG emissions reductions (55 percent compared to 1990 levels by 2030), leaving the executive branch to determine the means of meeting the target.[24] The European Convention on Human Rights recently ruled in favour of a group of older women, finding that Article 8 of the Convention encompasses a right to effective protection from the Swiss government from the serious adverse effects of climate change on their lives, health and well-being.[25]

---

22 While the District Court found the government responsible on the basis of a duty of care in its Civil Code, it also relied upon Article 21 of the Dutch Constitution which safeguards environmental protection. *Urgenda* District Court, above note 2. See also Gerechtshof Den Haag [The Hague Court of Appeal], The Hague, 9 October 2018, *The State of the Netherlands v Urgenda Foundation*, ECLI:NL:GHDHA:2018:2610, 200.178.245/01 (Netherlands), online (unofficial English translation): uitspraken. rechtspraak.nl [perma.cc/G6AS-XRK5]; Hoge Raad [Supreme Court of the Netherlands], The Hague, 20 December 2019, *The State of the Netherlands v Urgenda Foundation*, ECLI:NL:HR:2019:2007, 19/00135 (Netherlands), online: uitspraken.rechtspraak.nl [perma.cc/X698-TPGM] [*Urgenda* Supreme Court].

23 *Urgenda* Supreme Court, above note 22 at para 5.6.2.

24 Cour d'appel de Bruxelles [Brussels Court of Appeal], Brussels, 30 November 2023, *Klimaatzaak v Kingdom of Belgium and Others*, 2021/AR/1589 2022/AR/737 2022/AR/891 (Belgium), online: climatecasechart.com [perma.cc/4R5Q-TW4G] [*Klimaatzaak*]. Note that the Court of Appeal did not order the injunction against the Walloon Region which it found was meeting its human rights and civil obligations.

25 See *Verein KlimaSeniorinnen Schweiz and Others v Switzerland* [GC], No 53600/20, [2024], ECLI:CE:ECHR:2024:0409JUD005360020.

In *Friends of the Irish Environment CLG v The Government of Ireland & The Attorney General*, the Irish Supreme Court was invited to recognize a right to a healthy environment implicit in the Irish Constitution.[26] While it declined to do so in that case, the Court offered guidance about how such rights might be recognized.[27] It noted that rights which are not expressly stated in the Constitution can nonetheless be derived with reference to a constitution's values, structure, and reference to other rights.[28] The Court clarified that it was not advocating for a narrow textual approach but rather protecting against the possibility that judges could simply identify new unenumerated rights they approve of and determine them to be part of the Constitution, something the Court felt could bring the judiciary into the realm of the executive or legislature.[29] Importantly, the Irish Supreme Court observed that recognizing a derived right to a healthy environment may not be necessary given the possibility that it may not add anything new to the existing express rights to life and to bodily integrity in the Irish Constitution.[30]

There are examples outside of Europe of courts interpreting general rights, such as the right to life, as implicitly incorporating an environmental right. In *Leghari v Federation of Pakistan*, the Lahore High Court interpreted Pakistan's constitutional right to life and dignity as including a right to have the government take action on climate change.[31] Justice Shah described the serious concerns about food and water security created by heavy floods and droughts linked to climate change as a "clarion call for the protection of fundamental rights of the citizens of Pakistan, in

---

26   *Friends of the Environment CLG v The Government of Ireland & The Attorney General*, [2020] IESCDET 13 [*SC Ireland*].

27   *Ibid* at paras 8.1 to 8.17. The Court noted that the only other common law jurisdiction to do so is India.

28   *Ibid* at 8.6.

29   *Ibid* at paras 8.5 and 8.9.

30   *Ibid* at para 8.14. The Court also said that if the right to a healthy environment was meant to add something additional, that would need to be more precisely defined to avoid being overly vague. See also David R Boyd, "The Implicit Constitutional Right to a Healthy Environment" (2011) 20:2 Review of European Community and International Environmental Law 171–79.

31   *Leghari v Federation of Pakistan and Others* (2018), PLD 2018 Lahore 364 (Lahore High Court Pakistan), online: climatecasechart.com [perma.cc/2NLG-PD62] [*Leghari v Pakistan*].

particular, the vulnerable and weak segments of the society who are unable to approach this Court."[32] The Court held the government had an obligation to remedy the violations and used *mandamus* jurisdiction to constitute a Climate Change Commission comprised of the Joint Secretary for Climate Change and a point person from each government ministry and department. Justice Shah subsequently supervised a series of twenty-five further hearings involving the Commission, dissolving it in 2018 when he determined it had succeeded in sensitizing the government to the issue of climate change and achieved two-thirds of its goals.[33] Although the Commission was dissolved, the Court retained jurisdiction so it could be called upon if the government's climate policy regresses in such a way that may violate fundamental rights.[34] In 2024, Pakistan amended its constitution to include the right to a clean, healthy and sustainable environment.

In a ground-breaking decision in Latin America, the Supreme Court of Colombia recognized the role of deforestation in climate change and held the government's inaction in stemming deforestation to be a violation of Colombia's Constitution, which includes the right to a healthy environment. The Colombian Court directly linked a set of constitutional rights — including the right to life, health, minimum subsistence, freedom, and human dignity — to environment quality, stating that these rights are dependent upon, and determined by, the quality of the environment and ecosystems. The Court agreed with the twenty-five youth plaintiffs' claim that the government's failure to meet a net-zero deforestation target for

---

32   *Ibid* at para 11.

33   *Ibid*.

34   Collins, *The Ecological Constitution* above note 15 at 109, citing Emily Barritt & Boitumelo Sediti, "The Symbolic Value of *Leghari v Federation of Pakistan:* Climate Change Adjudication in the Global South" (2019) 30:2 King's L J 203. See Louis J Kotzé & Anel Du Plessis, "Putting Africa on the Stand: A Bird's Eye View of Climate Change Litigation on the Continent" (2019) 50:3 Envtl L 615; See also *Earthlife Africa Johannesburg v Minister of Environmental Affairs and Others*, [2017] ZAGPPHC 58 (SAFLII) [*Earthlife Africa*], in which the court held that the environmental assessment of a sizeable coal-fired power plant had to take climate change into consideration. While the case was based on statutory law, the law in question made no explicit reference to climate change. It's also likely that the court was influenced by the substantive environmental right in the South African Constitution, though it does not explicitly mention climate change.

the Colombian Amazon forest violated their constitutional rights, recognizing the Amazon forest as having legal personhood and holding that the environmental rights of future generations create legal obligations on the present generation to protect natural resources. It ordered the Presidency of the Republic along with several Ministries to develop an intergenerational plan to protect the Amazon forest, reducing deforestation to zero. It also ordered multiple levels of government to develop short, medium, and long-term plans to reduce deforestation, in collaboration with the plaintiffs, impacted communities, and the general public.[35]

Courts are increasingly finding that the right to a healthy environment includes the right to a safe, livable climate. For instance, the Indian Supreme Court observed that "[w]ithout a clean environment which is stable and unimpacted by the vagaries of climate change, the right to life is not fully realized."[36] Montana's Supreme Court stated that its constitutional "right to a clean and healthful environment and environmental life support system includes a stable climate system."[37]

These decisions showcase how courts have recognized the rights and duties of governments to address climate change in a variety of contexts, even where there is no explicit right to a healthy environment.

4.1.1.2 The Cases – Emerging Climate Change *Charter* Jurisprudence in Canada

Canada's climate litigation can be loosely divided into two waves. The first wave of climate litigation consisted of two cases launched within a few years of each other in the *Kyoto Protocol* era. In 2008, Friends of the Earth Canada sought judicial review of the federal government's climate change plan under the former *Kyoto Protocol Implementation Act (KPIA)*, which

---

35 Corte Suprema de Justicia [Supreme Court], Bogotá, 4 April 2018, *Future Generations v Ministry of the Environment and Others* [2018] STC4360-2018 (Columbia), online: climatecasechart.com [perma.cc/8SKE-AZQ5] [*Future Generations*]. See also Collins, *The Ecological Constitution*, above note 15 at 109.

36 *MK Ranjitsinh et al, v Union of India et al*, 2024 INSC 281, at para 25.

37 See, e.g., *Held v Montana*, 2014 MT 312, DA 23-0575, at para 30; See also *In Re Hawai'i Electric Light Company, Inc*, (2023) SCOT-22-0000418, Docket No. 2017-0122, at pg 16; *La Oroya v Peru*, Inter-American Court of Human Rights, online: www.corteidh.or.cr /docs/comunicados/cp_17_2024.pdf.

on its face would fail to achieve Canada's *Kyoto Protocol* commitments.[38] A few years later, in 2012, Member of Parliament Daniel Turp challenged the government's decision to withdraw from the *Kyoto Protocol*.[39] Both of these lawsuits were ultimately unsuccessful, largely because the courts held that they raised non-justiciable issues.[40] However, neither case raised *Charter* violations, which distinguishes them from the second wave of lawsuits which all raise *Charter* rights.[41]

At of the time of writing, there have been seven climate change cases filed in this second wave that engage the *Charter* in a climate context: *Raincoast Conservation Foundation v Canada (Attorney General)*, also known as *Youth Stop TMX*,[42] *Transmountain Pipeline ULC v*

---

38 *Friends of the Earth v Canada (Governor in Council)*, 2008 FC 1183 [*Friends of the Earth, FCTD*].

39 *Turp v Canada (Minister of Justice)*, 2012 FC 893 [*Turp*].

40 For a discussion of the two cases, see Nathalie J Chalifour & Jessica Earle, "Feeling the Heat: Climate Litigation under the Canadian Charter's Right to Life, Liberty and Security of the Person" (2018) 42 Vermont L Rev 688 at 24–25 and 63–69. The Federal Court of Appeal concluded that a contestation of the government's failure to enact a climate change plan in line with the requirements set out in the Kyoto Protocol Implementation Act was not justiciable. This conclusion has been criticized given that the legislation under review was drafted using expressly mandatory language (see *Friends of the Earth v Canada (Minister of the Environment)*, 2009 FCA 297, affg 2008 FC 1183). The Federal Court concluded that a contestation of the federal government's decision to withdraw from the Kyoto Protocol, in the absence of a Charter challenge, was not justiciable (see Turp, above note 40).

41 *Turp*, above note 39 at para 18.

42 This was a request by Youth Stop TMX for judicial review of the Governor in Council's decision to approve the Trans Mountain Pipeline's expansion project. Youth Stop TMX argued that the Government of Canada, as regulator and sole owner of the TMX Project, was obligated under sections 7 and 15 of the *Charter* to consider the climate change impacts of the pipeline's expansion on youth before approving the project. The Federal Court of Appeal dismissed the application, holding that the courts should defer to the GIC's finding that compelling public interest benefits associated with the pipeline outweighed the adverse environmental effects. The Court also found there was no evidence to support the applicants' *Charter* claims. This deferential approach may be explained in part by the judicial review context, where the court, based on the administrative standard at play, need only verify that the decision was correct or reasonable (see *Raincoast Conservation Foundation v Canada (AG)*, 2019 FCA 224 at paras 18, 19, 45, and 70, leave to appeal to SCC refused, 38892 (5 March 2020)). On judicial review and deference in the context of environmental decision-making, see, e.g., Lynda Collins & Lorne Sossin, "Approach to Constitutional Principles and Environmental Discretion in Canada" (2019) 52:1 UBC L Rev 293; Jason

Mivasair,[43] *Environnement Jeunesse v Canada (Attorney General)*, *La Rose v Canada (Attorney General)*, *Misdzi Yikh v Canada (Attorney General)*, *Mathur v Ontario*, and *Dykstra v Saskatchewan*.[44] *Youth Stop TMX* and *Mivasair* were efforts to raise *Charter* rights in the context of a judicial review and criminal proceeding, respectively. Both failed, and neither decision engaged in a *Charter* analysis. Therefore, they do not offer many insights about how courts interpret *Charter* rights in the climate context.[45] As such, I will not elaborate on these cases. The latter five cases all allege infringements of rights under sections 7 and 15 of the *Charter* by government, three by the federal government, one by the Ontario government, and one by a Saskatchewan crown corporation and the Saskatchewan government.[46]

These latter five cases are what can be described as "systemic" or "framework" cases that challenge the adequacy of the government's overall response to climate change, specifically the level of GHG emissions reductions committed to and achieved by the government in question. Each case is framed slightly differently but ultimately all argue that the government's decisions in relation to GHG emissions reductions threaten section 7's right to life, liberty and security of the

---

MacLean & Chris Tollefson, "Climate-Proofing Judicial Review after Paris: Judicial Competence, Capacity, and Courage" (2018) 31:3 J Envtl L & Prac 245.

43 *Trans Mountain Pipeline ULC v Mivasair*, 2019 BCSC 50 [*Mivasair* SC (TD)]. This case involved convictions of citizens trying to stop construction of the TransMountain pipeline expansion. The case was appealed to the BC Court of Appeal but not on the Charter issue. Two defendants sought leave to raise the common law defense of necessity and to argue that the government's conduct in relation to the pipeline, as owner and regulator, violated their section 7 rights. The case was dismissed (see *Trans Mountain Pipeline ULC v Mivasair*, 2020 BCCA 255).

44 *Environnement Jeunesse c Procureur général du Canada*, 2019 QCCS 2885 [*Enjeu* SC]; *La Rose v Canada*, 2020 FC 1008 [La Rose FCTD]; *Misdzi Yikh v Canada*, 2020 FC 1059 [Misdzi Yikh]; *Mathur v His Majesty the King in Right of Ontario*, 2023 ONSC 2316 [Mathur SC]; *Dykstra v Saskatchewan Power Corporation*, 2023 SKKB 118 [Dykstra].

45 See *Mivasair* SC (TD), above note 44 at paras 63–69.

46 The claimants in *Enjeu* also claimed a violation of section 46.1 of the *Charter of human rights and freedoms*, CQLR c C-12, which protects the right to a healthy environment.

person and the equality rights guaranteed under section 15(1) of the *Charter*.[47]

The only systemic case that has been finally decided at the time of writing was not a case on the merits but a request to certify a class action proceeding which was declined by the Quebec Court of Appeal in *Environnement Jeunesse v Attorney General of Canada*, or *Enjeu*.[48] Three cases (*Mathur, La Rose and Misdzi Yikh*) have been allowed to proceed to a hearing on the merits. *Mathur* is the only case to have already had a hearing and decision on the merits. The decision dismissing the case in *Mathur* was overturned by the Ontario Court of Appeal and the matter remitted back to the lower court to be reheard. Ontario filed leave to appeal the decision to the Supreme Court of Canada, and the youth claimants filed for leave to counter-appeal, however, the Supreme Court of Canada denied leave. A rehearing is scheduled for December 2025.[49] After being initially struck, *La Rose* and *Misdzi Yikh* have been allowed to proceed to a hearing on the merits for a portion of their claim (related to section 7) following amendment of their statements of claim to narrow the scope of the pleadings.[50] An eight-week trial is scheduled to begin in October 2026. *La Rose* and *Misdzi Yikh* (should the latter survive the motion to strike its amended claim) will be the first climate trial on the merits with live witnesses against a national government to go forward globally.[51] The five systemic cases are described in the section that follows and will be referenced throughout this chapter. They are presented mainly in chronological order (with the two Federal Court cases grouped together since the Federal Court of Appeal issued a joint decision for them).

---

47   Note that the liberty interest in section 7 has not been engaged in these cases despite the arguments that could be advanced. Consider the case of communities unable to leave or access their homes or traditional territories due to changes in permafrost and ice freeze and melt patterns or families unable to flee wildfires (see generally Chalifour & Earle, above note 40).

48   *Enjeu* CA, above note 6, leave to appeal to SCC refused, 40042 (28 July 2022).

49   *Mathur* CA, leave to appeal, above note 8.

50   Note that Canada filed another motion to strike in both the *La Rose* and *Misdzi Yikh* cases following filing of their amended pleadings. After further amendments to the *La Rose* claim, Canada abandoned its motion to strike. The motion to strike in the *Misdzi Yikh* case is proceeding as of the time of writing.

51   Note that the trial in *Held v Montana* included live witnesses.

TABLE 1   CANADIAN CLIMATE CHARTER CASES

Case Name	Against	Year launched	Year decided	Court	Status as of June 2025
Environnement Jeuneusse v Canada	Federal Government	2018	2021	Quebec Court of Appeal	Decided-Leave to appeal to SCC declined.
La Rose v Canada	Federal Government	2019	Ongoing	Federal Court	Trial scheduled to begin October 2026
Misdzi Yikh v Canada	Federal Government	2020	Ongoing	Federal Court	Trial scheduled to begin October 2026
Mathur v Ontario	Ontario Government	2019	Ongoing	Ontario Superior Court	Re-hearing scheduled for December 2025
Dykstra v Saskatchewan Power Corp	Saskatchewan Government and SaskPower Corp	2023	Ongoing	Saskatchewan Court of King's Bench	Pending decision on motion to strike claim

Environnement Jeunesse v Canada, Quebec Court of Appeal, 2021
Request to certify a class proceeding on behalf of residents of Quebec under age 35 declined
Commenced: November 2018
Status: Decided 2021

The claimants in this case (a non-profit environmental organization) filed an application in the fall of 2018 to authorize a class action against the federal government on behalf of all residents of Quebec under aged 35 on a given date. The claim alleged that the government had violated sections 7 and 15 of the *Charter*, as well as section 46.1 of the *Quebec Charter* (which protects the right to a healthy environment that respects biodiversity), by: (1) setting GHG emissions reduction targets at an insufficient level to avoid the harmful impacts of climate change and (2) failing to take sufficient measures to reach even the inadequate targets it did set — decisions that, the applicants argued, will disproportionately burden younger generations with future health, economic, and social costs.[52] As a remedy, the applicant sought a declaration that the government had failed to meet its obligations under the Canadian *Charter* and *Quebec Charter*, an order to cease these violations, and punitive damages of $100 per class member to be earmarked for climate-related remedial measures.[53]

The applicant's request to certify a class proceeding was first dismissed by Justice Morrison in the Superior Court of Quebec. Although Morrison J held that the claim was justiciable since it alleged *Charter* violations,[54] he found that he could not certify the class proceeding because the definition of the class was arbitrary and therefore inappropriate.[55] The Court held that the applicants had not provided a factual or rational explanation for the choice of age thirty-five as a cut-off, and that the decision to cap the age of members at thirty-five "excludes millions of other Quebecers because of their age." This choice, Morrison J found, was too subjective and limiting considering the type of case (a class proceeding) being put forward.

The applicants appealed the decision to the Quebec Court of Appeal, arguing that the Court should have waited to examine the evidence before deciding that the age of thirty-five was an arbitrary cut-off. The Quebec Court of Appeal upheld the dismissal but focused its reasons on a finding that the matter was not justiciable. Justice Vauclair pointed to the lack of a legislative anchor and characterized the case as one framed around inaction. The Court held that this was tantamount to asking the court to tell the legislature what to do and was therefore non-justiciable. The Court of Appeal also supported the lower court's finding that the choice of age thirty-five as a cut-off for the class was inappropriate. However, the emphasis at the appeal level was less on the irrationality of the choice of age thirty-five and more on the fact

---

52 *Environnement Jeunesse c Procureur général du Canada*, 2019 QCCS 2885 (Motion for Authorisation, Petitioner at paras 2.91–2.92 [*Enjeu* MfA]). The petitioners also argued the conduct of the federal government violated section 7 of the *Charter*, as well as sections 10 and 46.1 of the Quebec *Charter of human rights*, the latter of which protects the right to live in a healthy environment (see *ibid* at paras 2.82, 4.1, and 6).

53 *Enjeu* CA, above note 6 at para 23.

54 *Enjeu* SC, above note 44 at paras 69–71.

55 *Ibid* at para 123.

that the cut-off would exclude other residents who will also be harmed by climate change. Because class proceedings are meant to be an efficient use of judicial resources, the Court of Appeal found that limiting the proceeding to those under age thirty-five would exclude potential claimants and was not, therefore, an appropriate choice for a class action.[56] Leave to appeal the decision to the Supreme Court of Canada was declined.[57]

The next two decisions were bundled together at the Federal Court of Appeal, but their initial claims were separate and have some unique features, so I will introduce the claims separately first, then describe the Federal Court of Appeal's combined decision.

Cecilia La Rose v Canada[58] Federal Court
Systemic lawsuit against federal government – motion to strike initially successful, but overturned with leave to amend claim on section 7
Commenced: October 2019
Status: Proceeding to trial (scheduled for 8 weeks beginning October 2026).

La Rose was filed by fifteen children and youth plaintiffs, seven of whom are Indigenous, in October 2019 in the Federal Court.[59] The *La Rose* case is a classic example of systemic or framework litigation against the federal government for an inadequate response to climate change. There was no legislated climate target at the time it was filed, therefore the claim was deliberately framed around a combination of action and inaction. The original claim stated that the government knowingly allows, causes, and contributes to the dangerous destabilization of the climate, which deprives the plaintiffs of their section 7 and 15 rights and infringes the public trust.

---

56   *Enjeu* CA, above note 6 at para 44.
57   *Enjeu* CA, above note 6, leave to appeal to SCC refused, 40042 (28 July 2022).
58   *La Rose* FCTD, above note 44.
59   While each individual sought private interest standing, they also argued that they have public interest standing to represent the "rights of all children and youth in Canada, present and future" (*La Rose v Canada*, 2020 FC 1008 (Statement of Claim, Plaintiffs at para 27 [*La Rose* SoC])).

The claim pointed to a constellation of actions and government decisions that lead to a harmful level of GHG emissions.[60]

The plaintiffs argued that the government's response to climate change is "grossly insufficient" when measured against what scientists say is needed to protect the climate system and violates their section 7 and 15 rights.[61] The youth pointed to a range of harms, including impacts on their physical health, emotional health, culture, and food security, increased isolation of youth in Northern communities, displacement, property damage, and interruption to their day-to-day lives. The remedies sought by the plaintiffs included a declaration that the government is violating the plaintiffs' section 7 and 15 rights and an order requiring the government to conduct an accurate accounting of GHG emissions, adopt a climate recovery plan, and empowering the court to retain jurisdiction over the matter until the defendants have complied.[62]

The Federal Court granted the government's motion to strike the case in 2020.[63] The test on a motion to strike requires the court to accept the pleadings as true and consider whether it is "plain and obvious that the pleadings disclose no reasonable cause of action or that the claim has no reasonable prospect of success." Justice Manson found the claims under sections 7 and 15 to be non-justiciable and that even if the issues were justiciable, the sections 7 and 15 claims had no reasonable prospect of success.[64] The reasons turned on the finding that the claim targeted a broad range of actions and inactions that the Court characterized as being of undue breadth and of a diffuse nature. Justice Manson held that the way the claim was framed effectively put "the entirety of Canada's policy

---

60 Specifically, the *La Rose* plaintiffs allege that the federal government is: (1) causing, contributing to, and allowing a level of GHG emissions incompatible with a stable climate system; (2) adopting GHG emission targets that are inconsistent with the best available science about what is necessary to avoid dangerous climate change and restore a stable climate system; (3) failing to meet its own (inadequate) GHG emissions reductions targets; and (4) actively participating in and supporting the development, expansion, and operation of industries and activities involving fossil fuels that emit a level of GHGs incompatible with a stable climate system (see *Ibid* at para 5).

61 *Ibid* at paras 5–6.

62 *La Rose* SoC, above note 49 at para 222.

63 *La Rose* FCTD, above note 44.

64 *Ibid* at para 59.

response to climate change in issue," something he held was beyond the purview of the court.[65] He also found that the remedies requested were inappropriate and went too far.[66] The Federal Court of Appeal overturned the decision on section 7 and the plaintiffs have filed an amended Statement of Claim, as discussed below.

Misdzi Yikh v Canada, Federal Court
Systemic lawsuit against federal government – motion to strike initially successful, overturned with leave to amend claim on section 7, second motion to strike pending
Commenced: February 2020
Status: Motion to strike Amended Pleadings

Misdzi Yikh v *Canada*[67] is the second climate *Charter* case filed in the Federal Court. Because it raised similar issues to those raised in *La Rose*, the court heard the cases together. The case was brought by Dini Ze' Lho'Imggin (Alphonse Gagnon) and Dini Zé Smogilhgim (Warner Naziel) as individuals on behalf of two Wet'suwet'en house groups of the Likhts'amisyu Clans, Misdzi Yikh, and Sa Yikh. Like *La Rose*, this case is also systemic in nature and framed around government action and inaction. The claim alleges that the federal government has not implemented and enforced the laws and policies needed to ensure that Canada does its part to keep global warming below two degrees Celsius above pre-industrial levels.[68] The plaintiffs also challenged the authorization of several high-emission fossil fuel export projects.[69] The

---

65   *Ibid* at paras 26 and 43. The Court found the question relating to the public trust doctrine to be justiciable, though it ultimately found that the public trust claim had no reasonable prospect of success (see *ibid* at para 59).

66   *Ibid* at paras 40–41.

67   *Misdzi Yikh*, above note 45.

68   *Misdzi Yikh v Canada*, 2020 FC 1059 (Statement of Claim, Plaintiffs [*Misdzi Yikh* SoC]).

69   According to the plaintiffs, the permits that the government has issued to developments such as the liquified natural gas Canada Export Terminal Project in Kitimat and the Coastal GasLink Pipeline Project effectively guarantee that Canada will continue to emit "copious" amounts of GHGs for at least the next forty years (see *Ibid* at para 6).

Misdzi Yikh claimants argue that the government's inadequate climate change laws and policies disproportionately harm Indigenous Peoples, depriving the children, youth, and future members of their house group of their rights to "good health, to knowledge of their territories, fisheries, social relations and laws."[70] They link the deprivations to Canada's history of colonial racism, arguing that the government's disregard for their specific climate change vulnerabilities perpetuates their colonial trauma.[71]

The claimants based their claims on sections 7 and 15 of the *Charter* but also argued that the government's actions amounted to a failure to make laws in relation to peace, order, and good government (a novel argument that the government owes a duty based on the part of the Constitution allocating legislative authority between levels of government). In terms of remedies, the plaintiffs sought several declarations and orders including an order requiring the federal government to prepare an accounting of cumulative GHG emissions in a format allowing the emissions to be set in the context of a global carbon budget and asking the court to retain jurisdiction to ensure compliance with the other orders.[72]

Justice McVeigh for the Federal Court dismissed the case for lack of justiciability, holding that the "issue of climate change, while undoubtedly important, is inherently political, not legal, and is of the realm of the executive and legislative branches of government."[73] The Court pointed to the lack of specific laws as an anchor for the claim, finding that this rendered the matter "beyond the reach of judicial interference."[74] Justice McVeigh also found the remedies sought to be inappropriate, contributing to the

---

70   *Ibid* at para 92.

71   In addition to more general harms from climate change, the Indigenous plaintiffs point to harms that are culturally specific, such as the erosion of the Likhts'amisyu House's identity, culture, legal order, sustenance, and intergenerational knowledge transfer (see *Ibid* at paras 5 and 92).

72   The plaintiffs also argue that Canada has breached a duty under section 91 of the *Constitution Act, 1867* by refusing to respect international treaties that it has ratified (see *Misdzi Yikh*, above note 44 at para 5).

73   *Ibid* at para 77.

74   *Misdzi Yikh* SoC above note 68.

non-justiciability of the claim.[75] The Federal Court of Appeal overturned the decision, as discussed next.

> ***La Rose v Canada; Misdzi Yikh v Canada***, Federal Court of Appeal, 2023
> **Decision to strike systemic lawsuits against the federal government overturned, with leave to amend the pleadings on section 7 and proceed to a hearing**
> Status: Proceeding to trial October 2026.

The *La Rose* and *Misdzi Yikh* plaintiffs appealed the dismissals of their claims to the Federal Court of Appeal, which opted to hear their cases in parallel and issued one decision for both cases. In 2023, the Federal Court of Appeal granted the appeals in part, allowing the section 7 portion of both claims to proceed to trial after the filing of amended Statements of Claim narrowing the scope of their claims. The public trust arguments presented by the *La Rose* plaintiffs, the section 91 POGG arguments presented by the *Misdzi Yikh* plaintiffs,[76] and the section 15 claims in both cases remain dismissed.

Written by Rennie J, concurred in by Laskin and Leblanc JJ, the decision reiterated the test on a motion to strike, which is to determine whether it is plain and obvious that the claim has no reasonable prospect of success.[77] The Court then highlighted three related principles that inform the analysis. First, facts are assumed to be true unless they are patently incapable of proof. Second, the court must read the claim generously and err on the side of allowing novel but arguable claims to move to trial in recognition that law must evolve to address new and emerging issues. Third, the defendant bears the burden of proving there is no reasonable cause of action.[78]

Applying this test, informed by the aforementioned principles, Rennie J considered the justiciability of the claims, the substantive arguments, and the pleadings. He overturned the lower courts' findings on section 7, holding that the section 7 claims were justiciable and had a reasonable

---

75 *Misdzi Yikh*, above note 44 at paras 71–73.
76 *La Rose* FCA, above note 7.
77 *Ibid* at para 18.
78 *Ibid* at para 19.

prospect of success. However, Rennie J found the pleadings as framed too diffuse and expansive to sustain a proper *Charter* analysis. He granted leave to amend the claims to render them compatible with constitutional adjudication (by narrowing the scope of the impugned conduct).[79] The Federal Court of Appeal did not overturn the decision to strike the section 15 nor public trust claims, neither did it engage directly with the claims specific to the Indigenous plaintiffs.[80]

Mathur v Ontario, Ontario Court of Appeal, 2024
First climate hearing on the merits considering whether Ontario's climate target violates Charter rights dismissed by Superior Court, overturned on appeal and remitted for a re-hearing.
Commenced: November 2019
Status: Rehearing before Superior Court scheduled December 2025

Mathur v Ontario is the first climate *Charter* case in Canada to clear the preliminary hurdles of motions to strike and proceed to a hearing on the merits. While the case was dismissed by the Ontario Superior Court, the Ontario Court of Appeal allowed the applicants' appeal and ordered the case to be reheard by the lower court as outlined below. Ontario applied for leave to appeal this decision to the Supreme Court of Canada, and the youth applicants applied for leave to counter appeal the decision. However, the Supreme Court of Canada declined to hear the appeal. As such, the case is proceeding to a rehearing before Vermette J in the fall of 2025.

The applicants in *Mathur* consist of seven youth, some of whom are Indigenous, from Ontario who challenged the Ontario government's roll-back of the province's climate target as a violation of their section 7 and 15 *Charter* rights. Specifically, the applicants challenged the 2030 GHG emissions reduction target set by Ontario under section 3(1) of the *Cap and Trade Cancellation Act* and the related *Made-in-Ontario Environmental Plan*. In addition to removing the emissions trading program, the new target weakened the prior 2030 target and eliminated the provincial GHG

---

79   *Ibid* at para 22.
80   *Ibid* at para 21.

reduction targets for 2020 and 2050.[81] In terms of remedies, the applicants sought a number of declarations and an order directing Ontario to set a science-based GHG reduction target and revise its climate plan accordingly.[82]

In the first climate rights victory in Canada, the Superior Court of Justice of Ontario in November 2020 dismissed the government's motion to strike the claim, allowing the matter to proceed to a full hearing.[83] Justice Brown found that the claim was justiciable, that both the sections 7 and 15 claims have a reasonable prospect of success, and that a case for a positive obligation on the government could be made out in the context of climate change.[84]

After the hearing on the merits, the Ontario Superior Court of Justice dismissed the applicants' claim but made several important findings the youth applicants considered a success, despite their loss. For instance, Justice Vermette found that the application in general, including the section 7 and 15 claims, was justiciable.[85] Although this was a claim that Ontario's GHG reduction target was insufficient, like the other systemic cases, the *Mathur* claim was framed around a specific law and related policy. This helped Vermette J distinguish this case from *Enjeu*, *La Rose* and *Misdzi Yikh* (in which no legislative requirement for a climate target existed at the time) and other precedents such

---

81   *Ibid* at para 7. The former *Climate Change Mitigation and Low-Carbon Economy Act* called for three point-in-time GHG reductions relative to Ontario's GHG emissions in 1990: (1) a 15 percent reduction by 2020; (2) a 37 percent reduction by 2030; and (3) an 80 percent reduction by 2050 (see SO 2016, c 7, s 6(1)). The new *Cap and Trade Cancellation Act* sets a significantly weaker 2030 reduction target, allowing for thirty megatonnes more in annual GHG pollution by 2030 than permitted under the former *Climate Change Act* (see SO 2018, c 13). Further, the new legislation completely omits any GHG reduction target for 2050 (see Ministry of the Environment, Conservation and Parks, *Preserving and Protecting our Environment for Future Generations: A Made-in-Ontario Environment Plan* (Toronto: Queen's Printer for Ontario, 2018), online: ontario.ca [perma.cc/TZ2P-FYNR]). See also *Mathur v His Majesty the King in Right of Ontario*, 2023 ONSC 2316 (Factum, Applicants at 15–21).

82   *Mathur* SC, above note 44 at para 2.

83   *Ibid.*

84   *Ibid* at paras 132, 139, 140, 171, 189, 228, 233 & 234.

85   She held, however, that the question of determining Ontario's fair share of the global carbon budget was not justiciable (see *Ibid* at para 109).

as the Ontario Court of Appeal's decision in the housing rights case of *Tanudjaja*, discussed below.[86]

Having found the claim justiciable, Vermette J made several important findings of fact. For example, the Court held that Ontario's target falls severely short of the scientific consensus about what is required to limit warming to 1.5 degrees Celsius and that this deficiency contributes to an increased risk of death and harm faced by the applicants and all Ontarians.[87] She rejected the government's defence that the province's emissions were insignificant in light of the global nature of the problem, joining courts around the world in holding that every emission contributes to the problem and that jurisdictions are accountable for what they emit. Although the applicants took pains to frame their case as a negative rights case, challenging actions by the Ontario government that lead to emissions (versus inaction or insufficient action), Vermette J characterized the claim as one engaging positive rights. She broke new judicial ground in holding that the existential threat posed by climate change presented special circumstances that could justify the imposition of positive obligations under section 7.[88] However, she then held that the current section 7 test is not adapted to an analysis of positive obligations, specifically with respect to the analysis of principles of fundamental justice. Rather than adapt the test, Vermette J applied the standard analysis and concluded that the applicants had not shown that any deprivations of their rights were contrary to the principles of fundamental justice.[89] The claim thus failed. She also declined the invitation to adopt a new principle of fundamental justice of societal preservation or recognize an unwritten constitutional principle of ecological sustainability.[90]

---

86 *Ibid* at paras 33 and 103–4.

87 *Ibid* at para 147.

88 *Ibid* at para 138. Note that Rennie J in the *Kreishan* case, in *obiter dicta*, offered climate change as an example of the kind of circumstances that might qualify as special in the context of section 7 and positive obligations (see *Kreishan v Canada (Citizenship and Immigration)*, 2019 FCA 223 at para 139 [*Kreishan*]).

89 *Mathur* SC, above note 45 at para 171.

90 *Ibid*.

Justice Vermette also dismissed the section 15 claim. Following the approach taken by the majority in the Supreme Court of Canada decision in *R v Sharma*,[91] Vermette J held that courts are under no positive obligation to remedy inequalities and that leaving gaps between protected groups and others does not infringe section 15. While she held that young people are disproportionately impacted by climate change, she said this was not due to Ontario's target or plan but rather climate change as a whole.[92] Her analysis of causation under section 15 was more restrictive compared to her approach under section 7, leading to internal inconsistency in the decision. She also qualified the alleged distinction under section 15 as temporal (not a protected ground of discrimination) and not related to age.

The Ontario Court of Appeal allowed the appeal on the basis that the application judge had erred in characterizing the case as one seeking positive rights. By enacting legislation requiring a plan and target to be set, the Court of Appeal found that Ontario had voluntarily assumed a statutory obligation to combat climate change and in doing so, was obligated to produce a plan and target compliant with the *Charter*.[93] On section 15, the Court of Appeal noted that the judge had erred in her causation analysis and in holding that the applicants were seeking to impose a positive obligation. On remedies, the court clarified that ordering Ontario to establish a science-based target could be an appropriate remedy should a violation be found, since there are international standards for which there exists scientific consensus.[94]

The Ontario Court of Appeal also noted that the interveners had raised several relevant and important issues that were not addressed by the lower court, including whether Ontario's target breached the *Charter* rights of Indigenous peoples in Ontario and their section 35 rights under the *Constitution Act, 1982*; the application of international law, including

---

91   2022 SCC 39 [*Sharma*].
92   *Ibid* at para 178.
93   *Ibid* at para 5.
94   *Ibid* at paras 69–74.

international environmental law, in the interpretation of *Charter* rights; and the recognition and impact of unwritten constitutional principles, such as societal preservation and ecological sustainability.[95] The court remitted the matter for a new hearing before the Superior Court before the same or another justice, noting that the potential need for further evidence and/or amended pleadings justified the decision to remit.[96] As noted above, Ontario filed for leave to appeal the decision to the Supreme Court of Canada and the applicants filed for leave to counter-appeal to argue that the Court should have decided the case on the merits rather than remit for re-hearing.[97] Ontario's request for leave to appeal was denied, so the case is proceeding to a rehearing.

4.1.1.3 Conclusion

While each of the cases in the second wave of climate litigation in Canada is distinct, they are united by the fact that they are all aimed at government, allege violations of sections 7 and 15(1) of the *Charter* based on the government's insufficient response to climate change, and include claims by youth and Indigenous peoples based on disproportionate impacts. To date, the courts have all accepted the climate science evidence put forward by the claimants and the evidence of harm and disproportionate impact. The decisions have all turned on various inter-related issues common to systemic climate claims under the *Charter*, including justiciability, causation, the positive/negative rights framing, and remedies. In addition, the courts have grappled with specific issues in the section 7 and 15(1) claims, including the principles of fundamental justice in the former and the interpretation of age-related discrimination in the latter. While the courts have yet to discuss section 1 and have made only brief mention of unwritten constitutional principles, these may play a bigger role as the cases move forward.

---

95 *Ibid* at para 6.
96 *Ibid* at para 8.
97 *Mathur* CA, leave to appeal, above note 7.

4.2 JUSTICIABILITY

> "When the seas envelope our coastal cities, fires and droughts
> haunt our interiors, and storms ravage everything between,
> those remaining will ask: Why did so many do so little?"[98]
>
> *Justice Josephine Stanton*

Justiciability has been a significant barrier for climate cases around the world. Canada has been no exception, with *Enjeu* dismissed as non-justiciable and the lower courts in *La Rose* and *Misdzi Yikh* initially dismissing the cases as non-justiciable. The Federal Court of Appeal's decision to overturn the motions to strike in *La Rose* and *Misdzi Yikh* on section 7 and the Ontario Court of Appeal's confirmation of justiciability in *Mathur* for both the section 7 and 15 claims may signal a turning point in the Canadian judiciary's willingness to allow systemic climate claims to cross the justiciability threshold. In *La Rose*, the Federal Court of Appeal clarified that while the pleadings required refinement, this did not negate the fact that the section 7 claim was justiciable, as it was properly grounded in government conduct captured in legislation, Orders in Council, and executive action.[99] The Federal Court of Appeal also found that the motions judges had erred in striking the claims for being positive rights claims, stating that "[t]he claim of a right to a healthy and liveable environment, and the legislative sanctioning of something less, explores the scope of section 7 and tests its boundaries."[100]

This recent trend suggests that justiciability may not present the obstacle it once did, at least for section 7 claims. However, until these ongoing cases are finally decided and other decisions emerge, and until issues of justiciability in the section 15 context are further examined in

---

98 *Juliana v. United States*, 947 F.3d 1159 (9th Cir. 2020) at 64, Staton J, dissenting.

99 *La Rose* FCA, above note 7 at para 118.

100 *Ibid* at paras 105 and 109. While McVeigh J did hold that a positive rights framing was outside the scope of the court, Manson J specifically held that the plaintiffs in *La Rose* at the Federal Court were not barred from arguing a positive rights claim (see *La Rose* FCTD above note 44 at para 67).

the courts, justiciability remains a potentially significant obstacle for Canadian climate rights claims.[101]

4.2.1 What is Justiciability?

Justiciability is a necessary precondition for a court to exercise its jurisdiction.[102] To be justiciable, the issue presented must contain a sufficient legal component to warrant judicial involvement and be a question "that is appropriate for a court to decide."[103] In the context of a climate-based lawsuit, a finding of non-justiciability means the claimants will not have a chance to present a full evidentiary record and make their case before the court. It is, as such, a critical threshold question that can — if interpreted narrowly — limit access to justice for climate litigants.

Despite its significance, the courts offer surprisingly little practical or analytical guidance on how they should determine whether a matter is justiciable. The Supreme Court of Canada has observed that "[t]here is no single set of rules delineating the scope of justiciability."[104] As a result, "justiciability depends to some degree on context, and the proper approach to determining justiciability must be flexible."[105] To determine if a matter is justiciable, courts consider the suitability of the matter for adjudication in light of the legitimacy and institutional capacity of the courts.[106]

Legitimacy entails ascertaining the court's appropriate role given the separation of powers. In Canada's Parliamentary democracy, each of the

---

101 For a discussion of justiciability in the context of climate litigation in Canada, see Larissa Parker, "Let Our Living Tree Grow – Beyond Non-Justiciability for the Adjudication of Wicked Problems" (2022) 81:1 UT Fac L Rev 54 [Parker, "Beyond Non-Justiciability"].

102 Lorne Mitchell Sossin, *Boundaries of Judicial Review: The Law of Justiciability in Canada*, 2d ed (Toronto: Carswell, 2012).

103 *Highwood Congregation of Jehovah's Witnesses (Judicial Committee) v Wall*, 2018 SCC 26 at para 32 [*Wall*].

104 *Reference Re Secession of Quebec*, [1998] 2 SCR 217 at para 34 (SCC) [*Reference re Secession*].

105 *Ibid*. See also *Wall*, above note 103 at para 34.

106 Sossin, above note 102. See also Martha Jackman & Bruce Porter, "Socio-Economic Rights Under the Canadian Charter" in Malcolm Langford, ed, *Social Rights Jurisprudence: Emerging Trends in International and Comparative Law* (Cambridge: Cambridge University Press, 2008) 209.

three branches have distinct roles: the legislature establishes policies and laws, the executive implements them, and the judiciary maintains the rule of law by, *inter alia*, interpreting and applying laws and safeguarding the fundamental rights and freedoms guaranteed by the *Charter*.[107] Courts may hold a claim to be non-justiciable if it asks them to create policy or evaluate the wisdom of a government's policy choices. However, since legal issues are often intermingled with policy matters, delineating the bounds of justiciability can be challenging. This has certainly been the case for Canadian climate change cases so far, given the vast complexity of climate change.[108]

A crucial factor in assessing justiciability is the responsibility vested in courts to review government conduct for compliance with the *Charter*. The jurisprudence and scholarship are unified in the understanding that questions of potential *Charter* violations are almost always justiciable.[109] Courts have often held that the judiciary has an obligation to protect rights guaranteed by the Constitution. As such, courts should not decline to hear cases because *Charter* rights are entangled with political decisions.[110] The Supreme Court of Canada has noted that even when matters are complex or laden with social values, courts may

---

107 *Ontario v Criminal Lawyers' Association of Ontario*, 2013 SCC 43 at paras 28 and 30–31. See also *Mikisew Cree First Nation v Canada (Governor General in Council)*, 2018 SCC 40 at para 118 [*Mikisew Cree*]. In the context of a reference question, the SCC said that a matter may be non-justiciable if (1) addressing it would take the Court beyond its own assessment of its proper role in the constitutional framework of our democratic form of government or (2) if the Court could not give an answer that lies within its area of expertise: the interpretation of law (see *Reference re Secession*, above note 104 at paras 26 and 161).

108 The issue of justiciability has not been as significant a barrier in many other jurisdictions around the world that have had successful outcomes in climate litigation. This is likely due to a combination of factors including the recognition of an explicit right to a healthy environment, different judicial systems (e.g., civil versus common law) and different systems of government (e.g., federation versus other systems).

109 See, e.g., Wilson J's statement in *Operation Dismantle v The Queen* indicating that when a *Charter* challenge is involved, "it is not only appropriate that we answer the question: it is our obligation under the *Charter* to do so" (see [1985] 1 SCR 441 at 471–72 [*Operation Dismantle*]). See also Collins & Sossin, above note 24 at 308.

110 *Chaoulli v Quebec (AG)*, 2005 SCC 35 at para 107 [*Chaoulli*]. See also *Operation Dismantle*, above note 109 at 470–71; *Enjeu SC*, above note 44 ("[a]s the Supreme Court teaches in *Chaoulli*, courts cannot evade the exercise of judicial review only because the issue is complex or controversial or because 'it is laden with social values'" at para 70).

not "abdicate the responsibility vested in them by our Constitution to review legislation for *Charter* compliance" when challenged.[111] As famously stated by Wilson J in *Operation Dismantle* (in dissenting remarks, though not on this point), when asked whether "any particular act of the executive violates the rights of the citizens ... then it is not only appropriate that we answer the question; it is our obligation under the *Charter* to do so."[112] Who else but the courts (or perhaps federal and provincial human rights commissions) can decide if someone's human rights have been infringed by government? This is a legal question that courts are institutionally responsible to authoritatively resolve.[113] Indeed, the very purpose of the *Charter* is to hold the government accountable for respecting the rights enshrined in it. Why, then, has justiciability been such a significant hurdle in climate *Charter* litigation in Canada to date, at least until recently? The answer is multi-faceted, influenced by the unique features of climate change and especially the way these climate cases are framed (often of necessity). Justiciability has also been a challenge in other cases making broad/systemic claims, such as the efforts to invoke *Charter* rights in the case of homelessness.[114]

4.2.2   Justiciability in the Climate *Charter* Cases

Justiciability was the obstacle that proved fatal in *Enjeu* and continues to be a significant hurdle for section 15 climate claims (as discussed below in Section 4.4). In *Enjeu*, the Quebec Court of Appeal held that climate policy is a matter for the legislature, not the courts. Yet Canadian courts have a long history of decision-making in areas of complex

---

111 *Chaoulli*, above note 110 at paras 107–8.

112 *Operation Dismantle*, above note 109 at 472.

113 See *Reference Re Canada Assistance Plan (BC)*, [1991] 2 SCR 525 at 545–46.

114 *See Tanudjaja v Canada (AG)* 2014 ONCA 852 [*Tanudjaja* CA]; see also below note 151 (identifying sources of critiques of the approach to justiciability in *Tanudjaja*) and below note 172 and accompanying text (explaining the decision); see also Nathalie J. Chalifour, Jessica Earle & Laura Macintyre, "Coming of Age in a Warming World: The Charter's Section 15(1) Equality Guarantee and Youth-Led Climate Litigation" (2021) 17 JL & Equal 1 at 39 and 46–47 [*Coming of Age*] (discussing *Tanudjaja*).

public policy, especially in *Charter* cases.[115] As noted in the previous section, courts have held that cases engaging the *Charter* are generally justiciable, regardless of the nature of the government conduct.[116] It was also justiciability that led the Federal Court to strike the section 7 and 15 claims in *La Rose/Misdzi Yikh*. The Federal Court was concerned about the breadth of the claim which identified a constellation of government action and inaction as the impugned conduct. While the Federal Court of Appeal allowed the appeals in *La Rose/Misdzi Yikh*, it did so by granting leave to the claimants to amend their section 7 claim to address concerns about the pleadings being overly broad. The section 15 claims remain struck without leave to amend the pleadings. While the Court was not explicit about this, the section 15 claim likely remained struck due to overbreadth of the pleadings rather than justiciability. The Ontario Superior Court of Justice had no difficulty finding both the section 7 and 15 claims in *Mathur* justiciable, a finding upheld by the Ontario Court of Appeal.

What explains the different findings on justiciability in these cases? The answer is animated by three inter-related concerns: the prospect of violating the separation of powers, the framing of the impugned government conduct, and the spectre of positive rights. While each of these concerns was present in all of the cases, the separation of powers issue was dominant in *Enjeu*, the breadth of conduct was dominant in *La Rose/ Misdzi Yikh*, and the positive rights issue was prevalent in *Mathur*. The following section further explores each of these issues.

---

115   Courts have adjudicated matters related to health care, access to abortion, same-sex marriage, medically assisted death, safe injection sites, prostitution, secession of a province, and more (see *Chaoulli*, above note 111; *Morgentaler*, above note 10; *Marriage Reference*, above note 9; *Carter*, above note 101; *PHS*, above note 10; *Canada (AG) v Bedford*, 2013 SCC 72 [*Bedford*]; *Reference re Secession*, above note 104).

116   *Hupacasath First Nation v Canada (Foreign Affairs and International Trade Canada)*, 2015 FCA 4 at paras 61 and 70; *PHS*, above note 11 ("when a policy is translated into law or state action, those laws and actions are subject to scrutiny under the *Charter*" at para 105).

4.2.3 Challenges in the Justiciability Analysis in Climate Rights Cases

All five of the climate *Charter* cases discussed in this chapter are rooted in the same fundamental claim — that the total GHG emissions sanctioned (and often facilitated) by a given government exceed the scientific threshold necessary to avoid harm and therefore infringe the constitutional rights of the claimants. However, the specific way each of the cases is framed depends upon the government's approach to managing GHG emissions and is influenced by factors such as whether they set a GHG reduction target (a climate target) in legislation or as a matter of policy, and other decisions related to GHG emissions.

If a government sets a target, claimants can point to the target and argue it is insufficient relative to what science says is required to limit harm. If a government does not set a target, then claimants must either point to this fact — that no target has been set — or identify the constellation of government decisions that determine the level of emissions from that jurisdiction. The Canadian climate jurisprudence to date seems to suggest only the cases in the former category (where a target is set) are justiciable. This presents a legal conundrum that warrants reflection, since the distinction is one of form rather than substance. The following section unpacks the ways in which the framing of climate claims influences the justiciability analysis. Framing is again addressed in Section 4.3.2 when discussing positive obligations under section 7. The potential injustices arising from a focus on form, rather than substance, in determining justiciability is explored further in Section 4.2.4.

Framing is in the eye of the beholder

As already noted, when distilled to their essence, all of the climate *Charter* cases essentially boil down to the same allegation: the government in question is allowing a level of GHG emissions that is too high and this contributes to harm, infringing the claimants' rights. This same fundamental allegation can be framed in a number of ways:

- the government has set a GHG reduction target that is *insufficient* or *inadequate* relative to the science (inadequate target);

- the government is making decisions (such as approving or funding fossil fuel projects) that, together, allow a harmful level of emissions (allowing too many emissions);
- the government has failed to set a target or is otherwise not doing enough to reduce GHG emissions (failure to act or inaction);
- the government has failed to meet its prior targets and/or is not on track to meet existing targets (failure to enforce/implement commitments).

The claims are inter-related and, in some cases, are simply another way of saying the same thing. Yet the way in which the claims are framed has influenced their adjudication by courts.

How claimants choose to frame their case is necessarily influenced by what steps the government has (or has not) taken in response to the global imperative to reduce GHG emissions. The choice for claimants is relatively simple if there is a legislated target (or mandate to set a target), as there was in *Mathur*. Indeed, the Ontario Superior Court and Ontario Court of Appeal confirmed that the existence of legislation mandating the setting of a target (the *Cap and Trade Cancellation Act*) was central in holding that matter to be justiciable.[117]

When there is no legislated or policy target, claimants may identify the constellation of government decisions that together determines the aggregate level of GHG emissions for that jurisdiction ("allowing too many emissions"); frame the case as one of failure to act or insufficient action ("failure to act/inaction"); identify a single decision with great significance for GHG emissions (such as the approval of a major pipeline or pipeline expansion); or, some combination thereof.[118]

---

117 *Mathur* SC, above note 44 at para 106; *Mathur* CA, above note 7. Justice Vermette even noted that the applicants' *Charter* challenge was focused on the legislation and target to avoid the dismissal of their claim as non-justiciable (see *Mathur* SC, above note 44 at para 120).

118 By the end of 2023, only the federal government and six provinces and territories had climate legislation establishing a target or planning process for such a target. Four jurisdictions, which together represent 75% of Canada's emissions (Alberta, Saskatchewan, Manitoba and Nunavut) had no target or plan for 2030 (See Ross Linden-Fraser, "A Closer Look at the Varying Climate Targets of the Provinces and Territories" (5 October 2023), online: 440megatonnes.ca [perma.cc/74FX-V6PS]. See also Sarah

In the absence of a legislated target when the proceeding was filed,[119] the *Enjeu* applicants framed their complaint around both the inadequacy of the government's climate targets relative to the science and the government's failure to meet each of the targets whose deadline had passed ("inadequate target" and "failure to enforce/meet commitments"). The applicants referenced Canada's four climate targets, including the most recent target at the time of filing, and the *Greenhouse Gas Pollution Pricing Act*, in their claim.[120] While the lower court had found their claim to be justiciable considering it raised *Charter* issues, the Quebec Court of Appeal overturned this finding, rejecting the request to authorize a class proceeding on justiciability grounds. The Court of Appeal held the policy targets to be non-justiciable, characterizing the claim as one framed "in the absence of a statute" (therefore a "failure to act/inaction" framing) that sought "constitutional review of government inaction," something it held to be highly problematic.[121] While never explicitly calling it a claim for recognition of a positive right, Vauclair J stated that the "situation would be different if the Appellants were challenging the validity of a particular law enacting measures to address GHG emissions"[122] While he acknowledged the appellant's invocation of the *Greenhouse Gas Pollution Pricing Act*, he noted that the appellants did not challenge any of its provisions.

Similarly faced with a lack of legislated target when the original claim was filed, the *La Rose* claimants framed their claim broadly around a network of government actions that "cause, contribute to and allow a level of GHG emissions incompatible with a Stable Climate System" and the government's active participation in, and support of, the "development, expansion and operation of industries and activities involving fossil fuels" that emit a harmful level of GHGs ("allowing too many emissions").[123] The

---

McBain et al, *All Together Now: A Provincial Score Card On Shared Responsibility to Reduce Greenhouse Gas Emissions in Canada* (Calgary: Pembina Institute, 2024)).

119 Note that the federal government has since legislated a target (see *Canadian Net-Zero Emissions Accountability Act*, SC 2021, c 22 [*CNZAA*]).

120 *Enjeu* MfA, above note 53 at paras 2.26, 2.54, 2.56–2.59, 2.61, 2.65, 2.68, 2.70, and 2.77.

121 *Enjeu* CA, above note 6 at para 25.

122 *Ibid* at para 26.

123 See *La Rose* SoC, above note 59 at para 5. They also pointed to the setting of GHG targets that are too high and failure to meet past targets.

claim was initially struck and held to be non-justiciable on the basis of the breadth of conduct targeted. In justifying his decision, Manson J pointed to the diffuse nature of the claim that captures "all conduct leading to GHG emissions" as an attempt to seek "judicial involvement in Canada's overall response to climate change."[124] In *Misdzi Yikh*, the plaintiffs framed their claim as Canada allowing GHG emissions that cause global warming both directly (i.e., through regulations pertaining to road vehicles, fossil-fuel electrical generation, offshore fossil fuel developments and in the Arctic and Northwest Territories, and through federal impact assessment), as well as indirectly (i.e., through taxation and subsidies, approval of federally regulated oil and gas facilities and pipelines, and investments in fossil fuel infrastructure (thus also an "allowing too many emissions" framing).[125]

By the time the Federal Court of Appeal overturned the motion to strike and granted leave to amend the section 7 claims in *La Rose/Misdzi Yikh*, the situation had changed. In particular, the federal government passed in 2021 the *Canadian Net-Zero Emissions Accountability Act (CNZAA)* which establishes a national net-zero target for 2050 and requires the government to set national GHG reduction targets for milestone years by certain dates.[126] The Federal Court of Appeal referenced this law (and the target set under it) when it held the section 7 claim to be justiciable. While the Federal Court of Appeal did not observe that the legislated target was not in place when the original claim was filed, granting leave to amend the claim allowed the claimants to reframe their case around this legislation (which it did).[127]

Would the Federal Court of Appeal have still allowed the claim to proceed with leave to amend had the *CNZAA* not been enacted or if it were repealed (i.e., after a change in government)? The answer appears to be yes, since the Court's reasons did not turn only on the enacting of the *CNZAA* but also pointed to other government conduct. The Federal Court of Appeal noted that the federal government had made legislative

---

124 *La Rose* FCTD, above note 44 at para 44.
125 *Misdzi Yikh* SoC above note 68 at paras 36–39.
126 *CNZAA*, above note 120, ss 6–7.
127 *La Rose* FCA, above note 7 at para 114.

choices, pointing to the preamble of the *Greenhouse Gas Pollution Pricing Act* expressing Canada's commitment to meet (and increase) its national climate target. Distinguishing the *Tanudjaja* case (which it characterized as challenging the overall approach to homelessness rather than any law), Rennie J pointed to the government's Paris commitment (the Nationally Determined Contributions) which is a commitment ratified by Parliament and therefore "a legally defined, objective standards(s) against which the *Charter* claims can be assessed" rather than a challenge to policy alone.[128] The Federal Court of Appeal also left open the possibility that a claim could proceed on the basis of a purely positive right (if, for instance, the government were to repeal climate-related laws and programs, perhaps after a change in government).[129]

Although sub-national governments do not submit Nationally Determined Contributions, setting a policy target in the context of national and international climate commitments should be a sufficient basis upon which to find a provincial or territorial target justiciable. A key determinant might be whether the policy target is sanctioned in some way by the executive and/or legislative branches of government, rather than a mere electoral promise or broad political statement. If there is no sufficiently anchored climate target, there may still be legislation upon which to ground a claim.

The *Dykstra* case is framed around legislation establishing the Crown Corporation SaskPower and the latter's construction of new, unabated fossil fuel-based generation assets, including two specific power plants.[130] Specifically, the claimants allege that the decision to expand gas-fired electricity generation violates their section 7 and 15 rights. As such, it is framed as "allowing too many emissions," but unlike the original *La Rose* and *Misdzi Yikh* claims, *Dykstra* is anchored in a specific law and discernable government action. While the matter has not yet been decided, its framing is sufficiently narrow that it has a solid chance of clearing the justiciability threshold.

---

128 *Ibid* at para 38.
129 *La Rose* FCA, above note 7 at paras 96–99, 109, and 116.
130 *Dykstra*, above note 44 (Amended Originating Application, Applicants)

Reconcilable differences?

Are the differences in approach to justiciability — between the Quebec Court of Appeal and Federal Court of Appeal in particular — reconcilable? The federal government had made the same commitments at the time both original claims were filed, and they were both making fundamentally the same claim. Yet, the Quebec Court of Appeal interpreted justiciability more narrowly than the Federal Court of Appeal. One way to distinguish them is that these were different types of actions: *Enjeu* was a request to authorize a class proceeding, whereas *La Rose/Misdzi Yikh* are actions to review compliance with the *Charter*. Perhaps it is a reflection of this procedural difference that led the Quebec Court of Appeal to explicitly leave the door open to a future case, when it noted that courts may be called upon "in another context, to review the state's conduct with respect to global warming."[131]

The difference can also be explained by the nuance in framing between the two cases. *Enjeu* emphasized the inadequacy of the government's targets and action to reduce GHG emissions, including the failure to meet past targets. This emphasis on the failure to act led the Court to characterize the claim as one asking the court to tell the legislature what to do — something it held was not their role.[132] Though the Court never engaged with the positive rights debate that proved so central in *Mathur*, the *Enjeu* Court essentially characterized the case as one of positive rights (without using those words) and then held such to be non-justiciable.[133] In contrast, the Federal Court of Appeal in *La Rose/Misdzi Yikh* took pains to identify the government decisions being targeted by the claimants as leading to a harmful level of emissions. For example, the Court noted that the federal government had, at the time, signed and ratified the *Paris Agreement*, producing several targets including National Determined Contributions under the *Paris Agreement* and had passed the *Greenhouse Gas Pollution Pricing Act*.[134] While this slight

---

131 *Enjeu* CA, above note 6 at para 40.

132 *Ibid* at para 32.

133 The Court references positive obligations when summarizing the respondent's argument (see *Ibid* at para 19).

134 SC 2018, c 12, s 186.

difference in framing can distinguish the outcomes, it is a difference in form rather than substance.

One could also argue that it is possible to distinguish the precedents in terms of which is more recent. The Federal Court of Appeal decision in *La Rose/Misdzi Yikh* came later than the Quebec Court of Appeal's decision. The Quebec Court of Appeal relied upon the lower court's decision in *La Rose* to justify its finding of non-justiciability, and that decision has since been overturned.[135] However, the Supreme Court of Canada denied leave to appeal in *Enjeu*, so the precedent remains important. Interestingly, the Federal Court of Appeal made no mention of the *Enjeu* decision in its reasons in *La Rose/Misdzi Yikh*.[136]

Where does that leave us? It seems fair to conclude that challenges to legislation enshrining a GHG target or requiring the government to set targets will be found justiciable, at least for the section 7 claims. This was the case in *Mathur*, where the applicants grounded their claim in Ontario's *Cap and Trade Cancellation Act* which requires the government to establish a GHG reduction target and plan.[137] The Ontario Court of Appeal characterized this as the government having voluntarily assumed a statutory obligation that would then be subject to *Charter* scrutiny. What is less clear is how courts will treat a claim framed around an unlegislated policy target and/or network of decisions leading to a given level of emissions. The *La Rose/Misdzi Yikh* jurisprudence suggests that the network of decisions must not be too large or diffuse and also that climate targets set in relation to the *Paris Agreement*, such as the National Determined Contribution or the expressed commitment to adhere to the Paris goal in the Preamble of the *GGPPA*, are sufficient for justiciability (at least for section 7). The justiciability of section 15 claims and those framed entirely around a failure to act or inaction, in the absence of any relevant legislation or policy target, remain less certain. I return to this when discussing positive rights below. This uncertainty relating to the justiciability of section 15 claims is deeply concerning given that provision's legislative history and

---

135 *Enjeu* CA, above note 6 at para 35.
136 *La Rose* FCA, above note 7.
137 *Cap and Trade Cancellation Act*, SO 2018, c 13, s3(1).

its actual language, which suggest justiciability should not be the hurdle it is proving to be.[138]

The separation of powers

The need to respect the separation of powers is an important factor that arises in assessments of justiciability. This was one of the animating arguments used by the Quebec Court of Appeal to justify finding the claims in that case non-justiciable, since to find it justiciable would have, according to the Court, taken it beyond its judicial role into the purview of the legislature. The Quebec Court of Appeal noted that the response to climate change requires collaboration among governments in delicate negotiations which engage weighing health, transportation, economic, and regional development issues in coming up with solutions and then translating these solutions into budgetary priorities.[139] The Court held that it is "not the role of the courts to engage in such analysis"[140] nor is it the court's role to tell the legislation what do to.[141] In a similar vein, McVeigh J in the *Misdzi Yikh* decision held the claim was a "political one that may touch on strategic/ideological/historical or policy-based issues and determinations within the realm of the remaining branches of government."[142] Although McVeigh J acknowledged the existence of political elements of an issue like climate change does not mean that there cannot be sufficient legal elements to render something justiciable,[143] she ultimately concluded that it was "hard to imagine a more political issue than climate change."[144] She pointed to the remedies sought as ones that would be appropriate for the other branches of government rather than the judiciary.[145]

---

138 See Bruce Porter, "Expectations of Equality" (2006) 33 SCLR 23.

139 *Enjeu* CA, above note 6 at para 35.

140 *Ibid* at para 35.

141 *Ibid* at paras 24 and 32.

142 *Misdzi Yikh*, above note 44 at para 72.

143 *Ibid* at para 20.

144 Justice McVeigh's reasons were ultimately the third point — that policy choices must be translated into law or state action in order for them to be justiciable (see *Ibid* at paras 19 and 21).

145 *Ibid* at para 73.

In the lower court decision in *La Rose*, Manson J similarly found that some questions "are so political that the courts are incapable or unsuited to deal with them."[146] As examples, he offered "questions of public policy approaches - or approaches to issues of significant societal concern."[147] He characterized the plaintiffs' claim as effectively subjecting the holistic policy response to climate change to *Charter* review — something he held to be non-justiciable.[148] Justice Manson also pointed to the remedies sought by the plaintiffs as reason to hold that allowing the remedies would intrude on the policy-making functions of the legislature and the executive branches.[149]

Overturning these findings, the Federal Court of Appeal in *La Rose/Misdzi Yikh* clarified that matters are not non-justiciable because they reflect a political choice about how to address an issue such as climate change.[150] As long as some legislative choices or exercises of executive discretion have been made, it is appropriate for the courts to evaluate the constitutionality of those choices.[151] Justice Rennie defined the scope of government conduct needed to cross the justiciability threshold in broad terms, offering as an example the preambular language in the *Greenhouse Gas Pollution Pricing Act* and Canada's climate target under the *Paris Agreement* (the Nationally Determined Contribution) which was ratified by Parliament.[152] He contrasted this to the Ontario Court of Appeal's decision in *Tanudjaja* which challenged the government's overall approach to homelessness without any single legislative anchor.[153] The Federal Court of Appeal tied the justiciability finding to the existence of "legally defined, objective standards against which the Charter claims can be assessed," referring to the

---

146 *La Rose* FCTD, above note 44 at para 40.

147 *Ibid* at para 40.

148 *Ibid*.

149 *Ibid* at para 55.

150 *La Rose* FCA, above note 7 at para 32.

151 *Ibid* at para 33.

152 *Ibid* at paras 32–33 and 38.

153 *Ibid* at para 37. The approach of the Court in *Tanudjaja v Canada (AG)* on justiciability and separation of powers has been subject to widespread critique (see *Tanudjaja* CA, above note 115). See, e.g., Margot Young, "Temerity and Timidity: Lessons from Tanudjaja v Attorney General (Canada)" (2020) 61:2 C de D 469.

quantifiable GHG reduction targets developed internationally according to scientific consensus. In doing so, Rennie J joined courts around the world in pointing out the important difference between assessing whether the target infringes *Charter* rights (justiciable) versus telling the government how to fulfill its commitments (non-justiciable).[154] Justice Rennie also rejected the notion that a claim should be found non-justiciable because the remedies sought push the boundaries of a court's competence; rather, judges can exercise principled discretion to amend the scope of remedies sought when deciding a case.[155]

As such, while important to bear in mind, the potential of violating the separation of powers is not, in and of itself, a basis upon which to hold a climate *Charter* case to be non-justiciable. It may be, however, relevant in the choice of remedies. What plays a bigger role in justiciability is how the case is framed.

Lack of an anchoring statute or network of laws — the positive rights issue

The separation of powers issue is closely linked to the concern about the lack of an anchoring statute or set of laws to ground the claim and, relatedly, the prospect that a claim is aiming to impose a positive obligation on the government. The Quebec Court of Appeal in *Enjeu*, for instance, held that constitutional review of government inaction by courts is highly problematic in the absence of a statute.[156] In *Misdzi Yikh*, the Federal Court held there was "no impugned law or action to evaluate, there are no specific allegations of government actions, and the positive obligations (or limitations) sought by the Dini Ze' are vague."[157] In *La Rose*, Manson J for the Federal Court stated that "[p]olicy choices must be translated into law or state action in order to be amenable to *Charter* review and

---

154 *La Rose* FCA, above note 7 at para 38. See, e.g., *Urgenda* District Court, above note 3 at paras 4.94–4.102; *Klimaatzaak*, above note 24 at paras 153–56. Note that this issue could also be addressed as a matter of the appropriate remedy.
155 *La Rose* FCA, above note 7 at para 51.
156 *Enjeu* CA, above note 6 at para 25.
157 *Misdzi Yikh*, above note 44 at para 58.

otherwise justiciable."[158] While a law or a network of laws could be justiciable, the Federal Court in *La Rose* considered the "undue breadth and diffuse nature" of the network of actions and inactions identified by the plaintiffs to be effectively asking the court to evaluate "Canada's overall policy choices" for *Charter* compliance — something it held was non-justiciable.[159] At the heart of these reasons is a holding that matters framed around inaction or insufficient action — and thereby invoking a positive obligation — will be found non-justiciable.

In *Mathur*, the Ontario Court of Appeal drew a distinction between what it characterized as the imposition of a freestanding positive obligation (a possibility the Supreme Court of Canada in *Gosselin* left open in special circumstances under section 7) versus a voluntarily assumed statutory obligation (which is justiciable as an ordinary negative rights case).[160] The distinction seems to boil down to whether there is relevant legislation. What the Ontario Court of Appeal decision in *Mathur* leaves unanswered is whether government policy choices, such as a non-legislated target or a constellation of decisions that together lead to a harmful level of emissions ("allowing too many emissions") constitute a justiciable issue (albeit one that raises positive obligations in the second case). The Federal Court of Appeal decision in *La Rose/Misdzi Yikh* suggests the former does, since it pointed to a variety of government actions, including Orders in Council, which are not legislation.[161] Comparing *La Rose/Misdzi Yikh* to *Mathur*, the Federal Court of Appeal noted that the challenges are fundamentally the same, with both involving a "*Charter* challenge to the adequacy of a government's response to climate change."[162] Both claims were held to be justiciable.[163]

---

158   *La Rose* FCTD, above note 44 at para 38.

159   *Ibid* at para 46.

160   For a discussion on the positive/negative rights dualism, including a critique of it, in the context of section 7 of the *Charter*, see Section 4.3.2, below in this chapter.

161   *La Rose* FCA, above note 7 at paras 110 and 118.

162   *Ibid* at para 44. Justice Rennie could have included *Enjeu* here, since it was also fundamentally a challenge to the adequacy of a government's response to climate change, but as noted earlier, the Federal Court of Appeal decision makes no mention of *Enjeu*.

163   Only the section 7 portion of the claim was deemed justiciable in *La Rose/Misdzi Yikh*.

Both the Ontario Court of Appeal in *Mathur* and the Federal Court of Appeal in *La Rose/Misdzi Yikh* therefore suggest framing a matter around insufficiency is not, in and of itself, fatal in a justiciability analysis. The adequacy of a target relative to the science, and the adequacy of the actions to achieve that target, should both be justiciable. What is left unanswered is whether framing a case solely around a failure to act or inaction/insufficient action and seeking recognition of what the Ontario Court of Appeal qualifies as a freestanding obligation — is justiciable. I return to this issue in section 4.3.2 below.

Complexity

Complexity alone should not pose a barrier to justiciability. Even in the lower court decisions, Manson J in *La Rose* recognized that the complexity or novelty of an issue does not *per se* make it non-justiciable.[164] Drawing upon Supreme Court of Canada jurisprudence, Rennie J for the Federal Court of Appeal stated unequivocally that "claims are not rendered non-justiciable simply because they raise complex or controversial issues."[165] He cites several examples of complex, controversial public policy issues in which the Supreme Court of Canada has evaluated *Charter* rights, including access to private medical care (*Chaoulli*), wait times for surgery (*Cambie*), prostitution (*Bedford*), abortion (*Morgentaler*), injection sites (*PHS*), and medically assisted dying (*Carter*).[166] Similarly, in *Mathur*, the court cited *Chaoulli* and *PHS* in support of its finding that "the Constitution requires that courts review legislation and state action for *Charter* compliance when citizens challenge them,

---

164 See also McVeigh J in *Misdzi Yikh*, who recognized that "complexity itself does not mean that the Court cannot adjudicate an issue" though she notes in the same sentence that issues spanning various governments and involving issues of economics and trade, for instance, are non-justiciable (see *Misdzi Yikh*, above note 44 at para 56).

165 *La Rose* FCA, above note 7 at para 29.

166 *Bedford*, above note 115; *Morgentaler*, above note 10; *PHS*, above note 10; *Carter*, above note 10. See also *La Rose* FCA, above note 7 at para 30, citing *Cambie Surgeries Corporation v British Columbia (AG)*, 2022 BCCA 245.

even when the issues are complex, contentious and laden with social values."[167]

On this point, the *Enjeu* decision seems to be an outlier once again. The Quebec Court of Appeal found complexity to be a persuasive factor in its justiciability conclusion, holding that "[s]ome issues, because of their complexity, cannot be adequately dealt with by judicial orders."[168] With respect, this finding is at odds with Supreme Court of Canada jurisprudence and the other climate *Charter* cases; the Supreme Court of Canada has repeatedly held that neither complexity nor the novel aspect of an issue take it outside the scope of judicial intervention.[169] Most *Charter* cases that reach the Supreme Court of Canada involve public policy matters that are complex, controversial, and engage social values. It is likely that the Quebec Court of Appeal was primarily concerned with the claim's emphasis on inaction and the possibility of that leading to a finding of positive obligations and the implications for the separation of powers, rather than complexity itself.

4.2.4 Moving Forward on Justiciability in Climate Rights Cases – A Path and a Caution

This section identifies a key feature that supports a finding of justiciability in the climate *Charter* cases, even when framed around inaction or as a failure to act, and offers a critique about the implications of a narrow approach to justiciability in climate *Charter* cases.

---

167 *Mathur* SC, above note 44 at paras 100–1 and 106. The Ontario Court of Appeal characterized this as correct (see *Mathur* CA, above note 7 at para 36).

168 *Enjeu* CA, above note 6 at para 24.

169 For example, the Federal Court noted that novel, but arguable, claims should be permitted to proceed to avoid impeding the development of common law (see *La Rose* FCA, above note 7 at para 109). The Supreme Court of Canada also explained that difficult or ambiguous questions do not mean a question should not be answered but rather a Court is free to qualify and/or interpret the question (see *Reference re Secession*, above note 105 at para 31). See also *R v Imperial Tobacco Ltd*, 2011 SCC 42 at para 21.

The existence of measurable, objective legal standards is key to justiciability in climate cases

As noted earlier, the separation of powers is a fundamental and essential precept in Canada's Parliamentary democracy. But the line which courts should not cross is not easy to define with precision. Considering what would clearly fall outside the line can shed light on the zone of uncertainty. For instance, if courts were asked an open-ended question about what legislative tools the government should employ to respond to climate change, it would clearly be outside their ambit. However, the suggestion that an issue is non-justiciable simply because it is highly political or engages issues of public policy should be rejected. The issue for justiciability is not whether a question is too political but whether there is a legitimate legal question to be answered. It is entirely within the judiciary's ambit to determine whether government conduct unjustifiably infringes *Charter* rights, even if this conduct is part of a complex public policy issue like climate change.

Importantly, the determination of whether *Charter* rights are infringed can be made without wading into the merits of specific policy choices. Canadian courts have a long track-record of adjudicating alleged *Charter* violations in the context of highly politicized social policy issues, from health care and social housing to medically assisted dying and sex work.[170] There is no reason why the courts cannot evaluate alleged *Charter* violations in the context of climate change.

To the contrary, climate change offers something that many other complex public policy issues do not: objective, quantifiable standards — a barometer or measurement tool, if you will — that provide the judiciary with a scientifically valid standard against which to assess state conduct. The *Paris Agreement*, which Canada and some 194 other parties have ratified, sets two scientifically driven, quantifiable temperature thresholds beyond which lie increasingly catastrophic and irreversible impacts.[171]

---

170   See, e.g., *Carter*, above note 10; *PHS*, above note 10; *Bedford*, above note 115; *Chaoulli*, above note 110; and *Victoria* (City) *v Adams*, 2009 BCCA 563.

171   United Nations Climate Change, "Paris Agreement – Status of Ratification" (last visited 21 December 2024), online: unfccc.int [perma.cc/6JPB-6NED].

Science enables us to precisely quantify and measure the level of GHG emissions reductions required to avoid crossing that threshold, and several methodologies are available to determine how this global carbon budget can and should be allocated among jurisdictions.[172] The IPCC has reported that global GHG emissions must be reduced by 45 percent below 2010 levels by 2030 and reach net-zero by 2050. The international community agreed to these goals, as reflected in the *Glasgow Pact*.[173]

The existence of a quantifiable standard allows the courts to circumscribe their role to determining what the legal standard entails and the implications of breaching it for human rights, leaving it to the legislative and executive branches of government to determine *how* the standard should be met. Indeed, courts in other jurisdictions have relied upon scientifically generated and politically endorsed internationally GHG reduction targets to evaluate the adequacy of government decisions about GHG emissions and the implications for human rights.[174] In response to concerns about justiciability and the separation of powers, the Dutch Supreme Court in *Urgenda* explained that its role was to evaluate the Dutch government's climate target against the scientific standard to determine if the government was violating human rights. When it found the target wanting, the Court identified the minimum level of emissions reductions the government would need to achieve to safeguard human rights. The Dutch Court was careful to identify its role as limited to ruling

---

172   There is more than one barometer that courts can use as legal standards to evaluate government action in relation to climate change. The most common one is the temperature threshold of 1.5 degrees, embedded within the Paris Agreement and the Glasgow Pact. This temperature threshold has been translated by the IPCC into levels of GHG emissions reductions required to limit warming to 1.5 degrees. Other quantifiable methods for determining whether governments are doing enough to reduce GHG emissions is to consider the concentration of GHGs in the atmosphere in PPMs. This is the approach taken by the plaintiffs in *La Rose* and the American companion case *Juliana*, referenced as a Stable Climate System. Like with the temperature threshold, levels of GHG emissions allowed to stay within concentrations associated with a stable climate system can be calculated.

173   *Mathur* SC, above note 44 at para 20, citing *Glasgow Climate Pact*, UNFCCC, 2021, UN Doc FCCC/PA/CMA/2021/10/Add.1 FCCC Dec 1/CMA.3 art 22.

174   See, e.g., *Urgenda* Supreme Court, above note 22 at paras 5.2.3, 6.2, 7.2.11, 7.3.6, and 7.5.1; *Klimaatzaak* above note 24 at paras 282–87; *Earthlife Africa*, above note 34 at paras 16–17, 30, 35, and 119.

on the legal threshold the government was required to meet, and whether it had met that standard, without determining what policies should be implemented to achieve this standard. It was the role of the legislative and executive branches to determine *how* to meet that standard, as they have done in the years following the Supreme Court's decision.[175]

In a similarly framed case, the Belgian Court of Appeal in *Klimatzaak* ordered the government to reduce GHG emissions by 55 percent below 1990 levels by 2030, leaving the choice of the means used to achieve the target to the executive and legislative branches[176] To evaluate the adequacy of Germany's legislated climate targets, the German Constitutional Court (an Apex Court) in *Neubauer* engaged in a detailed carbon budget analysis and found the target insufficient. While the Court did not specify what the new targets should be, the government subsequently amended its legislation to strengthen its 2030 target and accelerate the net zero goal.[177]

The Ontario Superior Court of Justice in *Mathur* recognized the importance of metrics in evaluating Ontario's target when it referenced the *Glasgow Pact* and stated that it could "rely on the scientific consensus that GHG must be reduced by approximately 45% below 2010 levels by 2030, and must reach 'net zero' by 2050 in order to limit global average surface warming to 1.5 degrees C and to avoid the significantly more deleterious impacts of climate change."[178] Measured against this benchmark, the Court found Ontario's target severely deficient, a conclusion endorsed by the Ontario Court of Appeal.[179] The Ontario Court of Appeal went further in noting that the unchallenged international standards and scientific consensus on climate change and evidence about the remaining carbon budget provide clear standards that can inform the evaluation of a constitutionally compliant target and plan.[180] The Federal Court of Appeal in *La Rose/Misdzi Yikh* also identified

---

175 *Urgenda* District Court, above note 2 at para 4.101. The district court decision was upheld by the Supreme Court (see *Urgenda* Supreme Court, above note 22 at paras 5.7.9, 5.8, 7.5.1–7.5.3, and 8.2.7).

176 *Klimaatzaak*, above note 24.

177 *Ibid*.

178 *Mathur* SC, above note 44 at para 144.

179 *Ibid* at paras 20 and 147; *Mathur* CA, above note 7 at para 23.

180 *Mathur* CA, above note 7 at para 70.

Canada's climate target under the *Paris Agreement* as a "legally defined, object-ive standard against which the Charter claims can be assessed."[181]

The existence of scientifically driven, quantifiable thresholds in climate change offers a way to distinguish the Ontario Court of Appeal's decision in *Tanudjaja*, which was dismissed as non-justiciable. In that case, the applicants had argued that the Ontario and federal government's laws, policies, programs, and services led to inadequate housing and increased homelessness in violation of section 7 and 15 rights. The Court of Appeal characterized the claim as one seeking a positive right to housing and upheld the decision to strike the claim. In support of its finding of non-justiciability, the majority held that there was "no judicially discoverable and manageable standard for assessing in general whether housing policy is adequate."[182] Whether or not that was true in the *Tanudjaja* case, such a standard clearly exists in the case of climate change.

Implications of a formalistic approach to justiciability in the climate Charter cases — a false distinction

Finding cases against governments that have enacted a single climate target to be justiciable, while finding that claims framed around a broad constellation of government conduct or around a failure to act or inaction to be non-justiciable, is problematic for at least six inter-related reasons.

First, it is a distinction based on form rather than substance. Holding that the same state conduct is justiciable when the government has created a metric (a legislated target) but non-justiciable when the government has left its targets in the form of policy or opted not to act at all, is unsatisfactory. While a government's climate target is a helpful proxy that will guide government decision-making in relation to GHG emissions, there are still decisions being made behind that proxy that combine to determine the aggregate level of emissions for that jurisdiction. The distinction is one of form, not substance.

---

181 *La Rose* FCA, above note 7 at paras 38 and 45.
182 *Tanudjaja* CA, above note 114.

Second, the distinction creates a gap in *Charter* accountability. Unless there is a jurisdiction-wide GHG reduction target or law requiring one (as there was in *Mathur*), emission levels will be *de facto* dictated by a diffuse tapestry of government decisions. Allowing *Charter* scrutiny for governments that have set a target but withholding judicial scrutiny when governments fail to set a target ignores the crucial reality that omissions and systemic actions can lead to the same rights infringements, sometimes even more egregious ones.

Third, the distinction creates a problematic "catch-22" situation for claimants when there is no government target. They are faced with the choice of challenging the overall tapestry of government decisions that determines the level of emissions (as the claimants initially did in *La Rose/ Misdzi Yikh*), of targeting individual decisions (or small sets of decisions) (as the claimants did in *Youth Stop TMX* and *Dykstra*), of targeting inaction (as the claimants did in *Enjeu*), or of waiting for government to enact a target (as the claimants did in *Mathur*, when the government regressed its climate target by legislation, providing an anchor). If claimants opt to challenge inaction and/or the full swath of decisions that determine the level of emissions, the claims risk being ruled non-justiciable for venturing too far into the political arena (as was decided in *Enjeu*) or overly broad, as was the case initially in *La Rose/Misdzi Yikh*.[183] If the claimants opt to challenge only a single government decision (such as the authorization of a major fossil fuel development) or even a set of decisions (such as the authorizations of a series of GHG intensive industrial projects), these decisions risk not capturing a sufficient volume of emissions to meet the threshold of harm.

Holding cases where there is no legislated target to be non-justiciable forces claimants back to the drawing board to reframe their case in a piecemeal approach, challenging single laws or smaller sets of laws, or single major projects separately, possibly sequentially. This creates an injustice for those claimants given the existential threat posed by climate change and the urgency of the situation. The considerable time and resources

---

183　Recall that the Federal Court of Appeal is allowing the claimants to amend their pleadings.

required to mount a single challenge is already a significant obstacle for access to justice. Requiring claimants to mount multiple claims while running against the clock on the viability of the climate stability for the planet is unacceptable.

Fourth, and relatedly, denying systemic claims creates delays that risk exacerbating the infringement of claimants' rights. The window of opportunity to reduce GHG emissions to a level that avoids crossing dangerous thresholds is rapidly closing. The Canadian government has a perfect track record of failing to meet its climate targets.[184] It is impractical and unjust to require climate litigants to play "whack-a-mole" in defending their constitutional rights, when the justification for not allowing their claim to proceed to a hearing on the merits is one of form. Some courts have recognized the urgency of climate claims and expedited the hearing process to avoid delays. For instance, the Supreme Court of Ireland heard an appeal in a climate case directly from a lower court, on an expedited basis, leapfrogging over the traditional appeal route.[185] Similarly, the European Court of Human Rights fast-tracked the first three climate cases to reach it.[186]

Fifth, allowing claims to proceed only if a government legislates a target or plan may create a chilling effect on governments who may be incentivized to avoid setting climate targets to evade *Charter* challenges.

184　Auditor General of Canada, *Canadian Net-Zero Emissions Accountability Act – 2030 Emissions Reduction Plan*, (Reports of the Commissioner of the Environment and Sustainable Development to the Parliament of Canada), ISSN: 2561-1801 (Ottawa: Office of the Auditor General of Canada) online: oag-bvg.gc.ca [perma.cc/WS4T-XEGN].

185　The Irish Supreme Court justified its decision to hear the appeal directly due to the exceptional circumstances of climate change and the "degree of urgency in respect of the adoption of remedial environmental measures" (see *Friends of the Environment CLG v The Government of Ireland & The Attorney General*, [2020] IESCDET 13 at para 8 (SC Ireland)).

186　See *Verein KlimaSeniorinnen Schweiz and Others v Switzerland*, above note 25; *Carême v France*, No 7189/21, [2024], online: climatecasechart.com [perma.cc/4HV2-7Y3L]: *Duarte Agostinho and others v Portugal and 32 others* [GC], No 39371/20, [2024], online: climatecasechart.com ECLI:CE:ECHR:2024:0409DEC003937120. The European Court of Human Rights rejected a motion by the defendant governments to overturn the decision to fast-track the case (see *Duarte Agostinho and Others v Portugal and 32 Others* (preliminary objections), No 39371/20 (4 February 2021), online: climatecasechart.com [perma.cc/QLH5-RJKE]).

When the Ontario Court of Appeal held that Ontario's legislation represented a voluntary assumption of a positive statutory obligation, which opened the door to *Charter* scrutiny, it suggested that governments could shield themselves from *Charter* scrutiny by not enacting legislation to safeguard the climate (unless a positive obligation is found to exist). This form of conduct should not serve as an escape hatch through which governments can evade *Charter* scrutiny. When there is a clear, ascertainable standard against which emissions can be measured, the judiciary can (and should) assess the evidence and determine whether the standard is met, regardless of whether the government has set a target.

Sixth, focusing on form over substance creates a false distinction that places youth and Indigenous claimants in a passive, disempowered position. They must either wait for governments to enact a policy or legislated target or take specific legislative action before they can seek judgment from a court — based on a full evidentiary record — about whether these decisions infringe their *Charter* rights. This requirement compounds the powerlessness often experienced by the youth who have no right to vote and Indigenous claimants who face historic and ongoing discrimination.

Together, these arguments support the judiciary taking a broad approach to justiciability in the context of *Charter* allegations related to climate change.

4.2.5   Concluding Thoughts on Justiciability

The next few years will offer greater clarity on how the Canadian judiciary will interpret justiciability in the context of climate *Charter* cases. While it is impossible to generalize across individual cases, the preceding analysis suggests that courts are increasingly likely to find cases justiciable if they raise *Charter* challenges — at least under section 7 — as long as they are framed in a way that provides some legislative or policy anchor tied to government decision-making and do not target overly broad or diffuse conduct.

However, it is important that courts be flexible in identifying that legal component and not get misled by its form. Importantly, the approach to justiciability should answer the essential question of how a defendant government's decisions regarding the mitigation of GHG emissions impact *Charter* rights. Since the government's approach to GHG mitigation is necessarily determined either by a target or the suite of choices (including not to reduce emissions) that together dictate the level of emissions from the jurisdiction, a claimant will need to challenge one or the other. If the former is unavailable (because a target is not legislated or there is no target), then a reductionist approach to justiciability risks leaving claimants in a vacuum without the opportunity to have a court determine whether the government's choices infringe their rights.

Because it offers a single point of reference for courts to evaluate whether rights are infringed, the existence of a judicially manageable standard offers an answer to the concern about an issue being too political or not having a sufficiently circumscribed set of government decisions to evaluate. Either the government is or is not keeping its GHG emissions below the scientifically defined threshold required to avoid irreversible, catastrophic levels of harm. Precisely *how* the government is governing the level of emissions is a matter of form and should not dictate whether a matter is justiciable.

Larissa Parker convincingly argues that judges should resist approaching justiciability rigidly (something that "propels courts toward individual, narrow, and incremental thinking") and embrace a flexible approach that can handle the complexity of a wicked problem like climate change.[187] Indeed, as the Supreme Court of Canada has noted "justiciability depends to some degree on context, and the proper approach to determining justiciability must be flexible."[188] If courts take an overly cautious approach to justiciability for systemic climate *Charter* claims, they will miss the opportunity to develop the aspect of Canadian law dealing with environmental rights, which remains underdeveloped.[189] While there has been progress on section 7, this point is especially apt for section 15. The tendency of courts to strike public interest cases that claim environmental and climate

---

187 Parker, "Beyond Non-Justiciability," above note 101 at 89.
188 *Reference re Secession*, above note 104 at para 34.
189 Parker, "Beyond Non-Justiciability," above note 101 at 85.

rights prevents the living tree of Canadian constitutional law from evolving and growing as it is meant to do.[190]

4.3 THE SECTION 7 ANALYSIS

Section 7 of the *Charter* states that "[e]veryone has the right to life, liberty and security of the person and the right not to be deprived thereof except in accordance with the principles of fundamental justice."[191] To establish a violation, a challenger must first show that the impugned state conduct deprives them of life, liberty, or security of the person. In other words, does the impugned state conduct or law cause "a limitation or negative impact on, an infringement of, or an interference with those rights"?[192] Importantly, the claimant need not prove the deprivation but only show that there is a risk of such limitation, negative impact, or deprivation.[193] Second, the complainant must show that the interference or deprivation is not in accordance with the principles of fundamental justice.[194]

Breadth and scope of section 7 rights

The courts have given section 7 an expansive interpretation, holding that it confers both procedural and substantive rights.[195] The provision has been engaged in matters of criminal procedure but also in areas of great public policy importance, with the section having been successfully used in the context of laws that restricted abortion, prostitution, access

---

190 *Ibid.*

191 Above note 5 s 7.

192 *Canadian Council for Refugees v Canada (Citizenship and Immigration)*, 2023 SCC 17 at para 56 [*Refugees v Canada*], citing *Carter*, above note 10 at para 55.

193 *Refugees v Canada*, above note 192 at para 56. See also *Carter*, above note 10 at para 62; *R v Malmo-Levine; R v Caine*, 2003 SCC 74 at para 89 [*Malmo-Levine*]; *Suresh v Canada (Minister of Citizenship and Immigration*, 2002 SCC 1 at para 27 [*Suresh*].

194 Note that in Section 4.3.2 below, we discuss an alternative interpretation of section 7 which recognizes both a negative and positive right (see Section 4.3.2, below in this chapter).

195 *Reference Re Motor Vehicle Act (BC) S 94(2)*, [1985] 2 SCR 486 at 498–500 (SCC) [*Re Motor Vehicle Act*].

to private health care, physician-assisted dying, and more.[196] As such, section 7 has played an important role in "reflecting and safeguarding the public's evolving values."[197] However, despite many efforts, section 7 has never been successfully used in the environmental context. Most cases have failed for procedural or evidentiary reasons,[198] many of which we will return to in this section in the climate context.

Section 7 is broad in terms of who it protects and the scope of conduct from which individuals are shielded. It is enough that only one person's rights be violated. For example, in *Bedford* the Supreme Court of Canada held that a legislative "regime that is rational and non-arbitrary in almost all circumstances can nonetheless violate s. 7 if it is arbitrary, overbroad or grossly disproportionate for one individual."[199] Like other parts of the *Charter*, section 7 extends its scrutiny to actors outside government "if there is a sufficient causal connection between our government's participation and the deprivation ultimately effected."[200]

While the scope of the right to life and the right to liberty is fairly straightforward, security of the person has been subject to more judicial interpretation to define its contours. Courts have held that it protects against physical suffering or the threat of such, as well as serious and profound state-imposed psychological stress.[201] Most climate claims engage

---

196 See, e.g., *Chaoulli*, above note 110; *Morgentaler*, above note 10; *Carter*, above note 10; *PHS*, above note 10; *Bedford*, above note 115; *New Brunswick (Minister of Health and Community Service) v G(J)*, [1999] 3 SCR 46 (SCC) [*New Brunswick v G(J)*].

197 *La Rose* FCA, above note 7 at para 96, citing *Blencoe v British Columbia (Human Rights Commission)*, 2000 SCC 44 at para 188.

198 See David R Boyd, *The Right to a Healthy Environment: Revitalizing Canada's Constitution* (Vancouver: UBC Press, 2012); Lynda M Collins, "Safeguarding the Longue Durée: Environmental Rights in the Canadian Constitution" (2015) 71:20 SCLR at p 14 (footnote 71 cites a series of environmental claims under section 7 that have been dismissed).

199 *Refugees v Canada*, above note 192 at para 56, citing *Carter* above note 10 at para 76 and citing *Bedford*, above note 115 at para 123.

200 *Suresh*, above note 193 at para 54.

201 The security of the person was recently held to be engaged when someone experienced very cold conditions in detention (see *Refugees v Canada*, above note 192 at para 56, citing *Carter*, above note 10 at para 92).

the life and security of the person components of section 7, although there are situations where the liberty interest could be relevant.[202]

How section 7 applies in the context of climate change will necessarily be influenced by the unique features of climate change, including its global scope, the fact that harms are the result of the actions of multiple actors over time and across jurisdictions, the nature of the harms as existential, irreversible and pervasive on a human timescale, and the need for an urgent response that entails decarbonizing the economy.[203] Like justiciability, the section 7 analysis is also influenced by how a claim is framed, notably whether it is framed as a request for positive rights.

This section examines three issues that are central to climate claims under section 7 and considers how courts have interpreted them to date in the emerging climate *Charter* jurisprudence: causation, positive rights, and principles of fundamental justice. The discussion is by no means an exhaustive exploration of section 7 as applied to climate *Charter* claims, nor even exhaustive within each of these issues. I focus on the key points emerging from the jurisprudence to date.

4.3.1 Causation Under Section 7

Causation in the context of *Charter* claims is a different beast than it is in torts, where plaintiffs must prove, on a balance of probabilities, that "but for" the action of the defendants, the plaintiffs would not have sustained the injury. The causation standard in *Charter* cases is less onerous. In *Canada (Attorney General) v Bedford*,[204] a unanimous Supreme Court of Canada clarified that to engage section 7, there must be a "sufficient causal connection" between the state-caused effect and the prejudice suffered

---

202 Consider, for example, an Indigenous community that is no longer able to access or leave their community due to the melting of permafrost or ice that previously provided a reliable transportation route (see, e.g., Chalifour & Earle, above note 40 at 724–25).

203 See Chapter 1.

204 *Bedford*, above note 115.

by the claimant.[205] The onus rests on the claimant who must establish the connection on a balance of probabilities.

The issue of what constitutes the "state-caused effect" in climate cases is important for several reasons. Since climate change is a global phenomenon that results from the emission of GHGs from multiple sources over time, governments have defended themselves by claiming the prejudice suffered by claimants is not caused by their emissions but rather the broader problem of climate change. This *de minimis* defence has been squarely rejected by courts around the world, as well as Canadian courts (see Box 1). Identifying the "state-caused effect" also ties back to how the state conduct is framed, something discussed at length in the previous section on justiciability and to which I return in the context of positive rights below (Section 4.3.2).

The Supreme Court of Canada has clarified that the section 7 causation test is sensitive to the context of the particular case and does not require that the government conduct in question be the only or even the main cause of the harm suffered.[206] This nuance is especially helpful for climate litigants, since the GHG emissions from a given Canadian jurisdiction (federal or provincial/territorial) are not the only contributors to climate change. As long as the link between the state-caused effect and the prejudice is real, rather than speculative, the test can be met.[207] The *Bedford* approach to causation is therefore one that accounts for some of the unique features of climate change, including the global and collective nature of GHG emissions.[208] Justice Vermette of the Ontario Superior Court explicitly recognized this in *Mathur*, noting that the global nature and cumulative dimensions of emissions are key distinguishing features of climate change. She held that the "collective action problem" must be taken into consideration when assessing causation. Doing so led her to conclude that Ontario's severely deficient target contributes to an increased risk of death and risks to security of the person faced by the

---

205   *Ibid* at para 75.
206   *Ibid* at para 76.
207   *Ibid*.
208   *GGPPA References*, above note 11 at para 12.

appellants and others.[209] This approach is consistent with emerging jurisprudence addressing cumulative effects.[210]

The sufficient causal connection test in the Canadian climate cases

At the time of writing, the only case to have proceeded to a full hearing and have a decision rendered is *Mathur*. To defend the claim, Ontario argued that the province contributes only a small proportion of global GHG emissions and as such is not responsible for any resulting harms. Justice Vermette joined the chorus of other courts around the world and rejected Ontario's argument. She held that every ton of CO_2 emitted by Ontario "adds to global warming and leads to a quantifiable increase in global temperatures that is essentially irreversible on human timescales."[211] As such, Ontario's contribution to the problem was found by the Court to be real, measurable, and not speculative.[212] Even if the province's emissions are numerically small relative to the global scale, the court held that those emissions contribute to the applicants' increased risks of death and harm.

The Superior Court applied the *Bedford* causation standard[213] and found that Ontario's choice of an inadequate climate target (relative to

---

209 *Mathur* SC, above note 44 at paras 149–51.

210 As the British Columbia Superior Court recognized in the ground-breaking *Yahey* case, "[c]ausation in the context of a cumulative effects claim is something of a novel or currently developing issue at law" (see *Yahey v British Columbia*, 2021 BCSC 1287 at para 679). While *Yahey* was not a *Charter* case, Burke J's comment supports a flexible and context-specific analysis of causation in other cumulative effects contexts, including the climate cases. For commentary on cumulative effects in the context of the *Yahey* decision, see Martin Olszynski, "Counting Straws: Yahey v British Columbia and the Future of Cumulative Effects Management in Canada" (13 July 2023), online: ABlawg.ca [perma.cc/M7F3-YKXS].

211 *Mathur* SC, above note 44 at para 149.

212 See Section 4.3.1, below in this chapter.

213 In the *Miszhi Yikh* case at the Federal Court, the claimants proposed to use an attenuated version of the "material contribution" test sometimes used in negligence claims. The claimants may have been attempting to pre-empt challenges to the *Bedford* causation test due to the global nature of climate change. The Federal Court rejected the argument, pointing to the fact that the other emitters in a "material contribution" approach are outside of Canada and that there was no evidence provided that the other states are breaching their Paris commitments. The Federal Court of Appeal did not address this point (see *Misdzi Yikh*, above note 44 at paras 100–1).

the scientific benchmark) was sufficiently connected to an increased risk of death faced by the applicants and of harms and injuries affecting the applicants' security of the person.[214] Justice Vermette was clear that the fact that Ontario's target was not the only or even the dominant cause of the prejudice suffered by the applicants did not undermine the causal link, consistent with *Bedford*. She held that the context of climate change, namely its global, collective action nature, had to be taken into account in assessing causation.[215] The Court of Appeal confirmed the lower court's findings on causation, explicitly rejecting Ontario's argument that its actions did not cause or worsen climate change.[216]

The Federal Court of Appeal in *La Rose/Misdzi Yikh* addressed causation briefly when deciding whether the claim should proceed. Justice Rennie disagreed with the lower court's conclusion that a causal link could never be established,[217] holding that "[t]here is no reason to conclude that harms flowing from climate change and climate-related legislation are manifestly incapable of proof."[218] While the Federal Court of Appeal found the pleadings as such to be so broad that they lacked the focus needed for the Court in that case to analyze the *Charter* claims, it did not hold that causation could not be established.[219] Responding to the prospect that Canada might try to raise the *de minimis* defence, pointing to emissions from elsewhere to absolve itself of responsibility, Rennie J suggested there were answers to this concern.[220] First, returning to the sufficient connection test in *Bedford*, the existence of multiple causes does not constitute a barrier to constitutional challenges.[221] Second, Rennie J suggested this question would be best answered with the help of a full evidentiary record.[222] Having focused its analysis mainly on justiciability, the Quebec Court of Appeal in *Enjeu* did not address causation, nor the *de minimis* argument.

---

214   *Mathur* SC, above note 44 at para 147.
215   *Ibid* at para 149.
216   *Mathur* CA, above note 7 at para 33.
217   *La Rose* FCA, above note 7 at paras 91 and 113.
218   *Ibid* at para 114.
219   *Ibid* at para 133.
220   *Ibid* at para 134.
221   *Ibid*.
222   *Ibid*.

BOX 1: HOW THE GLOBAL NATURE OF GHG EMISSIONS IMPACTS
DOMESTIC RESPONSIBILITY – THE *DE MINIMIS* DEFENCE

Government defendants in climate lawsuits have often raised the "drop in the bucket" or "*de minimis*" argument, maintaining that they are not responsible for the impacts of climate change since their emissions contribute only a small proportion of the global emissions that influence the climate. Stated otherwise, they argue that climate change is a global problem caused by multiple factors and that the actions of one jurisdiction (especially if it is a small contributor on a global scale) are not going to make a dent in the otherwise massive global problem. As such, they argue there is no causal connection between emissions attributable to the jurisdiction and the harms suffered by claimants from climate change.

Courts around the world have squarely rejected this defence. The Dutch Supreme Court confirmed the District Court's finding in 2015 that "every emission of greenhouse gases leads to an increase in the concentration of greenhouse gases in the atmosphere" and thereby contributes to the problem, no matter how minor the emission.[223] The United States Supreme Court similarly rejected this argument in *Massachusetts v Environmental Protection Agency*, concluding that "[a] reduction in domestic emissions would slow the pace of global emissions increases, no matter what happens elsewhere."[224] The Montana Supreme Court similarly rejected the argument in 2024.[225]

The Supreme Court of Canada joined its counterparts around the world in unequivocally rejecting the *de minimis* argument in *GGPPA References*.[226] Chief Justice Wagner noted that accepting this line of reasoning must fail since it would apply equally to all individual sources of emissions everywhere.[227] The effect would be to let almost every jurisdiction — other than perhaps a handful of very large emitters — off the hook for contributing to climate change. For instance, Canada is among

---

223 *Urgenda* Supreme Court, above note 22 at para 4.6.
224 549 US 497 (2007) at 536.
225 *Held v Montana*, 2024 MT 312.
226 *GGPPA References*, above note 11 at paras 188–89.
227 *Ibid* at para 188.

the largest emitters globally, whether measured by historical, annual, or per capita emissions, yet we contribute under two percent of global emissions. Accepting a *de mimimis* defence would hamper the collective response needed to address this global problem, including the ability of countries to influence global negotiations to address climate change. As Wagner CJ in *GGPPA References* said in the context of Canadian emissions, "each province's emissions are clearly measurable and contribute to climate change."[228]

A second Canadian judicial decision has now effectively rejected the *de minimis* defence. In *Mathur*, the Ontario Court of Appeal endorsed the lower court's finding that Ontario's emissions, while relatively small compared to some large emitting jurisdictions, nonetheless make a "real, measurable, and not speculative" contribution to climate change.[229]

Unique temporal features of climate change and the section 7 analysis: present actions that create future harms

One of the unique features of climate change is its temporal scale.[230] While climate harms are already being experienced by claimants in these cases, as noted in Chapter 1, the climate destabilization experienced today is the result of past emissions. GHGs emitted today will exert their influence on the climate at some point in the future. In other words, actions today set into irreversible motion harms that will be experienced in the future.

Another aspect of the temporal scale is the length of time the harms will last. The scientific evidence is unequivocal in showing that most GHGs emitted today exert their influence for decades or longer.[231] Even if all

---

228 *Ibid*.

229 *Mathur* SC, above note 44 at paras 148–49.

230 For a detailed discussion of the temporal dimensions of climate change see Section 4.4.4.4, below in this chapter.

231 The long atmospheric life of gases like CO_2 and thermal inertia of oceans leads to this "warming commitment," as described by the IPCC (see Intergovernmental Panel on Climate Change, *Climate Change 2021: The Physical Science Basis*, Contribution of Working Group I to the Sixth Assessment Report of the IPCC, Valérie Masson-Delmotte et al, eds (New York: Cambridge University Press, 2021) at chs 4 and 7).

emissions were stopped immediately, the effect of increased concentrations of GHGs in the atmosphere will continue warming the ocean for centuries and the accumulating heat in oceans will persist for millennia.[232] This means that some of the prejudices claimed in climate lawsuits — including the most existential ones — may not occur until some point in the future. How does the fact that some of the harms from climate change will ensue in future impact the *Charter* analysis? How do the other temporal features of climate change (such as the delay between emissions and their effect) influence the analysis?

The extended temporal aspects of climate change are not, in and of themselves, a barrier to finding causation in Canadian *Charter* cases. Courts have held that claimants need only establish a real risk of harm. For instance, in *Khadr*, the Supreme Court of Canada clarified that *Charter* breaches in the past can impact rights in the present and future, and that this temporal quality does not preclude a remedy.[233] In *Chaoulli v Quebec (Attorney General)*, the Supreme Court of Canada held that provisions that had the effect of increasing the risk of health complications and death were contrary to section 7.[234] The majority in *Operation Dismantle* clarified that relief can still be available even when no harms have yet occurred, as long as the probability of future harm is not merely speculative.[235]

Courts in other jurisdictions have similarly found that future harms are a sufficient basis for present rights violations. The German Constitutional Court in *Neubauer*, for example, stated:

> [s]ince future impairments of fundamental rights could potentially be set into irreversible motion today, and given that lodging a constitutional complaint to address the ensuing restrictions on freedom might be futile by the time the impairments have arisen, the complainants already have standing to lodge a constitutional complaint at the present time.[236]

---

232   Expert evidence pertaining to this geological dynamic was cited by Staton J (see *Juliana*, above note 2 at 1176). See also IPCC, above note 231.

233   *Canada (Prime Minister) v Khadr*, 2010 SCC 3 at para 31 [*Khadr*]. See also *Canada (Justice) v Khadr*, 2008 SCC 28 at para 29.

234   2005 SCC 35 at para 124.

235   *Operation Dismantle*, above note 109 at 455–56.

236   Bundesverfassungsgericht [Federal Constitutional Court], Karlsruhe, 24 March 2021, ECLI:DE:BVerfG:2021:rs20210324.1bvr265618, 1 BvR 2656/18 at para 130 (Germany), [*Neubauer*].

Although these comments were made in the context of standing, the key point here are that the harms are set into irreversible motion today.

The Federal Court of Appeal in *La Rose/Misdzi Yikh* referenced *Bedford* and *PHS* as examples of cases based on government conduct that created or exacerbated a risk to life, liberty, or security or the person.[237] Although Rennie J did not address it explicitly, he seemed to understand the future orientation of risk in the climate cases, stating that "[j]ust as the prostitution provisions of the *Criminal Code* made the lives of the sex-workers more precarious, so too does the suite of federal laws challenged here."[238]

Instead of focusing on whether harms will ensue in some distant future the analysis should focus on whether the risks of harm are real and not speculative. As noted earlier, the *Bedford* causation standard requires risks to be real, but it does not require proving beyond doubt that the alleged harms <u>will occur</u>. Applicants must show (on a balance of probabilities) that the impugned state conduct contributes to an increased risk of death or harm impacting security of the person. In *Mathur*, Vermette J held that Ontario is contributing to this increased risk to human life, health and safety by not reducing GHGs in line with what the scientific consensus requires.[239] This finding was confirmed by the Ontario Court of Appeal.[240] The Federal Court in *La Rose* similarly agreed that the argument that Canada's role in GHG emissions that lead to harms under section 7 is not merely speculative.[241]

Identifying a benchmark upon which to assess harm or risk of harm

When determining whether section 7 rights have been violated, courts will benefit from having a benchmark or guardrail against which to make this assessment. There are several credible and quantifiable benchmarks against which to evaluate the potential for state-caused harms. The most obvious is the temperature threshold of 1.5 degrees Celsius to be achieved by 2030,

---

237 *La Rose* FCA, above note 7 at para 110.
238 *Ibid.*
239 *Mathur* SC, above note 44 at para 150.
240 *Mathur* CA, above note 7 at para 44.
241 *La Rose* FCTD, above note 44 at para 75.

as embodied in the Paris Agreement. This temperature threshold has been translated into a level of GHG emissions reductions required to avoid surpassing this benchmark, as reflected in the *Glasgow Climate Pact*: GHG emissions reductions of 45 percent below 2010 levels by 2030 and net-zero by 2050.[242]

In *Mathur*, Vermette J used the *Glasgow Climate Pact* as a benchmark against which to evaluate Ontario's target.[243] Justice Vermette found Ontario's target to fall severely short of this threshold.[244] To reduce emissions in accordance with the benchmark reflected in the *Glasgow Climate Pact*, Vermette J held that Ontario would need to reduce its emissions by 52 percent below 2005 levels by 2030.[245] The applicants had invited the Court to apply a carbon budget approach to determine Ontario's responsible share of emissions. Justice Vermette found the carbon budget issue to be non-justiciable, as she determined that the Court does not have the institutional capacity and legitimacy to determine Canada and Ontario's fair share of emissions, even under a number of difference scenarios presented.[246] However, since she relied upon the *Glasgow Climate Pact* threshold, she did not need to rely upon the carbon budget to assess the section 7 violation. In contrast, the Ontario Court of Appeal had no difficulty with the concept of the global carbon budget. Although it did not engage in an analysis of carbon budgets, the Court explained that the global carbon budget refers to the amount of carbon emissions that can be emitted before irreversibly locking in a harmful level of warming, which is currently defined as 1.5 degrees Celsius.[247]

Between the concept of a global carbon budget, scientific consensus on temperature thresholds, and agreement on what is required in terms of GHG emissions reductions to avoid crossing this threshold, courts are well equipped to evaluate the adequacy of a government's climate target.

---

242 *Mathur* SC, above note 45 at para 20. See *Glasgow Climate Pact*, above note 173 art 22.

243 While the Ontario Court of Appeal did not refer to the *Glasgow Pact*, Roberts J referenced the IPCC report upon which the *Glasgow* targets were established (see *Mathur* CA, above note 7 at para 23).

244 *Mathur* SC, above note 44 at paras 144 and 147.

245 *Ibid* at para 21.

246 *Ibid* at para 109.

247 *Mathur* CA, above note 7 at para 11.

Conclusion on causation

In *Mathur*, the only case in Canada that has benefited from a full hearing on the merits, the Ontario Superior Court concluded based on the evidence filed that "it is indisputable that, as a result of climate change, the Applicants and Ontarians in general are experiencing an increased risk of death and an increased risk to security of the person."[248] The Court further tied these harms to Ontario's insufficient target. Because Vermette J rejected the *de minimis* defence and took the global collective action nature of climate change into account, causation did not present a significant hurdle to the section 7 claim. In light of the Supreme Court of Canada's acceptance of the causal link between emissions and harm (albeit in a different context) and the global trend toward holding governments accountable to have targets that are aligned with what science shows to be required, causation should not be a significant hurdle for claimants under section 7. Section 15 claims are a different story, as explored in Section 4.4.

4.3.2   The Positive/Negative Rights Dichotomy

The question about the extent to which the *Charter* protects so-called positive rights is one that continues to spark debate and controversy in Canada. It is not nearly so controversial in other jurisdictions that routinely recognize positive human rights obligations on government.[249]

At first glance, section 7 of the *Charter* appears to explicitly confer both positive and negative rights, since it states that "[e]veryone has the right to life, liberty and security of the person *and* the right not to be deprived thereof except in accordance with the principles of fundamental justice" (emphasis added).[250] The use of the word "and" suggests that

---

248   *Mathur* SC, above note 44 at para 120.

249   See, e.g., Council of Europe, Directorate General of Human Rights, *Positive Obligations Under the European Convention on Human Rights: A Guide to the Implementation of the Eureopean Convention on Human Rights*, by Jean-François Akandji-Kombe, Human rights handbook, No. 7 (2007), online: rm.coe.int [perma.cc/SAD3-KLWF]. See also footnote 269 and accompanying text.

250   Above note 5 s 7.

there are two rights. However, the potential of two rights is less obvious in French, with the provision stating that "[c]hacun a droit à la vie, à la liberté et à la sécurité de sa personne; il ne peut être porté atteinte à ce droit qu'en conformité avec les principes de justice fondamentale."

Justice Arbour characterized section 7 as conferring negative and positive rights, based on this "two right" interpretation of section 7, in her dissent in *Gosselin v Quebec (Attorney General)*.[251] In that case, she aptly characterized the concept of positive rights as something that describes rights of "performance" that may be violable by a state's inaction.[252] The *Charter* includes many rights that impose some form of performance obligation on the part of the state. The right to vote (s 3), to trial within a reasonable time (s 11(b)), to trial by jury in certain circumstances (s 11(f)), to an interpreter in penal proceedings (s 14), and even the presumption of innocence (s 11(d)) are all rights that require a certain level of active, positive engagement by the government.[253] The minority language education rights enshrined in section 23 require execution by the state. In *Dunmore v Ontario (Attorney General)*, the Supreme Ccourt of Canada found there to be a positive obligation on the government in certain situations to ensure its labour legislation is appropriately inclusive, conferring a positive dimension to section 2(d)'s right to associate.[254]

Justice Arbour relied upon the wording of section 7 along with a purposive interpretation to arrive at her conclusion.[255] In the same decision, Justice McLachlin for the majority commented that a positive obligation to sustain section 7 rights could one day be made out in special circumstances, though not in that case.[256] In the more than twenty years since

---

251 *Gosselin v Quebec (AG)*, 2002 SCC 84 at paras 324–27 [*Gosselin*].

252 *Ibid* at para 319.

253 *Ibid* at para 320.

254 *Dunmore v Ontario (AG)*, 2001 SCC 94 [*Dunmore*], cited in *Gosselin*, above note 253 at para 320.

255 *Gosselin*, above note 251 at paras 324–27. See, e.g., *Re Motor Vehicle* Act, above note 197 at 499; *Irwin Toy Ltd v Quebec (AG)*, [1989] 1 SCR 927 at 95.

256 *Gosselin*, above note 251 at para 83. In fact, only one judge, Bastarache J, was dismissive of the idea that section 7 could include positive rights in *Gosselin*. See Martha

Gosselin, the judiciary has left open the possibility that section 7 may one day be interpreted to confer positive obligations, but that day has yet to come.[257]

Climate change has been a close contender for those special circumstances for some time now, as elaborated below. In *Kreishan v Canada,* Rennie J stated that he was cognizant that "section 7 is not frozen in time, nor is its content exhaustively defined, and that it may, some day, evolve to encompass positive obligations — possibly in the domain of social, economic, health or climate rights."[258] The debate has become central in Canadian climate rights cases, especially when the courts characterize the cases as ones complaining of inaction or inadequate government action.

This section begins by examining the problematic bifurcation of section 7 rights into negative and positive dimensions. Next, drawing upon the jurisprudence so far, it explores what the climate decisions have said about whether the claims are ones for positive rights and, if so, whether such a claim should be allowed and how it would be analyzed.

An unhelpful distinction

The bifurcation of rights into "negative" versus "positive" categories may appear straightforward. Negative rights imply that a government must refrain from doing certain things, whereas a positive right is said to impose some kind of obligation or duty on the government. As such, the right "not to be deprived of life, liberty and security of the person" other than in accordance with the principles of fundamental justice is said to confer negative rights, whereas the right to vote can be characterized as a positive right. A closer look at the bifurcation, however,

---

Jackman, "One Step Forward and Two Steps Back: Poverty, the *Charter* and the Legacy of *Gosselin*" (2019) 39 NJCL 85 at 100 [Jackman, "One Step"].

257　*Ibid* at para 82.

258　*Kreishan,* above note 88 at para 139, leave to appeal to SCC refused, 2020 CanLII 17609.

reveals that the distinction can be illusory and ultimately unhelpful.[259] As Wilson J pointed out in *Operation Dismantle*, "[a]ction by the state or, conversely, inaction by the state will frequently have the effect of decreasing or increasing the risk to the lives or security of its citizens."[260] What should matter is the link between government conduct and risk, not the form of that conduct.

The emphasis on the positive/negative dichotomy presents several problems for climate *Charter* rights. One of the concerns about interpreting section 7 rights as having "positive" dimensions is the possibility that constitutionally entrenched positive rights could erode the separation of powers by engaging courts in matters that involve the allocation of resources, bastions of the legislature.[261] While this is a fair concern, there are ways to safeguard against this possibility (such as having courts identify what is required to respect rights — such as establishing a science-based climate target — and leaving it to the other branches of government to determine how best to fulfill those rights). Canadian courts consistently interpret rights in a way that requires resource allocations by the state, whether in the administration of the criminal justice system, the protection of labour rights, or social policy. In *Canadian Doctors for Refugee Care v Canada (AG)*, Mactavish J noted that some section 7 claims may appropriately require government to spend money in a particular way.[262] Other jurisdictions that recognize positive obligations to protect human rights have been able to address these concerns.[263]

---

259 Martha Jackman & Bruce Porter, "Introduction: Advancing Social Rights in Canada" in Martha Jackman & Bruce Porter, eds, *Advancing Social Rights in Canada* (Toronto: Irwin Law, 2014) 1 at 14; See also Stephen Holmes & Cass R Sunstein, *The Cost of Rights: Why Liberty Depends on Taxes* (New York, W.W. Norton & Company, 1999) at 36–40 (questioning whether the distinction between positive and negative rights "helps illuminate reality").

260 *Operation Dismantle*, above note 109 at 488.

261 Colin Feasby, David Devlieger & Matthew Huys, "Climate Change and the Right to a Healthy Environment in the Canadian Constitution" (2020) 58:2 Alta L Rev 213 at 239.

262 *La Rose* FCA, above note 7 at para 104, citing *Canadian Doctors For Refugee Care v Canada (AG)*, 2014 FC 651 at para 520; see also *La Rose* FCA, above note 7 at para 104, citing *New Brunswick v G(J)*, above note 196.

263 See, e.g., Akandji-Kombe, above note 249.

Six of one, half dozen of the other — distinction without difference[264]

Critics of the binary approach point out that it is hard to conceive of a right that exists without some form of corresponding obligation to do or not do something.[265] Contrasting freedoms with rights, Chief Justice Dickson acknowledged that rights entail some form of corresponding positive obligation: "[r]ights are said to have a corresponding duty or obligation on another party to ensure the protection of the right in question whereas 'freedoms' are said to involve simply an absence of interference or constraint."[266] His caution reminds us that the distinction may not be only misleading, but could lead to miscarriages of justice. This is an important point, especially in the context of climate cases.

As noted earlier, the *Charter* includes many examples of rights that create obligations on the government, which could thus be characterized as "positive." Consider, for example, the right to vote and stand for election,[267] the right to be educated in the official language of choice,[268] the right to an interpreter in the criminal context,[269] the right to be tried

---

264 Justice Rennie used these words in rejecting the Dini Ze's argument that section 91 creates an obligation on Parliament (see *La Rose* FCA, above note 7 at para 69).

265 For scholarship addressing various aspects of positive rights, see Martha Jackman, "What's Wrong With Social and Economic Rights?" 2000 11 NJCL 235 [Jackman, "Social and Economic Rights"]; Martha Jackman, "The Protection of Welfare Rights under the Charter" (1988) 20:2 Ottawa L Rev 257 [Jackman, "Welfare Rights"]; Gwen Brodsky & Shelagh Day, "Beyond the Social and Economic Rights Debate: Substantive Equality Speaks to Poverty" (2002) 14 CJWL 185; Margot Young, "Section 7 and the Politics of Social Justice" (2005) 38:2 UBC L Rev 539 [Young, "Section 7"]; Emmett Macfarlane, "The Dilemma of Positive Rights: Access to Health Care and the Canadian Charter of Rights and Freedoms" (2014) 48:3 J Can Studies 49; Emmett Macfarlane, "Dialogue, Remedies, and Positive Rights: Carter v Canada as a Microcosm for Past and Future Issues under the Charter of Rights and Freedoms" (2018) 49:1 Ottawa L Rev 107.

266 In *Reference Re Public Service Employee Relations Act (Alta)*, for instance, Dickson CJ suggested that freedom of expression could require legislation preventing the monopolization of press ownership. He went on to offer an expansive interpretation of freedoms as well, noting that a narrow conceptualization could miss "situations where the absence of government intervention may in effect substantially impede the enjoyment of fundamental freedoms" (see [1987] 1 SCR 313 at para 77 [*Alberta Reference*]).

267 *Charter*, above note 5, s 3.

268 *Ibid*, s 23.

269 *Ibid*, s 14.

within a reasonable time,[270] the right to counsel, and the right to information in criminal proceedings.[271] All of these rights could fairly be construed as conferring positive rights.

The Supreme Court of Canada has also recognized that even rights framed in negative terms may require state action, thus conferring on them a "positive" dimension if we retain the dualistic terminology. For instance, in *Baier v Alberta*,[272] the Supreme Court of Canada explained how freedom of expression — a freedom traditionally described as ensuring the state does not interfere with one's expressive activities (a "negative" right) — could engage positive obligations. The *Baier* case involved a complaint that a law excluding school employees from accessing a platform for trustee candidacy infringed freedom of expression. Because the appellants were seeking access to the statutory platform, the Court held the case was a positive rights claim and that they needed to decide whether the claim qualified for an exception to the general rule that section 2(b) protects only against government interference.[273] Justice Bastarache defined the criteria specific to section 2(b) and concluded the conditions were not met, since the appellants were not deprived of their ability to express themselves but only deprived of access to a particular platform of expression.[274] In *Toronto (City) v Ontario (AG)*,[275] the Supreme Court of Canada affirmed the distinction between the tests for positive and negative freedoms in section 2(b): a positive claim for freedom of expression requires establishing substantial interference, whereas

---

270 *Ibid*, s 11(b).

271 *Ibid*, ss 10(a) and 11(a)–(b).

272 2007 SCC 31 [*Baier*].

273 *Ibid*.

274 *Ibid* at para 30. The court held that section 2(b) could confer positive obligations when: (1) the activity for which protection is claimed is a form of expression; (2) the claimant claims a positive entitlement to government action (as opposed to the right to be free from government interference); and (3) the three *Dunmore* criteria are met. The Dunmore criteria are: (1) the claim is grounded in a fundamental *Charter* freedom (as opposed to access to a particular statutory regime, for instance); (2) the claimant has demonstrated that exclusion from a statutory regime is a substantial interference or has the purpose of infringing freedom of expression; and (3) the government is responsible for the inability to exercise the fundamental freedom (see *Dunmore*, above note 254 at paras 24–26).

275 2021 SCC 34 [*City of Toronto*].

a negative claim requires consideration of whether the purpose or effect of government action restricts freedom of expression.[276]

The Supreme Court of Canada revisited the issue in *Société des casinos du Quebec inc v Association des cadres de la Société des casinos du Quebec* in the context of section 2(d).[277] Writing for the majority, Jamal J clarified that there is not a separate test for claims seeking positive intervention from the state and another for claims seeking negative protection from state interference for freedom of association. While the *Dunmore* factors may be relevant to the analysis, they do not create a separate test.[278] As such, the jurisprudence is mixed with respect to interpreting provisions traditionally deemed "negative" to have "positive" dimensions, and determining whether a claim bearing the hallmark of positive rights warrants a separate analysis.

The Supreme Court of Canada has also cautioned against simplistic dualisms, such as positive/negative and action/inaction, that risk masking rights violations. In *Vriend*, the majority characterized the attempt to create an artificial distinction between action and inaction as "very problematic."[279] Although this criticism was made in the context of section 32, which relates to the applicability of the *Charter*, the Court was clear in noting that a decision by a legislature not to act is relevant to *Charter* analysis, including for section 1 and when determining the appropriate remedy. A decision not to act in *Vriend* did not render such a choice outside the scope of *Charter* scrutiny. Rather, the decision not to act (specifically not to prohibit discrimination on the grounds of sexual orientation) was the basis of the violation. This substance-driven approach is preferable to one based on form.

The bifurcation of rights into positive and negative, with protections only for the former, has been especially criticized in the context of efforts to protect socio-economic rights under section 7.[280] Restricting section 7 to protect against interference with civil liberties leaves a gap

---

276 *Ibid* at paras 16–26.

277 2024 SCC 13 [*Société des casinos*].

278 *Ibid* at paras 33–37.

279 *Vriend v Alberta*, [1998] 1 SCR 493 at para 53 [*Vriend*].

280 See, e.g., Jackman, "Welfare Rights," above note 267; Gwen Brodsky & Shelagh Day, "Beyond the Social and Economic Rights Debate: Substantive Equality Speaks to Poverty" (2002) 14 CJWL 185; Young, "Section 7," above note 265.

for socio-economic rights, since such rights often require state action to protect them.[281] Professor Martha Jackman notes that the positive/negative distinction has long been abandoned in international human rights law and increasingly in other constitutional democracies.[282] However, efforts in Canada to interpret section 7 to create an obligation on government to provide basic levels of socio-economic rights have, thus far, failed.

For example, as mentioned earlier, the Ontario Superior Court rejected efforts to recognize a duty on government to address homelessness in the *Tanudjaja* case.[283] The Ontario Court of Appeal upheld the decision to strike the claim, holding that there was no sufficient legal basis upon which to find the case justiciable.[284] The Court deliberately chose to leave open the possibility that a positive obligation may be imposed on government to remedy a *Charter* violation, opting instead to rule on the lack of a legal anchor.[285] In another case, when considering whether changes to the Ontario school curriculum violated section 7, Lederer J for the Ontario Divisional Court considered whether it should impose a positive duty on the government. The Court concluded that no such duty was required. The applicants had cited *PHS* as a precedent, since the effect of that case was to provide a safe drug injection site for the applicants, a form of positive duty. However, the Court declined to qualify *PHS* as a "positive rights" case, characterizing the lack of availability of safe injection sites as an impairment of the claimant's ability to meet their health needs. While the remedy was an order to extend an exemption to allow

---

281 See, e.g., Martha Jackman, "Charter Remedies for Socio-Economic Rights Violations: Sleeping under a Box?" in Janice Payne, Kent Roach & Robert Sharpe, eds, *Taking Remedies Seriously* (Montreal: Canadian Institute for the Administration of Justice, 2009) 279 at 281–85 [Jackman, "Sleeping Under a Box"]. See also Jackman, "Social and Economic Rights," above note 267 at 242–43; Young, "Section 7," above note 267; Alison M Latimer, "A Positive Future for Section 7? Children and Charter Change" (2014) 67:14 SCLR 537.

282 Jackman, "Sleeping Under a Box" above note 281 at 281–85.

283 *Tanudjaja v Canada (AG)*, 2013 ONSC 5410 [*Tanudjaja* SC].

284 *Tanudjaja* CA, above note 114 at para 27.

285 *Tanudjaja* CA, above note 114 at para 37.

the continued operation of a safe injection site, the Court characterized this as a response to a rights infringement.[286]

In *Gosselin*, the Supreme Court of Canada considered the constitutionality of a provision that limited the level of welfare payments to recipients under age thirty unless they participated in workfare and learnfare programs. Holding that the claimants were not deprived of their section 7 rights, the Court also stated that section 7 did not place a positive obligation on the state to guarantee adequate living standards. The Court took pains, however, to note that the meaning of section 7 should be allowed to develop incrementally and that it was possible that the provision could someday be interpreted to confer positive obligations on the state in appropriate circumstances.[287] The "special circumstances" comment has been often repeated and has come to be especially relevant for the climate cases.

The special circumstances unlocking positive obligations under section 7

Justice McLachlin's comment in her majority reasons in *Gosselin* that section 7 could one day be interpreted to confer positive obligation on government in the right "special circumstances"[288] has generated both frustration and hope. It has led to subsequent efforts to have positive rights recognized in a variety of circumstances, albeit not yet in a successful claim.

Some cases have come close. In *Schulte v Alberta*, for instance, the Court dismissed a claim alleging inadequate workers compensation benefits but noted that interference with economic rights resulting in significant stress and anxiety could engage section 7.[289] As noted above, Rennie J noted that section 7 may someway evolve to include positive obligations and offered climate rights as an example.[290]

---

286 *ETFO v Her Majesty the Queen*, 2019 ONSC 1308 at para 146.
287 *Gosselin*, above note 251 at para 209.
288 *Ibid*.
289 *Schulte v Alberta (Appeals Commission for Alberta Workers' Compensation)*, 2015 ABQB 17.
290 *Kreishan*, above note 88 at para 139.

Although the applicants in *Mathur* took pains to frame their case as one challenging actions by the Ontario government that lead to emissions (i.e., a negative rights framing), Vermette J of the Superior Court characterized the claim as one engaging positive rights since the heart of the claim was that the target is insufficient. She went on to break new judicial ground in holding that the applicants made "a compelling case that climate change and the existential threat that it poses to human life and security of the person present special circumstances that could justify the imposition of positive obligations under section 7."[291] However, she then held the current section 7 test is not adapted to an analysis of positive obligations and chose not to attempt to modify the test. Instead, Vermette J applied the standard analysis and found the applicants had not shown that that any deprivations of their rights were contrary to the principles of fundamental justice.[292]

The Ontario Court of Appeal overturned the decision on this very point, holding that the application judge erred in characterizing the case as a positive rights case.[293] The Court drew a distinction between seeking to impose new obligations on the government to combat climate change (a positive rights claim) and evaluating the constitutionality of the statutory obligation voluntarily assumed by the government in the form of the *Cap and Trade Cancellation Act* (a negative rights claim).[294] The Court said that the appellants were not challenging the inadequacy of the target or Ontario's inaction but instead argued that the target commits Ontario to levels of emissions that violate their rights.[295] The distinction is subtle and one of form. The reasons suggest that courts continue to favour a negative rights framing and will lean toward such an interpretation when the

---

291 The Ontario Superior Court held in April 2023 held that the applicants made "a compelling case that climate change and the existential threat that it poses to human life and security of the person present special circumstances that could justify the imposition of positive obligations under section 7" (see *Mathur* SC, above note 44 at para 138).

292 *Ibid* at para 171.

293 *Mathur* CA, above note 7 at para 5.

294 *Ibid*.

295 *Ibid* at para 41.

circumstances allow. Ontario has applied for leave to appeal the decision to the Supreme Court of Canada but was denied, so it will return to the Superior Court for a re-hearing.[296]

The Ontario Court of Appeal's decision leaves ambiguous how a case seeking to impose positive obligations in the context of climate change could be made if there was no legislated target. Uncertainty also remains where there is a non-legislated target (e.g., a nationally determined contribution under the *Paris Agreement* or a target announced by government but not enacted into legislation). Would a soft target be a candidate for a positive rights framing where the case focused on the need for an overall jurisdiction-wide approach? Requiring claimants to frame their claim as a negative rights case targeting only whatever government action exists could create the same kind of catch-22 situation described earlier in the justiciability discussion, since positive claims remain as yet uncertain and, therefore, creating a possible vacuum for claims of failure to act or inaction.

In *La Rose/Misdzi Yikh*, where the claim is a hybrid of positive and negative rights, the Federal Court of Appeal addressed the positive rights issue in some depth even though it was deciding whether to overturn the motion to strike. Justice Rennie reiterated the point he made in *Kreishan* that section 7 remains open to positive rights in the right circumstances but provided numerous examples where courts have rejected efforts by claimants to have their case be the first one to hold that section 7 confers positive rights.[297] He nonetheless maintained that the door to positive rights under section 7 is not closed, pointing to the role of section 7 in protecting the public's evolving values. Justice Rennie surveyed arguments for and against the recognition of positive rights under section 7, then noted that there is agreement both in scholarship and judicial decisions that the line between positive and negative rights can be difficult to draw.[298]

Justice Rennie held that the motions judges had erred in striking the *La Rose* and *Misdzi Yikh* claims on the basis they were positive rights claims

---

296 *Mathur* CA, leave to appeal, above note 7.
297 *La Rose* FCA, above note 7 at paras 93–95.
298 *Ibid* at paras 98–104.

(even though Manson J did not, in fact, strike them for this reason).[299] Justice Rennie concluded that there were sufficient deprivations claimed tied to specific state action to construe the cases as negative rights claims. For instance, Rennie J pointed to permissive licensing of GHG emitting projects, deficient legislative standards, and the climate target set under the Paris Agreement.[300] Importantly, however, Rennie J also held that the question of whether section 7 may confer positive obligations in the case before it was not struck. He referenced the grave and widespread impacts of climate change and held that "[i]f these do not constitute special circumstances, it is hard to conceive that any such circumstances could ever exist."[301] However, he noted this would need to be determined by the trial judge based on a full evidentiary record.

Where does this leave us? It has been over twenty years that the Supreme Court of Canada left open the future possibility of recognizing positive rights under section 7. Judges have held more than once that climate change is the most likely candidate to present such special circumstances.[302] Yet, no court has — to date — held that the government has a duty to enact and respect a scientifically-driven GHG emissions reduction target in the context of the existential climate threat. Avoiding a positive rights framing in even the climate cases has the effect of rendering McLachlin J's words about special circumstances rather meaningless.

Although Vermette J came close, the Ontario Court of Appeal shut down the possibility, and the Supreme Court of Canada denied leave to appeal, signalling its agreement with the negative rights framing. The Federal Court of Appeal has left the possibility open in the *La Rose/Misdzi Yikh* case. The question of whether positive rights could one day be made out in the climate context remains unanswered. In the wake of this uncertainty, it's worth asking whether the extreme caution about positive rights

---

299 *Ibid* at para 105. See, e.g., *La Rose* FCTD, above note 44 at paras 65, 67, and 72.

300 *La Rose* FCA, above note 7 at paras 105–6. Later, Rennie J mentions legislation and Orders in Council permitting GHG emissions (see *La Rose* FCA, above note 7 at para 110).

301 *Ibid* at para 116.

302 *Kreishan*, above note 88 at para 139, leave to appeal to SCC refused, 2020 CanLII 17609.

is warranted. It's true that a positive rights framing would take Canadian courts out of their comfort zone and may require them to adapt some elements of the analysis (e.g., on principles of fundamental justice). It could, if applied broadly and carelessly, insert the judiciary in matters best left to the legislature. However, when proper guardrails exist to ensure a finding of positive obligation does not create unfettered judicial authority to intrude on the separation of powers, the caution may be unnecessary. Those guardrails exist in the climate context since there is robust science about required emissions reductions and metrics to measure progress toward those reductions. There are no guardrails on the climate.

BOX 2 - JUDICIAL RECHARACTERIZATIONS OF CLAIMS

Given the difficulty of recognizing positive rights, claimants in climate lawsuits have generally taken pains to frame their cases as negative rights claims. The *Enjeu* claimants, for instance, flagged a series of government targets and referenced the *Greenhouse Gas Pollution Pricing Act*. The *La Rose* plaintiffs initially framed their case in a hybrid format, including both a positive and negative claim: the negative rights framing was anchored around the government's climate target and a number of significant government decisions that influence the level of emissions (such as supporting fossil fuel intensive industrial developments).[303] After amendment, the *La Rose* plaintiffs now frame their negative rights claim around maintaining (i.e., through the *CNZEAA* and other state conduct) a GHG emissions trajectory that violates *Charter* rights and their positive rights claim around the failure to take action to maintain an emissions trajectory capable of safeguarding the right to life, liberty, and security of the person.[304]

In *Misdzi Yikh*, the claimants pointed to Canada's Nationally Determined Contribution (the federal government's GHG reduction goal under the Paris Agreement) and related laws and policies, such as the

---

303 *La Rose v Canada*, 2020 FC 1008 (Amended Statement of Claim, May 31, 2024) [*La Rose*, Amended SoC].

304 *Ibid.*

Pan-Canadian Framework on Clean Growth and Climate Change, and the assessments of several very large industrial developments with significant GHG emissions.[305] The claimants in *Mathur* pointed to the legislated requirement to establish a target and plan as the relevant state conduct.

However, some of the decisions recharacterized the claims into ones seeking positive rights.[306] The Quebec Court of Appeal in *Enjeu*, for instance, stated that the "[a]ppellants essentially ask the courts to find that the government has failed to act and to require it to legislate."[307] In *Misdzi Yikh*, McVeigh J characterized the claim as one about the alleged failure to enact adequate laws, and rejected the claimants' efforts to characterize their section 91 argument as a negative rights claim.[308] Justice McVeigh held that the positive obligations sought by the claimants were too vague to justify the court considering whether there was an appropriate case to impose a duty on the government (though it did not categorically close the door on that possibility).

The most overt recharacterization of a negative rights framing into a positive one was by the Ontario Superior Court in *Mathur*. Although the applicants challenged legislation as the source of the alleged rights violation, Vermette J stated that in her view, "the Application is seeking to place a freestanding positive obligation on the state to ensure that each person enjoys life and security of the person, in the absence of a prior state interference with the Applicants' right to life or security of the person."[309] Justice Vermette was influenced by the fact that the legislation at issue in fact repealed prior, stronger climate legislation

---

305 *Misdzi Yikh* SoC above note 68 at paras 38–71.

306 This also happened in *Gosselin*, where the claimants, contesting the reduction of their social benefits allocation, had initially framed their case as a negative rights violation. See Jackman, "One Step," above note 256 at 106–7.

307 *Enjeu* CA, above note 6 at para 32.

308 *Misdzi Yikh*, above note 44 at paras 44 and 58.

309 *Mathur* SC, above note 44 at para 132. See also *ibid* ("I disagree with the Applicants that this is not a positive rights case because Ontario's participation in creating the underlying harm, and its creation of the Target and the Plan pursuant to the CTCA triggers an obligation to ensure the resulting scheme is constitutionally compliant" at para 134).

(ratcheting down the province's target).[310] She relied upon the Supreme Court of Canada's decision in *Toronto (City) v Ontario (AG)*[311] that a "claim seeking to restore the status quo or an earlier statutory platform revealed 'a straightforward positive claim.'"[312] She then broke new judicial ground holding that the applicants made a "compelling case that climate change and the existential threat that it poses to human life and security of the person present special circumstances that could justify the imposition of positive obligations under section 7 of the *Charter*."[313] However, she subsequently rejected the claim for not violating the principles of fundamental justice. The Ontario Court of Appeal allowed the appeal of her decision on the basis that it was mischaracterized as a positive rights case, agreeing with the applicants' initial framing of the case as one of negative rights.

Justice Manson in *La Rose* was the only judge in the climate *Charter* cases to not recharacterize the claim, agreeing with the plaintiffs that the case disclosed both a negative and positive rights framing. He was also careful not to close the door on the positive rights argument.[314]

Positive rights in decisions outside Canada

The positive/negative dichotomy is more of an issue in Canadian courts than it is elsewhere, including in the context of climate rights cases.[315] Part of the reason for this is that in some jurisdictions, there exist legal instruments explicitly conferring positive obligations on governments. However, there are also examples of courts interpreting these obligations implicitly.

---

310 In *Misdzi Yikh*, the court noted that the issue is polycentric and international in nature and that it spans government departments and involves issues of economics and foreign policy, among other things, which rendered the claim non-justiciable and also one that required the court to tell the legislature what to do (see *Misdzi Yikh*, above note 44 at paras 56–57).

311 *City of Toronto*, above note 275 at para 30.

312 *Mathur SC*, above note 44 at para 136.

313 *Mathur SC*, above note 44 at para 138.

314 See *La Rose* FCTD, above note 44 at paras 67–72.

315 See Jackman, "Sleeping under a Box," above note 281.

For example, in the same year that the *Urgenda* trial-level decision was released, the Lahore High Court held that the Pakistani government's failure to implement a climate change framework offended the fundamental rights of its citizens.[316] Finding a positive obligation to exist on government to safeguard its citizens from the devastating impacts of climate change, the Court took a proactive approach to ensure the framework would be implemented by the government. First, it ordered the government to nominate a climate change focal person to ensure the framework's implementation and create a climate change commission to assist the court in monitoring progress on implementation. The Court dissolved the commission a few years later after it reported back to the Court that the government was on track to implement the rest of the framework policy. However, it ordered the creation of a Standing Committee to continue monitoring progress in implementing GHG reductions. While this level of judicial involvement is atypical of Canadian courts, the decision illustrates how seriously the Pakistani court took the violation of human rights, holding the government accountable for a positive duty to protect its citizens and providing additional accountability for citizens by imposing structured remedies with ongoing judicial oversight.

In the Colombian Supreme Court's 2018 ruling in *Future Generations v Ministry of the Environment and Others*,[317] the Court ruled in favour of twenty-five youth plaintiffs who claimed their rights to a healthy environment, life, health, food, and water had been violated by the government's failure to fulfill a target for zero-net deforestation in the Colombian Amazon by 2020. Like the *Leghari* case, the Colombian Supreme Court went beyond declaratory relief and ordered the government to formulate and implement action plans to address deforestation in the Amazon. In its reasons, the Court noted that the environmental rights of future generations create legal obligations on present generations to protect natural resources. The Court also held that the fundamental right of human dignity is substantially linked to and determined by the environment and

---

316 *Leghari v Pakistan*, above note 31.
317 *Future Generations*, above note 35 at 13–14.

the ecosystem — a finding that could be relevant in the context of a Canadian section 15(1) analysis.

There are several other examples emerging out of courts and tribunals ordering governments to act, with varying degrees of specificity regarding the actions to be implemented.[318]

Conclusion on positive rights

The climate *Charter* cases illustrate the problems with the trying to shoehorn claims into the positive or negative rights dualism. Instead of focusing on this dichotomy, courts should conduct a purposive, meaningful analysis of each *Charter* claim in its specific context. To do otherwise risks truncating *Charter* protections unnecessarily. Some rights and freedoms may require a positive dimension if they are to avoid being meaningless or lead to absurd results (such as the demise of safe living conditions for humanity). The *Charter* is a living tree that is meant to evolve with modern society and all its complexities. As Dustin Klaudt stated, referring to trees' function as carbon sinks, "[i]t appears prophetic that the *Charter*, rooted with the 'living tree' doctrine, should be interpreted to provide curative mechanisms to fight the global warming problem."[319] Faced with the existential and urgent threat of climate change, courts should move past semantics and focus on what is at the heart of these claims. The Supreme Court of Canada may have offered a hint of its views on environmental rights in the recent *Reference on the Impact Assessment Act* when the majority stated in its opening paragraph that: "[t]he Canadian judiciary, in tandem with the other branches of government, has an important role to play in protecting the 'right to a safe environment'" and the minority stated that "all levels of government bear an 'all-important duty' to use their powers to protect the environment."[320]

---

318   See, e.g., *PSB v Brazil (on Climate Fund)*, 2022 ADPF-708, online: [perma.cc/XYX7-WP4C]; *Klimaseniorinnen v Switzerland,* above note 25.

319   Dustin W Klaudt, "Can Canada's 'Living Tree' Constitution and Lessons from Foreign Climate Litigation Seed Climate Justice and Remedy Climate Change?" (2018) 31:3 J Envtl L & Prac 185.

320   *IAA Reference*, above note 1 at paras 1 and 219.

4.3.3 Principles of Fundamental Justice

Even if a court finds a violation of the right to life, liberty or security of the person, that violation may be acceptable if it is in accordance with the principles of fundamental justice. *Charter* rights are already subject to reasonable limits under section 1, but section 7 has an additional built-in limitation through the principles of fundamental justice. While the limitations share some similarities in that they are rooted in similar concerns, they are distinct inquiries in at least two ways. First, a claimant bears the onus of proof to establish a section 7 violation, whereas the government has the burden of justifying the infringement under section 1. Second, section 1 focuses on whether infringing an individual's rights may be justified in light of the broader public interest, whereas section 7 focuses on the impacts of an infringement on the claimant(s).[321] Courts are not required in the principles of fundamental justice analysis to consider the competing social interests or public benefits associated with the measure causing the infringement.[322] These broader issues are more appropriately considered in the section 1 analysis. Because of the double limitation, however, and the fact that both limits are rooted in similar concerns, infringements that violate the principles of fundamental justice will rarely be saved by section 1.[323]

Although the principles of fundamental justice were initially intended to provide procedural guardrails when section 7 rights were violated, they have been interpreted to be much more than that. In one of the first section 7 decisions, Lamer J wrote for a unanimous court in *Re BC Motor Vehicle Act* that fundamental justice principles were not limited to procedure but included substantive dimensions as well.[324] Since then, the courts have consistently interpreted the principles of fundamental justice as having a substantive as well as procedural component,[325] describing them as

---

321 *Bedford*, above note 115 at para 127.

322 *Carter*, above note 10 at para 79.

323 *Bedford*, above note 115 at paras 124–29.

324 *Re Motor Vehicle Act*, above note 195, cited in *ibid* at 264.

325 Martha Jackman, "Wizened Stump or Living Tree? Section 7 Principles of Fundamental Justice" in Howard Kislowicz, Kerri Froc & Richard Moon, eds, *The Surprising Constitution* (Vancouver: UBC Press, 2024) 260 [Jackman, "Wizened Stump"]; Hamish Stewart, *Fundamental Justice: Section 7 of the Canadian Charter of Rights and Freedoms*, 2d ed (Toronto: Irwin Law, 2019) [Stewart, *Fundamental Justice*].

"the minimum constitutional requirements that a law that trenches on life, liberty or security of the person must meet."[326]

Nadar Hasan suggests the principles of fundamental justice have moved through three iterations of interpretation in the *Charter*.[327] The first stage involved recognizing uncontroverted historical principles reflected in post-war human rights instruments, such as solicitor-client privilege.[328] The second stage involved recognizing principles deriving from the rights in sections 7–14 of the *Charter*, effectively filling gaps needed to give those rights meaning.[329] This third stage recognized that principles of fundamental justice could be recognized in other circumstances as well, reflecting a broader evolutionary stage of interpretation.[330] This third stage is anchored in respect for human dignity and individual autonomy.[331] An early example of this third stage is *R v Morgentaler*, where the provisions limiting access to abortion violated the principles of fundamental justice because the provisions were, as a whole, manifestly unfair.[332] In her concurring opinion, Wilson J recognized a new principle of fundamental justice that protects a woman's right to terminate her pregnancy free from state interference.[333] Justice Wilson also noted that "a deprivation of the s. 7 right which has the effect of infringing a right guaranteed elsewhere in the *Charter* cannot be in accordance with the principles of fundamental justice."[334] Justice L'Heureux-Dubé in *G(J)* also held that "systemically discriminatory decision-making is incompatible with the substantive requirements of section 7 principles of fundamental justice."[335]

---

326   *Carter*, above note 10 at para 72, citing *Bedford*, above note 115 at para 94.

327   Nadar Hasan, "Three Theories of 'Principles of Fundamental Justice'" (2013) 63 OHLJ 339. See also Erin Dobbelsteyn, "Fundamental (In)justice and the Existential Threat of Climate Change" (2025) CJWL (submitted).

328   Hasan, above note 326 at 346.

329   Hasan, above note 326 at 352. An example would be a pre-trial right to silence, extending into a right to silence at trial.

330   Hasan, above note 326 at 361.

331   *Ibid.*

332   *Morgentaler*, above note 10 at 72–73.

333   *Ibid* 10 at 166.

334   *Ibid* 10 at 175.

335   Jackman, "Wizened Stump," above note 325 at 18. See also *New Brunswick v G(J)*, above note 196 at 50.

The expansive approach to the principles of fundamental justice in the early *Charter* jurisprudence led to concerns by some about the potentially indeterminate nature of the principles and judicial overreach.[336] A retrenchment started taking shape in *Rodriguez v British Columbia (Attorney General)*[337] culminating in the Supreme Court of Canada articulating a test in 2003 to determine if a principle of fundamental justice exists.[338] This approach significantly narrowed the interpretation of fundamental justice, with the courts coalescing their interpretation around a triad of principles underpinned by proportionality: arbitrariness, overbreadth, and gross disproportionality.[339] This triad has been described as articulating principles of "instrument rationality," in the sense that they are designed to ensure the law meets its purpose without overstepping.[340] Since *Malmo-Levine*, the Supreme Court of Canada has been reluctant to recognize new principles of fundamental justice. After a brief overview of arbitrariness, overbreadth, and gross disproportionality, I consider the principles of fundamental justice analysis in the context of climate claims, including ones framed as positive rights.

Arbitrariness

The arbitrariness inquiry asks the question of whether there is a rational connection between the object of a law and its effects or between its purpose and the limit it imposes on life, liberty, or security of the person.[341] Does the law exact too great a constitutional price on rights without

---

336 See *Erin Dobbelsteyn*, "Fundamental (In)justice and the Existential Threat of Climate Change" (2025) CJWL (submitted) at 5 (surveying the evolution of the jurisprudence and critiques along the way).

337 [1993] 3 SCR 519 at 592 [*Rodriguez*].

338 *Malmo-Levine*, above note 193 at para 113. To quality as a principle of fundamental justice, the principle must be a "legal principle about which there is significant societal consensus that it is fundamental to the way in which the legal system out to fairly operate, and it must be identified with sufficient precision to yield a management standard against which to measure deprivations of life, liberty or security of the person." *Ibid*.

339 *Malmo-Levine*, above note 193 at para 104.

340 *Refugees v Canada*, above note 192 at para 124.

341 *Sharma*, above note 91; *Carter*, above note 10 at para 83.

justification?[342] Does the impact on the individual bear some relation to the law's purpose? The arbitrary effect does not have to be widespread — it can be on just one person.[343] Accurately identifying the purpose of the impugned law is key, since it is an important reference point.[344]

Turning to the climate cases and assuming a challenge is framed around a given climate target, a court must first ascertain its goal. In *Mathur*, Vermette J determined that the purpose of Ontario's target was "to reduce GHG in Ontario to address and fight climate change."[345] Justice Vermette concluded that the target — although deficient and contributing to increased risk of death and harm to security of the person — was not without any rational connection to its purpose.[346] While she acknowledged that the target is insufficient, she found that it "cannot be said that the effects of the target bear no connection to its objective."[347]

Following this line of reasoning, it would be difficult to find any climate target arbitrary, since it will always go some incremental distance toward the finish line. As per this approach, as long as the target does not counter the objective of GHG reduction (i.e., seek to increase emissions), it would not be arbitrary. This is unsatisfactory in the context of climate change, when rights infringements are effectively determined by the adequacy of targets. Adequate targets can help prevent cataclysmic changes to the planet's climate system; inadequate targets can precipitate changes that will make the climate system inhospitable to humans and other living beings.

In overturning the *Mathur* decision, the Ontario Court of Appeal held that the "application judge was unable to find that the Target was arbitrary" because of her mischaracterization of the issue as being that the target was inadequate, and that her analysis of gross disproportionality was similarly flawed.[348] Instead, the Court of Appeal said the court should have

---

342   *Mathur* SC, above note 44 at para 153.

343   *Ibid* at para 153, citing *Bedford*, above note 115 at paras 98–100, 108, 111,119, and 123; *Carter*, above note 10 at para 83; and *Sharma*, above note 91 at para 86.

344   *Sharma*, above note 91 at para 87.

345   *Mathur* SC, above note 44 at para 158.

346   *Ibid* at para 160.

347   *Ibid*.

348   *Mathur* CA, above note 7 at paras 51–52.

considered whether its target was compliant with the *Charter*.[349] Posed this way, the question for an arbitrariness analysis should be whether the target is meaningful and rational.

If Ontario's target has the purpose of "reduc[ing] GHG in Ontario to address and fight climate change," it would follow that the target would need to be set at a level that would be effective in addressing and fighting climate change. As already discussed, there exist quantifiable thresholds and scientific benchmarks upon which to evaluate the sufficiency of GHG reduction targets. The target guides decision-making toward a level of GHG emissions that the government will allow through its regulatory processes. If the target is set at a level that is fundamentally incompatible with the goal of meeting the scientifically justified thresholds, it should be considered arbitrary and counter to fundamental justice.[350] Justice Vermette already found the Ontario target to fall far short of what is known to be required to fight climate change. As such, it would follow that it is arbitrary to set the target at a level that will not meet the sufficiency benchmark, since it is not based on reason but perhaps on political considerations.[351] Determining if something is arbitrary when there exists a scientific benchmark of sufficiency entails assessing whether the target is set with consideration of the science and in a way that is objectively fair and reasonable.

This approach is consistent with the meaning of arbitrariness. Arbitrariness is defined in dictionaries as something that is random, aimless, casual, hit-or-miss, haphazard, or whimsical rather than based on reason or systematic thinking.[352] In other words, something that is arbitrary is illogical, nonsensical, or unpredictable. Arbitrariness can also be defined as arrogance — a "tendency to force one's will on others without any regard to fairness or necessity" and even as oppressive, with synonyms including unfair, unreasonable, erratic, and authoritarian.[353] The opposite

---

349 *Ibid* at para 53.

350 See Chalifour & Earle, above note 40 at 762.

351 Political considerations might come into play in the section 1 analysis.

352 Michael Proffitt, *Oxford English Dictionary* (Oxford: Oxford University Press, 2024), sub verbo "arbitrary"; Merriam-Webster, "Thesaurus: Arbitrary" (last modified 22 December 2024), online: merriam-webster.com [perma.cc/TB2B-6A7N].

353 Merriam-Webster, above note 352.

of arbitrariness is then something that is objective, reasonable, impartial, rational, fair, equitable, and just (antonyms to randomness), not to mention stable, constant, orderly, and systematic (antonyms to arrogance).

With the help of these definitions, a climate target could be considered arbitrary if it was not set in a way that is rational, fair, and objective. What is rational, fair, and objective? A target established based on scientific and international consensus. As such, a target not aligned with the science would be arbitrary. In oral argument in the *Mathur* case, Ontario characterized its climate target and plan as akin to a "glossy brochure," rather than a legal commitment. In doing so, it may have bolstered the argument that the target and plan was arbitrary.

Overbreadth or gross disproportionality

The principle of overbreadth or gross disproportionality is "infringed if the impact of the restriction on the individual's life, liberty or security of the person is grossly disproportionate to the object of the measure."[354] Are the effects of the government's conduct on life and security of the person so grossly disproportionate to the conduct's purpose that they cannot be rationally supported? The threshold for a finding of gross disproportionality has been set high by the courts, such that a disconnection between the impact of the government conduct and its purpose is not sufficient; the seriousness of the rights violation must be completely out of sync with the purpose.

In the Supreme Court of Canada's most recent (at the time of writing) decision interpreting this principle of fundamental justice, the court held that a legislative scheme that creates risks (e.g., of refoulement, or the forcible return of refugees to a jurisdiction where they may face the persecution they fled) but also provides safety valves to guard against those risks was not overbroad or grossly disproportionate.[355] In the context of climate litigation, the question to ask would be whether the impacts on claimants' rights are overly broad or grossly disproportionate given the purpose of the

---

354   *Carter*, above note 10 at para 89.
355   *Refugees v Canada*, above note 192.

legislation. The impacts on the applicants' rights are significant — we are talking about increased risk of death, as well as mental and physical harms. In their Amended Statement of Claim, the *La Rose* plaintiffs argue that the impact of the government's conduct is grossly disproportionate to any economic, security, or other objective in light of the severity and irreversibility of climate harms to the plaintiffs.[356] When asking whether the government conduct is adequate, the proper question would be to evaluate whether the target is set at a sufficiently robust level to be able to mitigate these grave risks. As discussed earlier, we know what is the threshold level of reductions for jurisdictions to avoid catastrophic levels of climate change.

While the existing principles of fundamental justice under the triad are ill-suited to a claim framed around inaction or insufficient action, they can be applied. Since inaction or insufficient action is the very thing that leads to grave harm in the case of climate change, inaction may simply be qualified as irrational and arbitrary. However, it may be that a different or adapted approach is needed if a claim proceeds as one for positive rights.

The principles of fundamental justice analysis in the context of positive rights

In a pure positive rights claim focused on a failure to act, there would be no state conduct or law to analyze through the principles of fundamental justice lens. As such, it would make sense that this kind of a positive rights claim does not require a principles of fundamental justice analysis, but simply proceeds to section 1.[357] The *La Rose* plaintiffs have advocated for this approach in their Amended Statement of Claim for the portion of their claim framed as a positive rights claim, arguing "there is no principle of fundamental justice that could justify continuing a trajectory of GHG emissions consistent with the grave and existential threats that climate change is causing the plaintiffs and all children and youth present and future."[358]

---

356 *Amended* SoC, above note 303 at para 77c.
357 See, e.g., Arbour J's dissent in *Gosselin*, above note 251 at para 386.
358 *Amended* SoC, above note 303 at paras 76 and 77a.

In the case of positive rights cases anchored around insufficient action, where there is still some government conduct at play, Erin Dobbelsteyn notes that courts could engage in a broader balancing exercise between the public purpose of the government conduct and the adverse effects.[359] This is the approach taken in extradition and deportation cases where courts are assessing the exercise of discretionary power rather than a law. This argument was also advanced by an intervener in the *Mathur* case.[360]

In the *Mathur* case, Vermette J for the Superior Court held that the analysis of principles of fundamental justice is not well adapted to cases claiming positive duties. Rather than eschewing the need for a principles of fundamental justice analysis altogether, she indicated that "if positive state obligations are to be recognized under section 7 of the *Charter*, it is very likely that a different framework of analysis would need to be adopted for such claims."[361] Justice Vermette recognized that courts have developed and applied a different test for positive claims under section 2(b), and held that while this modified approach may be helpful, it cannot simply be transposed to section 7.[362] In overturning the decision that this was a positive rights case, the Ontario Court of Appeal left the question of how the principles of fundamental justice analysis will unfold in a positive rights claim for another day.

Recognizing new principles of fundamental justice

Assuming the principles of fundamental justice analysis is not simply skipped over (such as in the context of a purely positive right) and given that the existing principles are ill-suited to the climate context, it is important to consider the potential that new principles need to be recognized. There are many candidates for new principles of fundamental justice. Some have argued that substantive equality could be recognized as

---

359   See *Dobbelsteyn*, above note 336 at 20.

360   *Ibid*. See also *Mathur v Ontario*, 2024 ONCA 762 (Factum of the Intervener, Canadian Association of Physicians for the Environment) at 29 (advancing the balancing approach in that case).

361   *Mathur* SC, above note 44 at paras 139 and 141.

362   See *Dunmore*, above note 254 at paras 24–26; *Baier*, above note 272 at para 30; For an explanation of the section 7 analysis for section 2(b) when framed as a positive claim, see *City of Toronto*, above note 270 at paras 16–26.

a new principle (either on its own or as part of gross disproportionality).[363] There are also several well-established legal principles in domestic and international law that could serve as principles of fundamental justice for the section 7 analysis in climate lawsuits. For instance, the principles of intergenerational equity, precaution, sustainable development, and polluter pays are all legal principles that are well-established in both Canadian and international law.[364]

In *R v Malmo-Levine*, the Supreme Court of Canada established a three-part test for recognizing new principles of fundamental justice. To be recognized, the proposed principle must be a: (1) legal principle; (2) about which there is significant societal consensus that it is fundamental to the way in which the legal system ought fairly to operate; and (3) it must be identified with sufficient precision to yield a manageable standard against which to measure deprivations of life, liberty, or security of the person.[365] The Supreme Court of Canada has distinguished between important public policy or state interests (which are not legal principles) and normative legal principles that are akin to recognized legal principles in domestic or international law.[366] The "best interests of the child" and the presumption of reduced moral culpability for children, for instance, have been accepted as principles of fundamental justice since they are recognized legal principles in both domestic and international law.[367] One of the indicia that a potential principle constitutes a legal principle is that it is used as a test or rule in common law, statutory law, or international

---

363 See, e.g., Suzy Flader, "Fundamental Rights for All: Towards Equality as a Principle of Fundamental Justice under Section 7 of the *Charter*" (2020) 25 Appeal 43; Kerri Froc, "The Meaning of Fundamental Justice" in Heather MacIvor & Arthur Milnes, eds, *Canada at 150: Building a Free and Democratic Society* (Toronto: LexisNexis Canada, 2017); Kerri Froc, "Equality's First Principles" in Heather MacIvor & Arthur Milnes, eds, *Canada at 150: Building a Free and Democratic Society* (Toronto: LexisNexis Canada, 2017).

364 The existence of societal consensus that the principles are fundamental to the way in which law ought fairly to operate is demonstrated by the prevalence of these principles in domestic laws.

365 *Malmo-Levine*, above note 193 at para 113; *Canada (AG) v Federation of Law Societies of Canada*, 2015 SCC 7 at para 87 [*Law Societies*].

366 *Law Societies*, above note 365 at paras 89–91.

367 *Ibid* at para 90.

law.[368] A lawyer's duty of commitment to a client's cause satisfied this part of the test since it is a well-entrenched legal principle.[369]

In *Mathur*, the applicants argued that "societal preservation" should be recognized as a principle of fundamental justice. They argued that engaging in conduct that will, or could reasonably be expected to, lead to future harm, suffering, or death of a significant number of its citizens would be contrary to the principle of societal preservation.[370] In rejecting societal preservation as a new principle, Vermette J pointed to the lack of any evidence of its use as a legal principle in common law decisions or in statutory or international law.[371] In overturning the decision, the Ontario Court of Appeal pointed to the potential relevance of this principle in informing whether the target and plan are *Charter* compliant, not necessarily as a principle of fundamental justice but as an unwritten constitutional principle which could help interpret the former.[372]

Several proposed principles of fundamental justice have been advanced in environmental rights cases as well as in the youth climate cases.[373] For example, interveners in *Mathur* argued in favour of a duty of care to limit GHGs to amounts needed to avoid the most catastrophic impacts of climate change, as per the science.[374] Another candidate for a principle in the climate cases is *pacta sunt servanda*, which refers to a country's obligation to "interpret and perform its international treaty obligations 'in good faith.'"[375] This obligation is clearly stated in the

---

368 *Ibid* at para 91.

369 *Ibid* at para 93.

370 *Mathur* SC, above note 44 at para 57.

371 *Ibid* at para 165.

372 The Court made a similar comment about the UCP of ecological sustainability, an argument made by Friends of the Earth Canada as an Intervener in the case (see Section 4.1.2, above in this chapter).

373 See also Dobbelsteyn, above note 336 at 22.

374 *Mathur v Ontario*, 2024 ONCA 762 (Factum of the Interveners, Environmental Defence Canada Inc and West Coast Environmental Law) at para 2. See also Dobbelsteyn, above note 336 at page 22.

375 Hat tip to Martha Jackman and Bruce Porter for proposing this idea. See also Jackman, "Wizened Stump," above note 327 at 19; *Vienna Convention on the Law of Treaties*, 23 May 1969, 1115 UNTS 331 art 26 (entered into force 27 January 1980, accession by Canada 14 October 1976) [*Vienna Convention*] ("[e]very treaty in force is binding upon the parties to it and must be performed by them in good faith").

Vienna Convention on the Law of Treaties.[376] The *pacta sunt servanda* principle is so fundamental to the international legal order that it can be considered a *jus cogens* norm, a subset of customary international law from which no derogation is permitted.[377] As such, the principle should have no difficulty meeting the criteria for being a legal principle and, in light of its recognition as a *jus cogens* norm, the criteria for societal consensus. In terms of its precision, the quantification of targets within the Paris Agreement (i.e., through the *Glasgow Pact)* would provide a manageable standard against which to measure deprivations of life and security of the person.

In the context of discussing international norms, LeBel J stated that *jus cogens* norms can be equated with principles of fundamental justice.[378] However, when examining the *jus cogens* prohibition against torture, LeBel J held that a provision in a treaty ratified by Canada does not transform into a principle of fundamental justice if it does not concomitantly meet the *Malmo-Levine* test.[379] An increasing number of states and jurists consider the right to a healthy environment as a norm of customary international law, as evidenced by the arguments presented by over sixty states before the International Court of Justice in December 2024.[380] To the extent that such a right includes the right to a stable climate system,[381] recognition of this as a principle of fundamental justice would strengthen section 7 climate claims.

Another factor that could influence the principle of fundamental justice analysis is the alignment between *Charter* values and international

---

376 Article 27 of the *Vienna Convention* adds that "[a] party may not invoke the provisions of its internal law as justification for its failure to perform a treaty" and that this rule is without prejudice to article 46, which pertains to provisions of internal law relating to the competence to conclude treaties (see *Vienna Convention*, above note 377 art 27).

377 John Currie, *Public International Law*, 2d ed (Toronto: Irwin Law, 2008) at 154. See also Jackman, "Wizened Stump," above note 325 at 20.

378 *Alberta Reference*, above note 266 at para 151.

379 *Ibid* at para 139. See also *Jackman*, "Wizened Stump," above note 327 at page 12.

380 See IISD, "Healthy Environment: A Human Right and Customary International Law, January 29, 2025, online: https://sdg.iisd.org/commentary/guest-articles/healthy-environment-a-human-right-and-customary-international-law/.

381 *Ibid*, see footnote 37 and accompanying discussion.

human rights norms, something that is recognized by a number of justices on the bench.[382] For example, the Supreme Court of Canada has recognized a "presumption of conformity" between the *Charter* and corresponding international human rights provisions in international treaties ratified by Canada.[383] Could compliance with commitments in the *UNFCCC* and/or the Paris Agreement help anchor a principle of fundamental justice?

Are there other principles of international law that could meet the *Malmo-Levine* test, such as the principle of intergenerational equity or precaution? Since GHG emissions are measurable, and there is scientific and global political consensus about the extent to which they must be reduced to avoid catastrophic climate change, there is a reasoned basis upon which each of these principles could be applied to climate change. For example, the principle of intergenerational equity refers to the idea that every generation needs to pass on the Earth and its resources (including a safe and stable climate system) in at least as good condition as that in which it was received.[384] The youth climate claims are anchored in intergenerational equity, since they are asking for emissions to be kept at a level that will not irrevocably damage the climate system so they have similar options as current generations. The precautionary principle protects against using scientific uncertainty to justify decisions that could lead to irreversible harms and favours adopting precautionary measures to safeguard against such harms, especially when the stakes are high (as they are in climate change). It is embedded in most multilateral environmental agreements and accepted as a principle of international law. Precaution would dictate emissions reductions that are sufficient to avoid crossing critical tipping points that trigger irreversible harms.

Sustainable development refers to development that meets the needs of the present without compromising the ability of future generations to meet their own needs. In the climate context, sustainable

---

382 *Quebec (AG) v 9147-0732 Quebec Inc*, 2020 SCC 32 at para 58.

383 *R v Hape*, 2007 SCC 26 at para 53.

384 Edith Brown Weiss, "Climate Change, Intergenerational Equity, and International Law" (2008) 91:3 VJEL 615 [Weiss, "Climate Change"].

development refers to the need to ensure energy production and development is managed in a way that does not compromise the ability of future generations to meet their needs, which include life and security of the person. The polluter pays principle imposes responsibility and accountability for the harms of pollutants such as GHGs on those who emit them.[385] It highlights the point of responsibility for harms, which should be borne by the polluters and regulators who allow the pollution and not the victims of the pollution who have little or control over the emissions.

The Ontario Court of Appeal noted that interveners had raised relevant and important issues that were not determined by the application judge, such as the application of international law including international environmental law.[386] The Court stated that the appellants may need to amend their notice of application and amplify the evidentiary record (despite its magnitude) if they wish to have these additional issues addressed.[387] This may be an opportunity to argue that recognized principles of international environmental law constitute principles of fundamental justice for the purpose of section 7 of the *Charter*.

An ecological baseline for constitutional rights – Unwritten constitutionalpPrinciples and the right to a healthy environment

Does the government have an obligation to establish or maintain the baseline conditions required to exercise the rights and freedoms protected by the *Charter*? Courts have recognized that certain principles, such as democracy and the rule of law, are so essential to a functional society that they are unwritten norms embedded within the constitution, even if they are not explicitly stated. This makes sense, as does the idea that air, water, and a stable climate system are prerequisites

---

385 Grantham Research Institute on Climate Change and the Environment, "What is the Polluter Pays Principle?" (18 July 2022), online: lse.ac.uk [perma.cc/5K4Q-9PVK].
386 *Mathur* CA, above note 7 at para 6.
387 *Ibid* at para 7.

to our Constitution.[388] Without these, the right to life, to security of the person, the freedom to associate, to exercise mobility rights, and to be free from discrimination, among many other rights, may be substantially compromised or even negated. Professor Lynda Collins has articulated this argument in her pioneering work advocating in favour of an unwritten constitutional principle of ecological sustainability and respect for nature in the Canadian Constitution.[389] She explains that "there is a clear need for an enforceable ecological "bottom line" if we are to ensure the sustainability of Canadian society."[390] Recognizing an unwritten normative principle of ecological sustainability in our Constitution would, she argues, help courts interpret legislation, the division of powers, and environmental claims under the *Charter*, as well as supervise the discretionary decisions of regulators"[391] in a way that would safeguard this critical bottom line.

Courts from countries around the world have interpreted the right to life as including, implicitly, the right to a healthy environment.[392] While Canadian courts have yet to join these other courts, the words of Justice Stanton in a dissenting opinion in the *Juliana* case offer an articulate and compelling justification for doing so. Justice Stanton observed that some rights — such as the right to a healthy environment — "serve as the necessary predicate for others."[393] She notes that civil

---

388   Natalie Kobylarz argues that the European system of human rights should integrate ecological minimum standards into the "fair balance" review (see Natalia Kobylarz, "Balancing Its Way Out of Strong Anthropocentrism: Integration of 'Ecological Minimum Standards' in the European Court of Human Rights 'Fair Balance' Review" (2022) 13 J Human Rights & Envt 16).

389   Lynda Collins, "Ecological Sustainability as an Unwritten Constitutional Principle: Why Canadian Courts should Recognize an Environmental UCP" (31 October 2017) at 3, online: ssrn.com/abstract=3061938 [perma.cc/VTN3-GNXH] [Collins, "Ecological Sustainability"]. See also Lynda Collins, "The Unwritten Constitutional Principle of Ecological Sustainability: A Solution to the Pipelines Puzzle" (2019) 70 UNBLJ 30; Lynda M Collins, "Constitutional Eco-Literacy in Canada: Environmental Rights and Obligations in the Canadian Constitution" (2022) 26:2 Rev Const Stud 227.

390   See Collins, "Ecological Sustainability," above note 389 at 2.

391   *Ibid*.

392   David R Boyd, "The Implicit Constitutional Right to a Healthy Environment" (2011) 20:2 *Rev of European Community & Int'l Env L* 171–79.

393   *Juliana*, above note 2 at 1177.

liberties imply the existence of an organized society and the state, and that without it, "all the liberties protected by the Constitution to live the good life are meaningless."[394] She added that the "blessings of liberty" secured by the Constitution are meant not only for one generation but for all future generations.[395] Justice Stanton also atply stated that the US Constitution "does not condone the Nation's willful destruction."[396] Surely, that statement should apply to any nation's constitution.

4.3.4 Conclusion on Section 7

The factual findings in *Mathur* were drawn from a voluminous evidentiary record and were largely uncontested by the government defendant. These findings included a conclusion that Ontario's climate target is severely deficient relative to the scientific consensus about what is required to avoid catastrophic levels of harm, and that this deficient target is causally linked to increased risks of death and harm to the applicants and all Ontarians. The Ontario Court of Appeal's decision that the lower court mischaracterized the case as a positive rights one and to remit it for a re-hearing has set the stage for the court to evaluate the sufficiency of Ontario's target against the scientific consensus. Similarly, the *La Rose/ Misdzi Yikh* plaintiffs have been given the green light to proceed to a hearing based on amended pleadings that present arguments under both a positive and negative rights framing. The analysis in this section suggests an increasing likelihood that an inadequate climate target could be found to be a violation of section 7 that is not in accordance with the principles of fundamental justice. Sections 4.5 and 4.6 below briefly discuss the analysis of section 1 and remedies if a section 7 infringement is found.

---

394 *Ibid* at 1178.
395 *Ibid.*
396 *Ibid* at 1175.

4.4   EQUALITY RIGHTS UNDER SECTION 15(1) OF THE *CHARTER**

> "The law is not static and unchanging – actions that were
> deemed hopeless yesterday may succeed tomorrow."[397]
>
> *Justice Rennie, Federal Court of Appeal*

4.4.1   The Equality Dimensions of Climate Change

Climate change is an equality issue of unprecedented magnitude. There is an ample and growing body of evidence showing that climate change is a multiplier of existing disadvantage.[398] As such, it will exacerbate existing inequalities and disproportionately impact certain groups and identities, including Indigenous Peoples, racialized communities, people living in poverty, women, girls and people with gender-diverse identities, young and old, and those with disabilities.[399] An intersectional approach to understanding climate injustice is therefore essential.

The Supreme Court of Canada offered the first Canadian judicial pronouncement of climate injustice in *GGPPA References* when it held that the irreversible harms of climate change will be "borne disproportionately by vulnerable communities and regions, with profound effects on Indigenous peoples, on the Canadian Arctic and on Canada's coastal regions."[400] As elaborated below, the Federal Court of Appeal in *La Rose*, the Quebec Court of Appeal in *Enjeu*, and the Ontario Court of Appeal in *Mathur* also recognized forms of climate injustice when noting that young people are disproportionately impacted by climate change both because

---

*   The author wishes to thank Professor Sheila McIntyre, Professor Sanda Rodgers, and Professor Martha Jackman for their insights about section 15(1) and its application to discrimination due to climate harms.

397   *La Rose* FCA, above note 7 at para 19, Rennie J.

398   See, e.g., Intergovernmental Panel on Climate Change, *Climate Change 2022: Impacts, Adaptation, and Vulnerability*, Contribution of Working Group II to the Sixth Assessment Report of the IPCC, Hans-Otto Pörtner et al, eds (New York: Cambridge University Press, 2022) at 8.

399   Nathalie Chalifour, "Equity Considerations in Loss and Damage" in Meinhard Doelle & Sara Seck, eds, *Research Handbook on Climate Change Law and Loss & Damage* (Cheltenham: Eward Elgar, 2021) 18 [Chalifour, "Equity Considerations"].

400   See *GGPPA References*, above note 11 at para 206.

of their young age and because they will live out their adult lives in an era when the effects of climate change are more severe and difficult to address.[401]

In addition to children, youth, and Indigenous Peoples, seniors, women and gender-diverse groups, people with disabilities, racialized groups, and those living in poverty are and will be disproportionately harmed by climate change.[402] For example, the British Columbia coroner's statement reporting on deaths during that province's extraordinary heat wave in June 2021 found that of the 619 people who died from the heat, 90 percent were over 60 years of age, and 67 percent were 70 years-old or more. Ninety-one percent had chronic health problems. Many were homeless.[403] Carter Vigh's family knows all too well the disproportionate impact of climate change on children and those with health vulnerabilities: the nine-year old boy died in 2023 in British Columbia from asthma exacerbated by wildfire smoke.[404]

While the disproportionate harms from climate change are clear, mounting a successful section 15(1) case in the context of climate change is no simple matter.[405] Equality rights claims based on the harms of climate change present a number of challenges for courts. One challenge relates

---

401 See *La Rose* FCA, above note 7; *Enjeu* CA, above note 6; *Mathur* CA, above note 7.

402 See, e.g., Sébastien Jodoin, Nilani Ananthamoorthy & Katherine Lofts, "A Disability Rights Approach to Climate Governance" (2020) 47:1 Ecology LQ 73; Nathalie Chalifour, "How a Gendered Understanding of Climate Change Can Help Shape Canadian Climate Policy" in Marjorie Griffen Cohen, ed, *Climate Change and Gender in Rich Countries: Work, Public Policy and Action* (Abingdon: Routledge, 2017) 233 [Chalifour, "Gendered Understanding"]; Larissa Parker et al, "When the Kids Put Climate Change on Trial: Youth-Focused Rights-Based Climate Litigation Around the World" (2022) 13:1 J Human Rights & Envt 64.

403 Death Review Panel, *Extreme Heat and Human Mortality: A Review of Heat-Related Deaths in B.C. in Summer 2021* (Burnaby: British Columbia Coroners Service, 2022) (chair: Michael Egilson).

404 Samantha Harrington, "Climate Change Played a Role in Killing Tens of Thousands of People in 2023," *Yale Climate Connections* (17 April 2024), online: yaleclimateconnections.org [perma.cc/6JMF-MM7N].

405 For an exploration of section 15 adverse effects claims for environmental innjustices, see Larissa Parker, "Not in Anyone's Backyard: Exploring Environmental Inequality under Section 15 of the Charter and Flexibility after Fraser v Canada" (2022) 27 Appeal 19 [Parker, "Backyard"].

to section 15(1) itself and its poor performance in safeguarding substantive equality. Another challenge relates to issues that arise when adjudicating equality claims in the unique context of climate change. After a brief introduction to section 15(1), this section explores both the general and specific challenges for climate equality claims, with an emphasis on the latter. Specifically, I address the way in which framing influences the analysis, including the potential framing of cases as seeking positive rights, causation, intersectionality, and the analysis of claims on the ground of age.

BOX 3   THREE LEVELS OF CLIMATE JUSTICE BURDENS

Individuals experiencing discrimination face several levels of burdens from climate change. The **first burden** is the disproportionate impact from the effects of climate change, which may exacerbate existing inequalities and create new disproportionate harms (such as the existential threat climate change poses for young people).

Equality-seeking groups may also be disproportionately impacted by the policies enacted to address climate change, creating a **double burden** of harm. Unless mitigation, adaptation, and loss and damage policies are crafted intentionally to ensure they do not further entrench or worsen inequalities, they may impose an additional burden on those already facing systemic discrimination. For instance, carbon pricing can be an effective GHG mitigation strategy and can be neutral or even equality promoting *if designed with this purpose in mind* (i.e., with exceptions, credits or rebates for those in lower income brackets, and funding from the revenues directed toward equality-promoting programs).[406] But

---

406 See, e.g., Nathalie Chalifour, "A Feminist Perspective on Carbon Taxes" (2010) 22:1 CJWL 169; Karen-Bubna Litic & Nathalie Chalifour, "Are Climate Change Policies Fair to Vulnerable Communities? The Impact of British Columbia's Carbon Tax and Australia's Carbon Pricing Policy on Indigenous Communities" (2012) 35:1 Dal LJ 127. As it is currently designed, the fuel charge portion of national carbon pricing system was estimated to have in 2030–31 a progressive impact on households — relative to household disposable income and the fiscal-only impact of the charge — in provinces where it applies. Canada, Office of the Parliamentary Budget Officer, *A Distributional*

without careful and inclusive design, GHG mitigation policies can also serve to exacerbate inequality, such as electric vehicle rebates that target expensive cars or public transit tax credits that are not refundable, and thus fail to reach low income Canadians. Adaptation policies can be designed in a way that supports equality, such as by ensuring services like cooling centres and air filtration are accessible and affordable to all, and by designing evacuation plans in a way that proactively assists those with disabilities.[407] Policies that respond to the inevitable losses and damages of climate change that are not avoided through mitigation or adaptation may similarly be designed fairly to reach equality-seeking groups first.[408] Policy-makers must be aware of, and actively work to counter, this potential for a double burden of injustice.

There is also the possibility of a **intersecting and compounding burdens** of injustice due to the undermining of geopolitical stability and the economic problems that could result from the ongoing and worsening emergencies that result from climate change.[409] Climate change may lead to the dismantling of social institutions (e.g., in the face of budgets diverted to address the crises, for instance), increased displacement from affected communities, food and water insecurity, and climate-related violence and conflicts.[410] Those with socio-economic disadvantages, health inequities, and who face systemic discrimination will be disproportionately impacted by such disruption. This multilayered burden of injustice is compounded by the fact that those facing systemic inequality have contributed the least to the problem. Consider the plight of those

---

Analysis of the Federal Fuel Charge – Update (Ottawa: Office of the Parliamentary Budget Officer, 2024), online: pbo-dpb.ca [perma.cc/AW4U-T2CN].

407　See, e.g., Jodoin, Ananthamoorthy & Lofts, above note 404.

408　See, e.g., Chalifour, "Equity Considerations," above note 398 at 41–42; Chalifour, "Gendered Understanding," above note 401.

409　The Ontario Court of Appeal referenced the increase in probability of large-scale displacement, food insecurity at the regional level, and climate-related violence and conflict linked to increased warming (see *Mathur* CA, above note 7 at para 12).

410　See, e.g., United Nations Security Council, Press Release, SC/15318 "With Climate Crisis Generating Growing Threats to Global Peace, Security Council Must Ramp Up Efforts, Lessen Risk of Conflicts, Speakers Stress in Open Debate" (13 June 2023), online: press.un.org [perma.cc/HW5A-CWFD].

> living in small island developing states who are already being devastated
> by the climate crisis in part due to Canadian GHG emissions.
>
> These climate justice burdens address mainly intragenerational jus-
> tice and justice among human communities. Though not discussed here,
> intergenerational justice and inter-species justice are equally important.

4.4.2 The Section 15(1) Test

Section 15(1) of the *Charter* states that:

> (e)very individual is equal before and under the law and has the right to
> the equal protection and equal benefit of the law without discrimina-
> tion and, in particular, without discrimination based on race, national
> or ethnic origin, colour, religion, sex, age or mental or physical disability.

When section 15(1) came into force in 1985, it held great promise as a means
of advancing substantive equality in Canada.[411] However, the provision
has yet to realize its promise. The Supreme Court of Canada has openly
struggled to determine a definitive framework for section 15(1), describing
equality as "the most difficult right,"[412] an "elusive concept,"[413] and "perhaps
the *Charter*'s most conceptually difficult provision."[414] Ongoing attempts
by the Supreme Court of Canada to overhaul the section 15(1) test over

---

411 Substantive equality requires paying attention to the full context of a person or
group's situation and the impact of government conduct on that situation. *Fraser v
Canada (AG)*, 2020 SCC 28 at para 42 [*Fraser*], citing *Withler v Canada (Attorney General)*, 2011 SCC 12 [*Withler*] at para 2; *Fraser*, above note 413, citing *R v Kapp*, 2008 SCC
41 at paras 15–16; *Fraser*, above note 409, citing *Quebec (AG) v Alliance du personnel
professionnel et technique de la santé et des services sociaux*, 2018 SCC 17 at para 25
[*Alliance*]. The inquiry in section 15 is therefore meant to always be contextual, since
substantive equality is concerned with real-world effects (see *Sharma*, above note 91
at para 196, Karakatsanis J, dissenting, citing *Withler* above note 373 at para 93. See
also Jonnette Watson Hamilton & Jennifer Koshan, "Sharma: The Erasure of Both
Group-Based Disadvantage and Individual Impact" (6 July 2023) at 8, online: ssrn.com/
abstract=4491798 [perma.cc/67F3-7HS3] [Hamilton & Koshan, "Erasure"]. Context is
especially important in adverse effects cases (see *Fraser*, above note 409 at paras 42
and 57, citing *Withler*, above note 409 at para 43).

412 Beverley McLachlin, "Equality: The Most Difficult Right" (2001) 14 SCLR 17 at 17.

413 *Andrews v Law Society of British Columbia*, [1989] 1 SCR 143 at 164 [*Andrews*].

414 *Law v Canada (Minister of Employment and Immigration)*, [1999] 1 SCR 497 at para 2
[*Law*].

the decades have created uncertainty for plaintiffs about what is needed to establish a violation of their rights,[415] with a new variation on the test emerging every decade or so.[416]

In its most recent form, the analysis proceeds in two parts, though the Court has noted that these are not "impermeable silos" and there may be overlap between the two.[417]

1. Does the law or state action, on its face or in its impact, create a distinction based on an enumerated or analogous ground?
2. Is the distinction discriminatory, meaning that it imposes a burden or denies a benefit in a way that has the effect of reinforcing, perpetuating, or exacerbating disadvantage?[418]

The phrase "on its face or in its impact" in the first prong of the section 15(1) test recognizes that discrimination can be direct or indirect. Direct discrimination occurs when a law or government decision makes an overt distinction based on a characteristic that is protected (an enumerated or analogous ground).[419] Indirect discrimination, referred to as "adverse impacts," "adverse effects," or "systemic" discrimination, arises when an otherwise neutral law or government conduct has a disproportionate impact on members of a protected group.[420] In other words, the impugned

---

415 For critiques on the Supreme Court of Canada's equality jurisprudence, see generally Bruce Ryder, "The Strange Double Life of Canadian Equality Rights" (2013) 63 SCLR 261; Daphne Gilbert, "Time to Regroup: Rethinking section 15 of the 'Charter'" (2003) 48:4 McGill LJ 627; Jennifer Koshan & Jonnette Watson Hamilton, "The Continual Reinvention of Section 15 of the Charter" (2013) 64 UNBLJ 19; Hart Schwartz, "Making Sense of Section 15 of the Charter" (2011) 29:2 NJCL 202; Daphne Gilbert, "The Silence of Section 15: Searching for Equality at the Supreme Court of Canada in 2007" (2008) 42 SCLR 497.

416 For a brief summary of the changes to the section 15 test over its three-decade history, see Chalifour & Earle, above note 40 at 33–36.

417 *Fraser*, above note 411 at para 82; *Sharma*, above note 91 at para 30.

418 *Fraser*, above note 411 at para 27; *Sharma*, above note 91 at para 28.

419 For examples of Supreme Court of Canada cases addressing claims of direct discrimination, see *Law*, above note 416; *Hodge v Canada (Minister of Human Resources Development)*, 2004 SCC 65 [*Hodge*].

420 See Jonnette Watson Hamilton & Jennifer Koshan, "Adverse Impact: The Supreme Court's Approach to Adverse Effects Discrimination under Section 15 of the Charter" (2015), 19:2 Rev Const Stud 191 [Hamilton & Koshan, "Adverse Impact"]; Margot

law or conduct does not directly single out a particular group but indirectly places them at a disadvantage.[421] Because of their indirect nature, adverse effects cases have been difficult to establish. In fact, only three adverse effects cases under section 15(1) have succeeded at the Supreme Court of Canada.[422] I return to this issue in Section 4.4.3 below.

Against the backdrop of section 15's ongoing underperformance, the likelihood of a climate change equality rights case succeeding may be low. On the other hand, the dramatic way in which climate change will impact equality rights could offer courts an opportunity to reset and refocus section 15(1) on the core values of the *Charter's* struggling equality guarantee. This section examines a set of issues central to climate claims under section 15(1) and considers them through the lens of the emerging case law. These include how the issue is framed, causation, positive rights, and the analysis of the ground of age considering the temporal dimensions of climate change. It also begins with a brief consideration of two obstacles which relate more generally to section 15(1). While this is not by any means an exhaustive list of barriers faced by climate *Charter* equality claimants, these issues seem to be the ones causing difficulty for the courts to date in the jurisprudence.

4.4.3 General Obstacles to Section 15(1) Claims

Second fiddle

A general hurdle climate equality claims face relates to section 15 itself, which has underwhelmed expectations that it would play an important

---

Young, "Unequal to the Task: 'Kapp'ing the Substantive Potential of Section 15" (2010) 50 SCLR 183 [Young, "Unequal to the Task].

421 *Fraser*, above note 411 at para 30. See also Hamilton & Koshan, "Adverse Impact," above note422; Colleen Sheppard, *Inclusive Equality: The Relational Dimensions of Systemic Discrimination in Canada* (Montreal: McGill-Queen's University Press, 2010).

422 See *Eldridge v British Columbia (Attorney General)*, [1997] 3 SCR 624 [*Eldridge*]; *Vriend*, above note 281; *Fraser*, above note 411. Note that the first SCC case to interpret section 15 recognized that section 15(1) includes protection against adverse effects discrimination, even though it was not an adverse effects case (see *Andrews*, above note 413). For an explanation of the difference between direct and indirect discrimination, see also *Withler*, above note 411 at para 64.

role in advancing substantive equality in Canada.[423] While both sections 7 and 15 offer critical protections for vulnerable members of society, section 15 has seen less success in *Charter* litigation. There are several elements to this phenomenon. For instance, when claims include both section 7 and section 15(1) arguments, the former are more likely to succeed.[424] The Supreme Court of Canada is granting fewer leaves to appeal for section 15 cases compared to other *Charter* claims.[425] Empirical analysis has shown that there is a pattern whereby the Supreme Court of Canada dismisses section 15 claims at a greater rate than other *Charter* claims, including those under section 7.[426] Sometimes the Court has disposed of analyzing the section 15 claim entirely.[427] Together, these factors create a pattern that scholars have criticized as weakening substantive equality in Canada.[428]

In the recent *Canadian Council for Refugees* case, the Supreme Court of Canada acknowledged this phenomenon, stating that "claims based on

---

423 For a variety of critiques of section 15, see Hamilton & Koshan, "Adverse Impact," above note422; Sonia Lawrence, "Critical Reflections on Fraser: What Equality Are We Seeking?" (2021) 30:2 Const Forum Const 43; Young, "Unequal to the Task," above note 422; Joshua Sealy-Harrington, "The Alchemy of Equality Rights" (2021) 30:2 Const Forum Const 53; Mary Eberts & Kim Stanton, "The Disappearance of Four Equality Rights and Systemic Discrimination from Canadian Equality Jurisprudence" (2018) 38:1 NJCL 89; Terry Skolnik, "Expanding Equality" (2024) 47:1 Dal LJ 195; Jennifer Koshan & Jonnette Watson Hamilton, "Women's Charter Equality at the Supreme Court of Canada: Surprising Losses or Anticipated Failures?" in Howard Kislowicz, Richard J Moon & Kerri Anne Froc, eds, *Canada's Surprising Constitution* (Vancouver: UBC Press, 2024) 235; Bruce Porter, "Expectations of Equality" (2006) 33 SCLR 23.

424 See Jennifer Koshan, "Redressing the Harms of Government (In)Action: A Section 7 Versus Section 15 Charter Showdown" (2013) 22:1 Const Forum Const 31.

425 Bruce Ryder & Taufiq Hashmani, "Managing Charter Equality Rights: The Supreme Court of Canada's Disposition of Leave to Appeal Applications in Section 15 Cases, 1989-2010" (2010) 51 SCLR 505 at 506, 517–18.

426 In fact, in cases where section 15(1) is claimed and the SCC declines to rule on other *Charter* claims, the court has declined to rule on section 15 in 85.7 percent of cases (see Cheryl Milne & Caitlin Salvino, "Analyzing the Treatment of Multiple *Charter* Claims: Judicial Restraint and the Case for Section 15" (last modified 22 September 2023) at 16–17, online: ssrn.com/abstract=4557943 [perma.cc/JAS6-X8MN]).

427 The SCC is granting fewer leaves to appeal for section 15 cases compared to other *Charter* claims.

428 See, e.g., Martha Jackman, "The Unfulfilled Promise of s. 15" in Heather MacIvor & Arthur Mills, eds, *Canada at 150: Building a Free and Democratic Society* (Toronto: LexisNexis Canada, 2017) 195.

s. 15 are not secondary issues only to be reached after all other issues are considered" and that there is no "hierarchy of rights in which s. 15 occupies a lower tier."[429] The Court ordered the matter back to trial for the section 15 issues to be properly adjudicated. It remains to be seen whether this acknowledgement will translate into more robust application of section 15(1) in future cases. For now, climate cases are replicating this trend, with the equality portions of the claims either remaining dismissed (e.g., *La Rose* and *Misdzi Yikh*) or being subject to a more stringent analysis (e.g., *Mathur* SC) as discussed below.

In the lower court decision in *La Rose*, Manson J disposed of the section 15 analysis in five short paragraphs. While he acknowledged that section 15 applies to government action in a variety of forms, he nonetheless held the lack of an "impugned law (that) creates the claimed distinction, whether on its face or in its impact," to be fatal to the claim.[430] In *Misdzi Yikh*, the Federal Court skirted the section 15 question, concluding that there is no cause of action with regards to both the sections 7 and 15 alleged breaches after noting that no specific law or government action had been specified and following an analysis of only the section 7 causation test.[431] The Quebec Court of Appeal in *Enjeu* spent even less time on equality, noting only that it is difficult to establish a breach of equality in the absence of a specific decision by the state and that climate change will affect all Canadians.[432] I return to this below.

Justice Vermette's analysis of the section 15 claim in *Mathur* was a bit more detailed (which is to be expected since this was a hearing on the merits versus a motion to strike). However, the Ontario Superior Court's analysis of section 15 still comprised a relatively small portion of the judgment — twelve paragraphs as compared to fifty-three paragraphs for the section 7 analysis — and she applied a stricter causation standard

---

429 *Refugees v Canada*, above note 192 at para 180. The SCC cited the intervention of a group including the David Asper Centre for Constitutional Rights and the Women's Legal Education and Action Fund Inc (see *ibid*).

430 *La Rose* FCTD, above note 44 at para 79.

431 *Misdzi Yikh*, above note 44 at paras 91–104.

432 *Enjeu* CA, above note 6 at para 43.

for section 15 as compared to section 7.[433] The Ontario Court of Appeal decision in *Mathur* may signal a change, since it noted several errors in the lower court's section 15 analysis and remitted the matter for re-hearing on both sections 7 and 15. I return to the substance of the analysis further below.

Adverse effects cases

The five rights-based climate lawsuits filed to date in Canada are adverse effects cases.[434] They do not allege that the GHG target or network of laws and policies that dictate GHG levels are directly distinguishing between protected groups but rather that the effect of these policies is to impose a disproportionate burden on people based on their young age and (in some of the cases) based on their Indigeneity.[435] This presents a second general obstacle for climate litigants, since adverse effects cases are more difficult to establish than cases of direct discrimination.

The section 15(1) analysis is the same whether the claim is for direct or adverse effects discrimination. However, the proof required to establish an adverse effects case is less direct.[436] When applying step one of the test in adverse effects cases, the claimants must show that the challenged law or government conduct has a disproportionate impact on members of the protected group. This might happen if members of a protected group are denied benefits or required to take on burdens more frequently than others. The impacts of the law and the specific circumstances of the group are relevant. There is no need to prove that all members of the group are similarly impacted nor that the law is the cause of the group's social situation. In other words, there is no need to show the government intended

---

433 Her reasons devote twelve paragraphs to section 15 compared to fifty-three devoted to section 7 (see *Mathur* SC, above note 44 at paras 118–71 and 172–83).

434 *Dykstra*, above note 44; *Mathur* SC, above note 44; *Enjeu* CA, above note 6; *La Rose* FCTD, above note 44; *Misdzi Yikh*, above note 44.

435 *Dykstra* is the only case to date to specifically raise the ground of race in its pleadings, though *La Rose*, *Misdzi Yikh*, and *Mathur* all provide evidence of the intersectional harms faced by the Indigenous claimants.

436 For a discussion of adverse effects discrimination in the context of environmental racism, see Parker, "Backyard," above note 405.

to create a disparate impact. In the most recent section 15(1) decision by the Supreme Court of Canada at the time of writing (*Sharma*), the majority made a number of changes to the test that have been characterized (both by scholars and the dissent in that case) as a "wholesale revision" that makes it more difficult for equality claims to succeed.[437] I return to the decision's approach to causation and positive rights below.

The first successful adverse effects case before the Supreme Court of Canada was *Eldridge v British Columbia*.[438] In that case, the Court held that a hospital's decision not to provide sign-language interpreters discriminated against people with hearing disabilities.[439] A year later, the Court held in *Vriend v Alberta*[440] that provincial human rights legislation that did not include sexual orientation as a ground of discrimination violated section 15(1).[441] It was more than twenty years later that the next adverse effects case succeeded in a case involving pension benefits. *Fraser v Canada* involved three retired female members of the RCMP who were excluded from pension benefits because they had participated in a job-sharing program. The program was used primarily by women trying to balance child-care with work.[442] The Supreme Court of Canada's holding that the exclusion of those who participated in the job-sharing program violated section 15(1) (since the effect of this was to exclude mainly women) provided an important precedent and a significant development for claimants advancing adverse effects claims.[443]

To illustrate how the adverse effects test works, the Supreme Court of Canada in *Fraser* cited one of the first cases to apply indirect discrimination (an American example). In that case, an employer used high school education as a criterion for employment, which was found to have a

---

437 *Sharma*, above note 91 at para 206. See also Hamilton & Koshan, "Erasure," above note 411 at p 6.

438 *Eldridge*, above note 422.

439 *Ibid*.

440 *Vriend*, above note 279.

441 *Ibid*.

442 *Fraser*, above note 411 at paras 3–5.

443 See Jonnette Watson Hamilton, "Cautious Optimism: *Fraser v Canada (Attorney General)*" (2021) 30:2 Const Forum Const 1; Carissima Mathen, "Equality before the Charter: Reflections on Fraser v. Canada (Attorney General)" (2022) 30:2 SCLR 104.

disproportionate impact on African Americans.[444] It was unnecessary for the claimant to show that his exclusion from employment was directly based on race or lack of education — instead, the emphasis was on the experience of disproportionate impact.[445] In *Fraser*, the Court held similarly that there is no requirement to show that the protected characteristic "caused" the disproportionate impact.[446] In other words, it is not necessary to decide if the law or government conduct itself was responsible for creating the situation that led to the disadvantage.[447] The focus of the inquiry is on whether there is a disproportionate impact. This flexible approach to causation in the section 15 analysis is important for capturing systemic discrimination.

There has been, however, some tension in the Supreme Court of Canada about how to prove the disproportionate impact, resulting in some inconsistency within the jurisprudence.[448] The more flexible approach, reflected in *Fraser*, is animated by the norm of substantive equality and emphasizes the experience of disproportionate impact by claimants. In contrast, the stricter approach reflected in *Sharma* brings forward a critique of substantive equality expressed by Brown and Rowe JJ in their dissent in *Fraser*. In that dissent, Brown and Rowe JJ critiqued the notion of substantive equality, claiming it has a "slippery quality" that creates an "unbounded, rhetorical vehicle by which the judiciary's policy preferences and personal ideologies are imposed piecemeal upon individual cases."[449] They brought this perspective to their majority reasons in *Sharma*.

At the heart of the difference in approach is the extent to which pre-existing disadvantage through ongoing systemic inequality is relevant to the analysis in the first prong of the test. According to the majority's view in *Sharma*, the key point is whether the distinction in impact is explained by pre-existing disadvantage or something the state has done

---

444 See *Griggs v Duke Power Co*, 401 US 424 (1971), cited by *Fraser*, above note 413 at para 32.
445 *Fraser*, above note 411 at para 70.
446 *Ibid*.
447 *Ibid* at para 71.
448 See, e.g., *Fraser*, above note 411 as compared to *Sharma*, above note 92.
449 *Fraser*, above note 411 at paras 216, 219, and 221.

to contribute to it. According to Brown and Rowe JJ's stricter interpretation in *Sharma*, the first prong of the section 15(1) test is met only in the latter case. While the impugned conduct need not be the only or even the dominant cause, it must be *a* cause since there is, as per the majority, no positive obligation on government to remediate pre-existing social inequalities. When applying the test in *Sharma*, the majority did not allow the claimant to use evidence of historical disadvantage, such as the overincarceration of Indigenous Peoples. As such, the claimant was not able to establish disproportionate impact.[450] The dissent noted that this created an unfair situation for the claimant, since the government was the gatekeeper of data on incarceration by race and Indigeneity.[451]

The majority in *Sharma* renewed an emphasis on evidentiary requirements in the causation analysis, requiring proof that the impact of the impugned law or conduct creates or contributes to a disproportionate impact. Some interveners had attempted to supplement the record by providing such evidence, but the majority rejected this effort.[452] The dissent criticized the majority's approach to causation as being unnecessarily onerous, with the effect of raising the bar for equality claimants and eroding the foundational premises of decades of equality jurisprudence.[453] Penned by Karakatsanis J, the dissent critiqued the majority's emphasis on causation which eschews the test's language of "creates a distinction," and characterized the majority's reasons as a wholesale revision that dislodges the foundational premises of equality jurisprudence.[454]

The majority in *Sharma* also declined to recognize the role of section 15(1) in remedying broader social inequalities (in this case, the

---

450 *Sharma*, above note 91 at paras 74 and 76. See also Hamilton & Koshen, "Erasure," above note 410 at 7.

451 *Sharma*, above note 91.

452 See *Sharma* above note 91 at para 75.

453 See *Sharma*, above note 91 at para 205. Justice Karakatsanis notes that broad the framework she describes is settled law, affirmed five times in the last four years with roots that extend even deeper in the case law (see, e.g., *Alliance*, above note 413 at para 25; *Centrale des syndicats du Quebec v Quebec (Attorney General)*, 2018 SCC 18 at para 22; *Fraser*, above note 411 at para. 27; *Ontario (AG) v G*, 2020 SCC 38 at para 40; *R v CP*, 2021 SCC 19 at paras 56 and 141).

454 *Sharma*, above note 91 at para 206.

discrimination stemming from patterns of intergenerational incarceration for Indigenous women), holding that section 15(1) does not confer a positive obligation on government to remedy social inequalities.[455] In their dissent in *Fraser*, Brown and Rowe JJ had acknowledged that Parliament has a constitutional obligation to take affirmative remedial measures when the disadvantage is created or exacerbated by state conduct.[456] However, they did not reference this proviso in *Sharma*.

Should it remain in place, the stricter approach represented by Brown and Rowe JJ's majority reasons in *Sharma* means that adverse effects cases will continue to be more difficult to establish. I return to the influence of this test in the following section which examines the specific context of climate cases.

4.4.4 Hurdles Specific to Climate Change

Section 15(1) claims related to climate change present a set of hurdles specific to that context. One of the challenges presented by climate equality claims (which is also an issue under section 7) is identifying the state conduct at issue. How a claim is framed (e.g., around insufficient targets, inaction or affirmative state conduct) will have implications not only for justiciability (as discussed earlier) but also the section 15 analysis. A second challenge relates to causation. Climate equality challenges require courts to determine what discriminatory conduct means in the context of climate change which involves harms at a global scale, some of which will occur in the future, and an indirect causal link between emissions and harms. A third challenge stems from the unique temporal aspects of climate change which are causing confusion in the analysis of age-related discrimination. To the extent that climate equality claims are framed around age, they raise questions about the interpretation of that ground and related questions about whether future generations and/or birth cohorts can be considered nested within the ground of age or recognized as analogous grounds. I also offer some brief comments about how the intersecting grounds of inequality within some of the claims, such as

---

455 *Ibid* at para 42.
456 *Fraser*, above note 411 at para 207.

the experience of Indigenous youth, have been almost completely ignored by the courts in the climate cases to date. First, though, I discuss the scope and breadth of section 15.

4.4.4.1 Framing Is in the Eye of the Beholder

Recall that section 15(1) is phrased in a way that protects individuals' rights to equality *before and under the law* and safeguards the right to *equal protection and benefit of the law* without discrimination. The Supreme Court of Canada has interpreted the word "law" in section 15(1) very broadly. While it clearly includes laws and regulations, it has also been held to include government decisions and administrative action,[457] the common law,[458] collective agreements with governments,[459] decisions of government delegated decision-makers such as health services,[460] and programs developed in partnership with third parties.[461] Part of the rationale for doing so has been to ensure section 15 lives up to its purpose of redressing discrimination. Since discrimination is frequently perpetuated through informal practices, the Supreme Court of Canada recognized that "it would be altogether inconceivable that [such informal practices] should be treated as insufficient to trigger the application of s. 15."[462]

It follows that the form of government conduct that dictates the level of GHG emissions should not be the basis to determine whether section 15(1) applies to youth climate claims. Regardless of whether a violation arises from one decision, a series of decisions, or a broad set of choices over a period of time — including a decision not to limit emissions — it remains government conduct that could constitute a violation. In other

---

457 *Little Sisters Book and Art Emporium v Canada (Minister of Justice)*, [2000] 2 SCR 1120 [*Little Sisters*].

458 *R v Swain*, [1991] 1 SCR 933.

459 *Douglas/Kwantlen Faculty Assn v Douglas College*, [1990] 3 SCR 570 [*Douglas College*].

460 *Eldridge*, above note 422.

461 *Lovelace v Ontario*, [2000] 1 SCR 950 [*Lovelace*]; See, e.g., *McKinney v University of Guelph*, [1990] 3 SCR 229 at 276–77. For a detailed discussion of the *Charter's* application to government conduct, see also Nathalie J Chalifour, "Environmental Discrimination and the Charter's Equality Guarantee: The Case of Drinking Water for First Nations Living on Reserves" (2013) 43 RGD 183 at 196–202; Chalifour & Earle, above note 40 at 730–44.

462 *Douglas College*, above note 458 at 613.

words, the courts should not remove government decisions from *Charter* scrutiny based on a narrow interpretation of their form but should focus on asking whether the government conduct — whatever its form — violates the plaintiffs' *Charter* rights.[463] As discussed below, the climate decisions to date suggest form is playing an influential role in determining the outcome of section 15 claims.

Six of one, half dozen of the other

As noted earlier, one of the distinct features of climate change is that it is caused by the accumulation of GHG emissions from a diverse range of sources around the world over time. This creates challenges with respect to how *Charter* claims are framed.[464] At their core, all of the climate *Charter* cases are fundamentally about the same claim: that the government is not reducing GHG emissions fast enough, thus contributing to catastrophic harm. However, claims can be framed in multiple ways depending, *inter alia*, on whether there is a legislated target.[465] When there is no legislated target, claimants are forced to frame the case either around a failure to act (or inaction) or a network of government conduct. The former may create a problem with justiciability, as was the case in *Enjeu*. The latter, where claims are framed around a network of government laws and decisions that together lead to excessive GHG emissions, create problems of scope as illustrated in the lower court decisions in *La Rose* and *Misdzi Yikh* where the Courts found the impugned conduct too broad and diffuse to support a *Charter* analysis.[466]

When a target is legislated, as it is now federally under the *Canadian Net-Zero Emissions Accountability Act*, or when there is a legal requirement to establish a target, as was the case in *Mathur*, there is a legislative anchor

---

463 See *Vriend*, above note 279.

464 This feature of climate change also creates issues related to causation (see Section 4.3.1, above in this chapter; Section 4.4.4.2, below in this chapter).

465 See earlier discussion in the context of justiciability, also titled "framing is in the eye of the beholder" (see Section 4.2.3, above in this chapter).

466 Note that the Federal Court of Appeal overturned the decision to strike the section 7 claim and allowed the plaintiffs to modify their claim to narrow the scope of government conduct targeted.

for a claim. However, even with this anchor, the root of the claim may still be characterized as one of inaction or insufficiency, which in the absence of a right to a healthy environment or stable climate, might be characterized as seeking to impose a positive obligation (as illustrated by the lower court decision in *Mathur*). The Ontario Court of Appeal overturned the application judge's finding that the application was seeking to impose a positive obligation, holding the matter to be a negative rights claim since the government had voluntarily assumed a positive statutory obligation to address climate change.[467] While subtle, this distinction has a significant impact on the ensuing *Charter* analysis.

While the Ontario Court of Appeal's decision narrows the scope of claims that could be characterized as seeking positive obligations, it has not eliminated it. For example, in the absence of any statutory requirement to establish a climate target or plan, a court might conclude the case is one framed around a positive right. In *La Rose/Misdzi Yikh*, the Federal Court of Appeal did not explicitly characterize the section 15(1) claim as one seeking imposition of a positive obligation, but its reasons suggest it was implicitly characterized as such. For example, the Federal Court of Appeal justified its decision to maintain the lower court's decision to strike the section 15(1) claim based in part on the separation of powers. The Court characterized the claim as one focused on future harm and intergenerational equity, something the court held was outside the scope of section 15 in part because it would require courts to participate in policy choices around resource allocation.[468] This suggests it was understood as a case seeking a positive obligation to allocate resources fairly. While the Federal Court of Appeal allowed the claim to be amended for section 7 (to reduce the scope of challenged government conduct), the claimants were not given the opportunity to reframe their section 15(1) claim (perhaps to more clearly emphasize present versus future harms, though these were well-established in the claim). The door on the equality portion of the claim is therefore closed.

---

467 *Mathur* CA, above note 7 at para 5.
468 *La Rose* FCA, above note 7 at paras 82–83 and 123.

Ultimately, the end result is that the remedies available to claimants vary based on the form of government conduct (whether and how it legislates GHG reduction targets and plans), rather than the substance (what level of GHG emissions are condoned by the jurisdiction). The difference is dictated by a distinction of form versus substance.[469]

Positive obligations

As discussed earlier, the positive/negative dualism is an ongoing tension in the context of section 7. The prospect of establishing climate change as the "special circumstances" referred to in *Gosselin* has made distinguishing between positive and negative claims particularly important. This positive/negative dualism has been less prevalent in the context of section 15, though it has become important in recent decisions.

There is no parallel obligation under section 15(1) to establish climate change as a special circumstance. In fact, in the early cases interpreting section 15, the Supreme Court of Canada characterized the provision as a "hybrid of positive and negative protection"[470] and held that "a government may be required to take positive steps to ensure the equality of people or groups who come within the scope of s. 15."[471] In *Eldridge v British Columbia*, the Supreme Court of Canada held that deaf patients had the right under section 15(1) to access sign language interpretation services at hospitals.[472] The Court held that section 15(1) does not distinguish between laws that impose unequal burdens ("equal protection" of

---

469　Nathalie Chalifour, Jessica Earle & Laura Macintyre, "Detrimental Defense," *CBA/ABC National* (18 November 2020), online: nationalmagazine.ca [perma.cc/THW4-X6MQ]; Nathalie Chalifour, Jessica Earle & Laura Macintyre, "Coming of Age in a Warming World" (2021) 1:17 U of T J of L & Equality at 47, 49 and 53 [Chalifour, Earle & Macintyre, "Coming of Age"].

470　*Haig v Canada (Chief Electoral Officer)*, [1993] 2 SCR 995 [*Haig*].

471　*Haig*, above note 472 at 1041, citing *Schacter v Canada*, [1992] 2 SCR 679. See also *Doucet-Boudreau v Nova-Scotia (Minister of Education)*, 2003 SCC 62 at para 28.

472　*Eldridge*, above note 422. However, some express caution against too much optimism about the *Eldridge* and *Vriend* decisions (see Margot Young, "Change at the Margins: *Eldridge v British Columbia (AG)* and *Vriend v Alberta*" (1998) 1 CJWL 244.

the law) and those that deny equal benefits.[473] When facially neutral laws and actions (or the absence thereof) are discriminatory in their impact, the government may be required to provide additional resources to the affected cohort to level the playing field.[474] In *Eldridge*, this meant that hospitals would need to provide deaf patients with sign language interpreters to ensure that these patients were able to receive the same level of care as hearing patients.[475]

In *Vriend v Alberta*,[476] the Supreme Court of Canada held that provincial human rights legislation violated section 15(1) because it failed to include sexual orientation as a ground of discrimination.[477] Justice Cory's observations in that case are of particular relevance to the youth climate lawsuits, since he emphasized that it is only within the context of a section 15 analysis that the courts can interpret whether the legislature's silence on a particular question is neutral as opposed to discriminatory. Responding to the argument that a political choice not to legislate is outside judicial purview, Cory J emphasized that there is nothing in the wording of section 15(1) "to suggest that a *positive act* encroaching on rights is required."[478] He concluded that, if omissions were not subject to the *Charter*, "the form, rather than the substance, of the legislation would determine whether it

---

473 *Eldridge*, above note 422 at para 77. See also *Lovelace*, above note 460 ("the purpose of s. 15(1) encompasses both the prevention of discrimination and the amelioration of the conditions of disadvantaged persons" at para 60).

474 *Eldridge*, above note 422 at para 62.

475 The Supreme Court in *Eldridge* noted that the duty to take positive action is subject to the principle of reasonable accommodation but that this should be left to the section 1 analysis (see *Ibid* at para 79). The principle of reasonable accommodation comes from human rights jurisprudence. The duty of reasonable accommodation "envisions a dynamic process" that "takes into account the specific details of the circumstances of the parties and allows for dialogue between them" and ultimately "enables them to reconcile their positions and find common ground tailored to their own needs." The principle is most often invoked in an employment context, wherein the employer and employee must "adjust the terms of their relationship in conformity with the requirements of human rights legislation, up to the point at which accommodation would mean undue hardship for the accommodating party" (see *Alberta v Hutterian Brethren of Wilson Colony*, 2009 SCC 37 at para 68 [*Hutterian Brethren*]).

476 *Vriend*, above note 279.

477 *Ibid*.

478 *Ibid* at para 60 [emphasis in original].

was open to challenge," the results being "illogical and more importantly unfair."[479] While the *Vriend* claimants were challenging inaction within the context of provincial human rights law, Cory J's language and reasoning offers a compelling justification for courts to consider whether the legislature's silence on a matter (such as the absence of an enforceable GHG reduction target), or an inadequate target that leads to harm, is discriminatory.

In *Haig v Canada (Chief Electoral Officer)*, L'Heureux-Dubé J recognized that "a situation might arise in which, in order to make a fundamental freedom meaningful, a posture of restraint would not be enough, and positive government action might be required."[480] In a case that challenged the repeal of employment equity legislation, the Ontario Court of Appeal confirmed its view that *Eldridge* stands for the idea that section 15(1) "may oblige the state to take positive actions to ameliorate the symptoms of systemic or general inequality."[481]

In *Sharma*, however, the Supreme Court of Canada held that section 15(1) creates no general obligation to remedy social inequalities. Writing for a four-judge dissent, Karakatsanis J critiqued this approach, stating that the Court's long-standing approach, grounded in substantive equality, requires a broad, remedial and generous approach that accommodates new or different understandings of equality and new issues raised by varying fact situations.[482] Justice Karakatsanis stated that such flexibility "gives life to s. 15(1)'s constitutional directive to continually scrutinize the existing order of things to see how Canadian law shapes up. To the extent the state's laws or acts are unjustified, equality may well unsettle the *status quo*. Section 15(1) demands nothing less."[483] In the context of climate *Charter* cases that may be characterized as framed around inaction or insufficient action, the question of whether silence or inaction can be discriminatory is paramount.

---

479 *Ibid* at para 61.
480 *Haig*, above note 469 at 1039.
481 *Ferrell v Ontario (AG)* (1998), 42 OR (3d) 97 at para 43 (CA).
482 *Sharma*, above note 91 at para 203, citing *Law* above note 413 at para 3.
483 *Sharma*, above note 91 at para 203.

Can silence or inaction be discriminatory?

Is section 15 a right not to be discriminated against or a right to substantive equality? Notwithstanding the clear language and history that suggests the latter, section 15 has sometimes been considered to confer only negative rights.[484] However, as noted earlier, the characterization of rights as either negative or positive is fraught with linguistic and legal difficulty. The positive/negative rights dichotomy was initially less fraught in the context of section 15, in part because the wording of the section itself references the right to equal benefit of the law. As noted earlier the Supreme Court of Canada has stated that[485] "[i]n some contexts it will be proper to characterize s. 15 as providing positive rights."[486] In the unanimous *Eldridge* decision, LaForest J noted that the discriminatory impacts on deaf persons stemmed "not from the imposition of a burden not faced by the mainstream population, but rather from a failure to ensure that they benefit equally from a service offered to everyone."[487] Once the state provides a benefit, "it is obliged to do so in a non-discriminatory manner ... In many circumstances, this will require governments to take positive action; for example, by extending the scope of a benefit to a previously excluded class of persons."[488] This is similar to the Ontario Court of Appeal's reasoning in the *Mathur* case,

---

484 See, e.g., *Tanudjaja* SC, above note 283; *Sharma*, above note 91. See also Kerri Froc, "A Prayer for Original Meaning: A History of Section 15 and What It Should Mean for Equality" (2018) 38:1 NJCL 35; Bruce Porter, "Beyond Andrews: Substantive Equality and Positive Obligations after Eldridge and Vriend" (1998) 9:3 Const Forum Const 71.

485 *Schacter v Canada,* above note 470.

486 *Ibid* at 721. See also *Vriend*, above note 279; *Eldridge*, above note 422.

487 *Eldridge*, above note 422 at para 66.

488 *Eldridge*, above note 422 at para 73. In *Auton (Guardian ad litem of) v British Columbia (Attorney General)*, 2004 SCC 78, the SCC rejected a claim that the government's refusal to fund an intensive behaviour therapy for autistic children was discriminatory. The case has been distinguished from *Eldridge* on the basis that the latter was concerned with unequal access to a benefit conferred by law, versus the former was seeking access to a benefit that had not been conferred. The *Auton* decision has been criticized on other bases, including its use of a narrow comparator group (see Emmett Macfarlane, "Positive Rights and Section 5 of the *Charter: Addressing a Dilemma*" (2016) 38 NJCL 147 at 165.

whereby it held that once the government enacts a climate target, such target must conform to the *Charter*.

It may also be possible to characterize government action to reduce climate change (such as a climate target and related plan) as a government benefit since it addresses an existential threat to humanity. Following this reasoning, the target would need to be applied in a non-discriminatory way (i.e., does not place a disproportionate burden for GHG reductions on younger people by delaying action). This is similar to what the German Constitutional Court concluded in *Neubauer* when it held that the government's choice not to specify reductions for the 2030–50 period created too great a burden for younger generations who would have to shoulder a more radical burden in later years.[489]

However, the issue of positive obligations remains contentious in the context of section 15, with conflicting approaches exemplified in two recent cases leaving this part of the law in flux. In *Quebec (Attorney General) v Alliance du personnel professionnel et technique de la santé et des services sociaux*, the Supreme Court of Canada (in a 6:3 split) found that a law aimed at improving pay equity was discriminatory because it did not go far enough, an interpretation that can be construed as conferring a positive obligation.[490] The *Pay Equity Act* was aimed at addressing systemic wage discrimination against women, but it still created inequities by denying women retroactive compensation for wage disparities that occurred in the time period between audits.[491] The *Pay Equity Act* narrowed the wage gap, but not as well as it should have since it allowed wage inequities to persist between five year audits without requiring retroactive pay. The fact that the government did not cause or create the pay discrimination did not exclude the legislation from *Charter* scrutiny; the legislation perpetuated disadvantage for a protected group and that was sufficient.[492] The Supreme Court of Canada found this to be in violation of section 15.

---

489 *Neubauer*, above note 235 at para 243.
490 *Alliance*, above note 410.
491 *Ibid* at para 37.
492 *Ibid*.

In a dissenting opinion, Côté, Brown and Rowe JJ argued that Abella J's approach was effectively conferring positive rights in section 15 of the *Charter*, something they stated was beyond the judiciary's purview: "An interpretation of the *Charter* that imposed a *positive* obligation such as that would change its nature and confer on the courts the unusual responsibility of overseeing compliance with it."[493] Justices Brown and Rowe had a chance to reassert this view, this time in a majority opinion in *Sharma*.[494] They distinguished *Alliance* by noting Abella J's decision did not mean the government had a "freestanding positive obligation" to enact schemes to redress social inequalities.[495] Rather, they held the obligation on government is to take whatever action it *does* take in a non-discriminatory way.

While the distinction is subtle, it is important and will continue to have repercussions for climate equality arguments. In *Mathur*, the Court relied upon the majority's reasons on this point in *Sharma* to dismiss the claim by the youth. Although Vermette J held that "young people are disproportionately impacted by climate change" and found that Ontario's target was severely deficient, she held that the deficient target did not create or contribute to the disproportionate impacts experienced by young people. She then disposed of the claim on the basis that the *Charter* "does not impose a general, positive obligation on the state to remedy social inequalities or enact remedial legislation," observing that leaving a "gap between a protected group and non-group members unaffected does not infringe section 15(1)."[496] She stated that the impacts of climate change are not worsening because of the current target or plan.[497] However, she failed to complete her logical sequence by also noting that the impacts would be lessened by a more stringent target aligned with the science. The Ontario Court of Appeal overturned the decision, characterizing the claim as a negative rights claim and outlining the lower court's errors stemming from Vermette J's mischaracterizing of the claim as one for positive rights. As such, the equality portion of the *Mathur* case remains open.

---

493 *Ibid* at para 65.
494 *Sharma*, above note 91 at para 42.
495 *Ibid* at para 63, citing *Alliance*, above note 411 at para 42.
496 *Mathur* SC, above note 44 at para 178.
497 *Ibid* at para 179.

The Ontario Court of Appeal decision in *Mathur* narrowed the scope of positive claims considerably by holding that any legislation establishing a climate target or plan (or requiring the establishment of such) is a voluntary assumption of a statutory obligation that must, therefore, be *Charter* compliant. In other words, it seems that if a government has taken some legislative step that creates some obligation toward addressing climate change, that step can be challenged as a negative claim. In *Mathur,* that legislative step was the *Cap and Trade Cancellation Act,* which required setting a target and establishing a plan. As long as the government has enacted some sort of target, then, it opens the door to *Charter* scrutiny for that target. What is less clear is whether a non-legislated policy target, such as the National Determined Contribution under the *Paris Agreement,* would meet the Ontario Court of Appeal's test. Although the Federal Court of Appeal shut down the section 15 portion of the claim in *La Rose,* it held that the Nationally Determined Contribution was an example of government conduct that could be subject to *Charter* scrutiny under section 7. The precise contours of what qualifies a claim as one for "positive rights" is still somewhat unclear. It seems that as long as the government has enacted some form of statutory obligation — even a weak target — it qualifies as a negative rights claim. A claim grounded solely in the absence of a climate target or plan or system-wide GHG reduction legislation would likely qualify as a positive rights claim. A claim based on a policy target such as the Nationally Determined Contribution would seem to qualify as a negative rights claim as per the reasoning of the Federal Court of Appeal, though recall that the section 15 claim was not allowed to proceed in *La Rose.*

The way forward

The Supreme Court of Canada's decision in *Vriend* offers a way through the trappings of a formalistic approach to equality claims. In *Vriend,* the Court stated that the threshold test for determining whether section 15 applies in a given case is whether there is "some matter within the authority of the legislature" as per section 32 of the *Charter.* In that case, the respondents had argued that courts must defer to a decision of the legislation not to enact a particular provision, but the Court rejected this argument. It held that "[a]side from the very problematic distinction it draws between legislative action and inaction, this argument seeks to substantially alter the nature of considerations of legislative deference in Charter analysis." Considering whether there is "some matter within the authority of the legislature" offers a constructive way to avoid the problems that are arising in the climate cases where some cases are being held to be non-justiciable or having no reasonable prospect of success on section 15 based on how they are framed, rather than the substance and impact of government conduct.

4.4.4.2 Causation

The causation tests for sections 7 and 15(1) are similar in that they both seek to establish a causal link between government conduct and the alleged *Charter* violation. Both require a connection between the impugned conduct and its impact on rights, and neither requires the government conduct to be the only (or even the dominant) cause of harm. For section 7 claims, there must be a real (not speculative) link which can be satisfied by a reasonable inference, on a balance of probabilities.[498] The burden of proof is on the claimant.

The burden of proof in a section 15(1) claim is also on the claimant(s) and established on a balance of probabilities. The first part of the section 15(1) test, as articulated in *Fraser*, asks whether the government conduct, on its face or in its impact, "creates a distinction" based on an enumerated or analogous ground. As such, section 15(1) claims emphasize the claimant's experience of discrimination or disadvantage.[499] Claimants may rely upon evidence of the claimant group's situation and/or evidence of the practical outcomes of the law or conduct to establish an adverse effects distinction.[500] However, the question of what evidence is required to establish distinction has been the source of some debate at the Supreme Court of Canada.

In a divided judgment in *Sharma*, the majority made the evidentiary burden to satisfy this part of the test more onerous by requiring claimants

---

498 *Bedford*, above note 115 at para 76.

499 For a discussion of causation in the context of environmental justice, see Nathalie J Chalifour, "Environmental Justice and the Charter: Do Environmental Injustices Infringe Sections 7 and 15 of the *Charter?*" (2015) 28:1 J Envtl L & Prac 89.

500 *Fraser*, above note 410 at para 56.

to produce evidence showing the impugned conduct "creates or contributes" to the disproportionate impact.[501] In previous decisions, the Court took judicial notice of disadvantage, or the systemic discrimination faced by some groups, and therefore did not require overt evidence of the distinction. By holding that section 15 does not confer a positive obligation to address pre-existing disadvantage, the approach in *Sharma* made it more onerous for claimants to meet the first prong of the section 15 test.

As such, equality claimants must now, according to the approach in *Sharma*, produce evidence to show on a balance of probabilities that government conduct creates or contributes to a distinction based on an enumerated or analogous ground, though — as with section 7 — the government conduct need not be the "only or the dominant cause of the disproportionate impact"[502] and the causal connection may be established by a "reasonable inference."[503]

The only climate *Charter* case to undergo a causation analysis to date has been *Mathur*, since the other cases have all been decided on preliminary matters.[504] In the context of the section 15(1) claim, Vermette J held that young people are disproportionately impacted by climate change because of their young age and because of worsening impacts of climate change.[505] However, she then went on to hold that there is no causal link between this disproportionate impact and Ontario's target.[506] She held that the disproportionate harms from climate change would continue irrespective of Ontario's actions. Recall that the Supreme Court of Canada has been consistent in

---

501 *Sharma*, above note 91 at para 42. The incorporation of this causal language harkens back to the 1993 decision in *Symes v Canada*, [1993] 4 SCR 695. For a critique of this shift, see Caitlin Salvino, "R v Sharma's 'Clarification' of the Section 15 Framework and its Creation of Unique Barriers for Disability-Based Equality Claims" (2024) 32:4 Const Forum Const 1.

502 *Sharma*, above note 91 para 45; See also *Fraser*, above note 413 at para 70.

503 *Sharma*, above note 91 at para 49.

504 In *La Rose*, the Federal Court of Appeal held that there was enough of a potential causal link between the impugned conduct and the prejudices claimed for the case to go ahead, contrary to the lower court's decision to strike the claim on the basis that such a causal link could never be established (see *La Rose* FCA, above note 7 at para 91.

505 *Mathur* SC, above note 44 at para 178.

506 *Ibid*.

holding that the law or state action need not be the sole contributor to the disproportionate impact; it only has to contribute to the disproportionate impact.[507] Further, "[t]he causal connection may be satisfied by reasonable inference."[508]

Applying the more onerous causation standard led the court in *Mathur* to reject the section 15 claim. This likely stemmed from an error in the court's application of the test, not the test itself. Specifically, Vermette J failed to apply a contextual analysis to the causation test in section 15, even though she did so for section 7, producing an internally inconsistent decision. In her section 7 analysis, Vermette J held that we have to take the global, collective nature of climate change into consideration when assessing causation.[509] In doing so, she joined courts from around the world (including the Supreme Court of Canada) in squarely rejecting the idea that one jurisdiction's emissions are not causally related to the harms of climate change simply because other jurisdictions also contribute to the problem.[510] She rejected Ontario's argument that its GHG emissions do not count simply because they are not the only source of emissions. Every single GHG emission contributes to the problem; even if Ontario's contribution may be relatively small, it is "real, measurable, and not speculative."[511] She concluded that Ontario's severely deficient target is part of the causal chain between climate change and the harms disproportionately experienced by young people; Ontario's actions contribute to an increased risk of death and risks to security of the person faced by the youth appellants.[512] In overturning the *Mathur* decision, the Ontario Court of Appeal noted that Vermette J erred in failing to apply this logic to section 15 as well as section 7.[513]

---

507   *Sharma*, above note 91 at para 45. See also *Fraser*, above note 413 at paras 63 and 70.

508   *Sharma*, above note 91 at para 49.

509   *Mathur* SC, above note 44 at para 149.

510   *Ibid*. See Box 1 on *De Minimis* emissions (see Section 4.3.1, above in this chapter).

511   *Ibid* at para 148.

512   The Application Judge found that even if Ontario's contribution to climate change may be relatively small, it is "real, measurable, and not speculative" and is causally connected to the harms alleged by the appellants (see *Mathur* SC, above note 45 at paras 147–49).

513   *Mathur* CA, above note 7 at paras 64–65.

The lower court's approach in *Mathur* offers another example of section 15(1) playing second fiddle to section 7, setting a higher bar for equality claimants contrary to the Supreme Court of Canada's direction in *Canadian Council for Refugees v Canada (Citizenship and Immigration)*.[514]

Section 15 jurisprudence does not require a direct and singular causal link between government conduct and harm. Indeed, section 15 is not concerned with causation as such but rather whether government conduct imposes a burden or denies a benefit to some based on an immutable personal characteristic. A government's choice of an insufficient target — which Vermette J held would increase the risk of death and harm for all Ontarians — imposes a greater burden on young people who will bear the brunt of worsening climate instability. A weak target also places a proportionately greater burden on young people who will bear the brunt of having to undertake more radical mitigation efforts at greater cost, will need to take more drastic adaptation measures, and who will live the bulk of their lives on a planet whose ability to sustain life as we know is increasingly uncertain.

4.4.4.3 Intersectionality

Although the Ontario Superior Court in *Mathur* highlighted the evidence submitted that the discriminatory impacts of the climate target would be amplified for the Indigenous claimants,[515] it did not address this intersectionality at all in the rest of its analysis. Similarly, the Federal Court of Appeal did not address the specific claims of the *Indigenous* claimants in *Misdzi Yikh* nor of the Indigenous youth in *La Rose* in its discussion of the equality claim. While Rennie J acknowledged that Indigenous communities will be disproportionately impacted by climate change, he characterized the section 15(1) claim as being one of intergenerational equity and analyzed it on this basis. He only briefly referenced the harms claimed by the *Misdzi Yikh* plaintiffs in the context of section 7, noting the claims about climate change's impacts on food security, culture and economies.[516]

---

514 *Refugees v Canada*, above note 192 at para 180.
515 See, e.g., *Mathur* SC, above note 44 at para 74.
516 *La Rose* FCA, above note 7 at para 105.

The Ontario Court of Appeal recognized this omission in the *Mathur* case, noting that the application judge had not addressed several important issues including whether Ontario's target breached the *Charter* rights of Indigenous Peoples in Ontario.[517] By drawing attention to this specific question, the Court opened the door to a more fulsome (and hopefully intersectional) analysis of the rights of young Indigenous claimants who face discrimination on more than one protected ground.

The lack of engagement by any of the climate decisions to date with the compounding, intersectional harms faced by Indigenous claimants in the *La Rose*, *Mathur*, and *Misdzi Yikh* cases was influenced by how the cases were pleaded and framed (i.e., around the ground of age, and not race), and the evidence provided. However, there was sufficient material put forward by the claimants to warrant some discussion of this intersectionality in the decisions. The lack of attention to intersectionality is a recurring phenomenon in equality cases.[518] However, it is especially problematic given the significant threat posed by climate change to substantive equality for these claimants. This is an extraordinarily important issue that warrants further unpacking, especially in light of the likelihood that more intersectional claims could be filed.

4.4.4.4 How the Temporal Dimensions of Climate Change Are Muddling the Analytical Waters of Age-Related Discrimination

> "A child born today will experience a world that is more than four degrees warmer than the pre-industrial average, with climate change impacting human health from infancy and adolescence to adulthood and old age. Across the world, children are among the worst affected by climate change."
>
> *Nick Watts et al,* 2019 Report of the Lancet Countdown on Health and Climate Change[519]

---

517 *Mathur* CA, above note 7 at para 6.

518 Grace Ajele & Jena McGill, *Intersectionality in Law and Legal Contexts* (Toronto: Women's Legal Education and Action Fund).

519 Nick Watts et al, "The 2019 Report of the Lancet Countdown on Health and Climate Change: Ensuring That the Health of a Child Born Today Is Not Defined by a Changing Climate" (2019) 394:10211 Lancet 1836 at 1836.

The main ground of discrimination alleged in the climate *Charter* cases to date has been age. While the potential for discrimination based on race or Indigeneity and the analogous ground of future generations is also relevant, this section is focused on the ground of age to show how the temporal dimensions of climate change have led to an impoverished analysis of the discrimination faced by young people in the face of this existential threat.

Temporal dimensions of climate change

The temporal dimensions of climate change arise from the fact that the phenomenon has a directional timeline — a trajectory, per se — where the impacts of climate change worsen over time as GHG emissions rise, concentrations in the atmosphere increase, and tipping points are crossed.[520] There are several important things for courts to understand about this timeline. First, the timeline is not reversible on a human timescale.[521] Even if all industrial GHG emissions were to cease immediately (which is practically impossible), concentrations of emissions in the atmosphere would take time to decrease and many of the biophysical changes would already have been set into motion.[522] Due to built-in inertia associated with thermal energy, oceans will continue warming for centuries even after concentrations of GHGs stabilize.[523] Also, importantly, the effects of climate change will not occur in a linear fashion. While emissions lead to a gradual warming of average temperatures, some climate impacts result from the passing of tipping points that set in motion a self-perpetuating cycle of collapse when triggered (such as the rapid melting of ice sheets in Greenland

---

520 See Chapter 1 for a more fulsome discussion of climate science.

521 While there are discussions of geoengineering solutions that could possibly impact the concentrations of GHGs in the atmosphere, which are worth exploring, none are viable at this juncture, and they all have significant uncertainties associated with them that raise important ethical concerns.

522 There are many complexities in the scientific modelling of emissions scenarios. For a discussion that clarifies some of the recent science, see Zeke Hausfather, "Explainer: Will Global Warming 'Stop' as Soon as Net-Zero Emissions Are Reached?" Carbon Brief (29 April 2021), online: <arbonbrief.org [perma.cc/D39G-DFDA].

523 Intergovernmental Panel on Climate Change, Press Release, 2019/31/PR, "Choices Made Now Are Critical for the Future of Our Ocean and Cryosphere" (25 September 2019), online: ipcc.ch [perma.cc/T7MV-3KHW].

and Antarctica and the collapse of the Atlantic current).[524] Recent research suggests we are, at current levels, perilously close to triggering some of these tipping points which will have ripple effects hundreds of years (or more) into the future.[525]

The temporal aspect of climate change means humanity is on a path to catastrophic destabilization of the planet's life-support systems. It is unsurprising, then, that this has spurred claims of disproportionate impact by (or on behalf of) children and youth, as well as future generations, and on the basis of intergenerational equity.[526] The element of time in climate equality cases is central to the claim, yet it poses a number of practical and conceptual challenges to the *Charter* analysis.[527]

One might think that framing a case around age-discrimination for young people in the face of a vanishing future might be straightforward. It is not. To start, a proper analysis of age-related discrimination in the climate context requires some precision of definitions. The precise contours of categories such as "child" or "young people" are not always clearly defined in pleadings and cases, resulting in some uncertainty.[528] There is also uncertainty over whether "future generations" are part of the ground of age or need to be recognized as an analogous ground and how intergenerational equity fits into the analysis. Professor Aoife Daly, for instance, remarks that the relationship between intergenerational equity and children's rights in the context of climate change has not been

---

524 Tessa Möller et al, "Achieving Net Zero Greenhouse Gas Emissions Critical to Limit Climate Tipping Risks" (2024) 15:6192 Nature Communications 1.

525 *Ibid.*

526 The seminal book by Edith Brown Weiss has provided a foundation for the development of the idea of intergenerational equity as both a theoretical concept and a principle of international law. See Edith Brown Weiss, *In Fairness to Future Generations: International Law, Common Patrimony, and Intergenerational Equity* (Tokyo: United Nations University, 1989) [Weiss, *In Fairness*]). See also Weiss, "Climate Change," above note 386.

527 For a fascinating exploration of time in the context of environmental law, see especially Benjamin J Richardson, *Time and Environmental Law: Telling Nature's Time* (Cambridge: Cambridge University Press, 2017) at ch 2.

528 In *Enjeu*, for instance, the court expressed concern about the choice of age thirty-five as a cut-off for the class, noting that the claimants had not provided adequate justification for this choice (see *Enjeu*, above note 6 at para 44).

elaborated in the cases nor in the academic literature.[529] Professor Aoife Nolan offers a thoughtful critique about how the lack of rigorous definition of "future generations" in climate litigation risks obscuring the rights of children.[530] The Federal Court of Appeal's decision in *La Rose/ Misdzi Yikh*, which conflated the intergenerational aspects of the claim with the discrimination faced by young people today, illustrates the need for clarity on this point.

a.   *Three aspects of age-related discrimination claims by children and young people in the Canadian climate cases*

Many section 15 cases consider discrimination based on age.[531] For example, in *Law v Canada*, Nancy Law was denied pension benefits because of her young age. While the Court had no trouble finding that the legislation drew a distinction based on age and treated her differently, the Court ultimately found that the law did not promote the notion that younger people (in this case those under age 45) are less capable or less deserving of concern, respect, and consideration.[532]

BOX 5   DIFFERENT ASPECTS OF AGE-RELATED HARMS IN CLIMATE CASES:

1.   Unique physiological and psychological vulnerabilities of children.
2.   Inability to vote or influence decisions.
3.   Temporal aspects of harms:

---

529   Aoife Daly, "Intergenerational Rights Are Children's Rights: Upholding the Right to a Healthy Environment Through the UNCRC" (2023) 41:3 Nethl QHR 132 at 4. Children can be defined as those under eighteen, and youth as those between fiftheen and twenty-two years of age (see also Aoife Daly, "Intergenerational Rights are Children's Rights: Upholding the Right to a Healthy Environment through the UN Convention on the Rights of the Child" (23 June 2022) at 2 online: ssrn.com/abstract=4141475 [perma.cc/PT7S-2LM8]).

530   Aoife Nolan, "Children and Future Generations Rights Before the Courts: The Vexed Question of Definitions" (2024) 13:3 Transnational Envtl L 1.

531   For, e.g., *Withler*, above note 411; *Gosselin*, above note 251; *R v CP*, above note 452.

532   *Law*, above note 413.

> A. Will experience harms for longer duration of time / greater proportion of lives.
> B. Will experience increased severity / magnitude of harms over time.
> C. Will bear greater burden of responsibility to address climate change (mitigation, adaptation, loss & damage).

Canadian courts to date have identified at least three ways in which children and youth face disproportionate harms from climate change.[533] First, they are vulnerable to the effects of climate change because they are **uniquely vulnerable physically and mentally as children.** Children experience distinct physical and mental health harms as a result of climate change because of their age, which dictates their physiology and stage of life that studies have shown render them especially susceptible.[534] Developing organs and growing tissues are uniquely vulnerable to things like heat waves and wildfire smoke.[535] In the decision rejecting the motion to strike in the *Mathur* case, Brown J of the Ontario Superior Court observed that young people will suffer the most of all generations and that the impacts of climate change will "exacerbate their pre-existing vulnerability and disability."[536] The Ontario Court of Appeal and Federal Court of Appeal both recognized the disproportionate impact of climate change on young people.[537]

---

533 For a summary of the three categories, see *Mathur* SC, above note 40 at paras 177–78.

534 For instance, the developing organs of children increase their vulnerability to heat waves and air pollutants linked to climate change (see, e.g., Nathalie J Chalifour & Jessica Earle, above note 40; Melissa Hornbein et al, "Held v State of Montana: Plaintiff's Proposed Final Findings of Fact and Conclusions of Law" (5 July 2023) at paras 8, 104–7, 111, 119–24, 127, and 133–36, online: climatecasechart.com [perma.cc/ D775-7UFT]).

535 A British Coroner's Court in the verdict of an inquest into the death of a child recognized the causal link between air pollution and respiratory symptoms in a young person. The case involved Ella Adoo-Kissi-Debrah, a nine-year-old child from southeast London who died of an asthma attack causally linked to air pollution. See The Ella Roberta Foundation, "Ella's Life and Her Legacy" (last visited 28 December 2024), online: ellaroberta.org [perma.cc/SAS7-KYF6].

536 *Mathur* SC, above note 44 at para 178.

537 *Mathur* CA, above note 7 at para 13; *La Rose* FCA, above note 7 at para 82.

In the face of the planetary crisis, children and youth are also uniquely vulnerable to anxiety about their futures. The United Nations Committee on the Rights of the Child recently issued a report urging states to fulfill their human rights obligations by taking immediate steps to address climate change given the impacts on children's right to life, survival, development,and a healthy environment, recognizing the impacts on their mental and physical health.[538] The youth litigants in the relevant cases have provided details of these physiological and psychological harms in their materials.[539]

Second, children and youth may be disproportionately impacted because they are **not of the age to vote** and thereby cannot influence the choice of politicians who will prioritize climate action. Since the federal voting age in Canada is 18, this category could be defined as under 18 years of age. This line of argument was rejected by the Ontario Superior Court in *Mathur*, with Vermette J characterizing the claims as a challenge to the right to vote and that the record on this issue was "woefully" inadequate.[540] Similarly, in *La Rose*, the Federal Court of Appeal held that considering the lack of a vote or voice in significant climate decisions that impact them would be "an unprecedented application of section 15(1), and not the kind of gradual, incremental change by which the law evolves."[541]

The third category relates to the **disproportionate harms experienced by younger people due to the temporal aspects of climate change.** There are at least three related components to this third category. First (A), because the impacts of climate change worsen and compound over time, the younger a person is, the longer they will be exposed to the impacts of climate change ("**duration of harm**"). As a result of their comparatively longer period of time to live, they will experience harms for a longer period and a larger proportion of their lives, relative to adults. For instance,

---

538 *La Rose* FCA, above note 7 at para 87, citing *General Comment No. 26 (2023) on Children's Rights and the Environment, with a Special Focus on Climate Change*, CRC, 2023, UN Doc CRC/C/GC/26 [*General comment*].

539 See, e.g., *La Rose* SoC, above note 59 at paras 78–89, 100, 112, 140, 149, 203, and 225; *Mathur* SC, above note 44 (Notice of application, Applicants at paras 46–47); Dykstra, above note 44 (Origination application, Applicants at paras 36–37).

540 *Mathur* SC, above note 44 at para 181.

541 *La Rose* FCA, above note 7 at para 84.

for children born in the last few years, their entire lives will be impacted by climate change (100 percent of their lives). For someone born in 1950, only a small proportion of their lives will be spent experiencing the effects of climate change (perhaps 15 percent or 20 percent of their lives).

Second (B), children and youth will be proportionately subject to the increased severity and worsening effects of climate change ("**compounding magnitude over time**"). Especially as we reach certain tipping points that may set in motion accelerated and intensified changes, the effects of climate change will worsen and compound with time. As such, children and youth will experience more extreme storm surges, wildfires, heat waves, floods, droughts and more unlike anything experienced by modern humanity in the past. The younger a person is, the more likely they will experience these more extreme, unprecedented and often life-threatening events. These worsening effects will create great uncertainty and likely lead to political and socio-economic destabilization as governments try to cope.[542] Children and youth will be denied the benefit of enjoying life and growing old in a safe, stable climate. Justice Vermette recognized both the duration of harms (A) and their compounding nature (B) in the *Mathur* decision when she noted that the "catastrophic impacts of climate change will worsen over time" and that "by virtue of their age, youth and future generations will bear the brunt of these impacts as they live longer into the future."[543] The Ontario Court of Appeal recognized that young people are disproportionately impacted by climate change but did not discuss the different types of impacts.[544]

Third (C), children and youth will be faced with a greater burden of responsibility to address climate change, which includes mitigating GHG emissions, implementing adaptation measures, and compensating for the inevitable loss and damage that will arise as climate destabilization unfolds. This "**burden of responsibility**" will fall disproportionately on

---

542 See, e.g., "How Climate Events Can Lead to Social and Political Stress" in John Steinbruner & Jo Husbands, eds, *Climate and Social Stress: Implications for Security Analysis* (2013, Washington DC, National Academies Press) 75.

543 *Mathur* SC, above note 44 at para 177. As discussed later, however, Vermette J held that the disproportionate impact is not causally tied to Ontario's climate target.

544 *Mathur* CA, above note 7 at para 64.

the shoulders of younger people, something the German Constitutional Court recognized in the *Neubauer* decision.[545]

These categories apply to children and young people that are already born. However, the second and especially the third categories invite consideration of future generations, since they involve comparisons between younger and older generations. I will return to this point in the section below.

b. *How these categories of harms fit into the analysis of age-discrimination*

In *La Rose*, Rennie J commented that the ground of age has a unique status because it is universal (someone at a given age has experienced all earlier ages and expects to experience later ages).[546] He also pointed to the fact that government choices will often affect different generations differently.[547] While both of these points are true, they are not a full answer to the question of whether (and, if so, how) government conduct in relation to climate change creates a distinction on the ground of age. It's helpful to unpack the way age is conceptualized in the case law before proceeding to its consideration in the climate cases.

Conceptualizing age

The concept of "age" can be defined in various ways in age discrimination cases. The most obvious definition is the length of time (number of years) a person has lived (i.e., a person's "age") (**numerical age**). Age may also be defined as the "stages of life" that people progress through, such as infancy, childhood, adolescence, youth, middle age, and old age) (**life stages**). It could also refer to a distinct period, such as birth cohorts (i.e., Generation Z,

---

545 See the text accompanying note 453; *Neubauer*, above note 235 at para 243. See also National Human Rights Commission of the Republic of Korea, "Submission of Opinion to the Constitutional Court of Korea Regarding Article 8, Section 1 of the Framework Act on Carbon Neutrality and Green Growth for Coping with Climate Crisis and Article 3, Section 1 of the Enforcement Decree of the Act" (2023), online: climatecasechart.com [perma.cc/H9SW-9MJC].

546 *La Rose* FCA, above note 7 at para 85.

547 *Ibid* at para 86.

representing people born between the years of 1997 to 2012) or periods in history, such as the "industrial age" or "internet age" (**time periods**).[548]

Canadian age-discrimination cases usually rely upon the first definition — numerical age — when considering whether distinctions amount to discrimination. This is likely the definition Rennie J was referring to when he said that age was universal. For example, age may be a prerequisite or a cut-off to attain certain benefits or privileges (e.g., mandatory retirement, denying unemployment insurance to those over a certain age). Cases also consider numerical age in the context of rights tied to presumed levels of maturity (e.g., voting, marriage) or ability (e.g., to provide medical consent).[549] Climate equality claims do not fit neatly into any of these categories.

Age in the section 15(1) cases to date

It is not difficult to see that the first and second categories described above (**unique physiological and psychological vulnerabilities of children** and **lack of voting rights**) are directly related to numerical age and, as such, fit within the enumerated ground of age. Both relate to specific ages or spans of ages (e.g., under 18).[550] The evidence to establish present physical and mental health harm in *La Rose* and *Mathur* is robust. Yet, curiously, none of the courts have squarely addressed the physiological or mental health harms on children and youth in the context of section 15(1). The courts in *Mathur* and *La Rose* declined to pursue an analysis of the voting ground for different reasons (lack of sufficient evidence in the former and too unprecedented an approach in the latter).

---

548 The generational categories are as follows: Baby Boomers (born 1946-64); Generation X (born 1965-80); Millennials or Generation Y (born 1980-96); Generation Z (born 1997-2012); Generation Alpha (born 2013-25). Will those born in 2026 or later be Generation Beta or perhaps Generate Climate Catastrophe? (see Chalifour, Earle & Macintyre, "Coming of Age," above note 114 at 73, n 355).

549 See Chalifour, Earle & Macintyre, "Coming of Age," above note 468.

550 While there may be some debate about the cut-off age for "children," there are existing laws and international agreements that provide principled definitions. Similarly, the right to vote is legislated in Canada to be eighteen years of age, so anyone under eighteen years of age on voting date would meet the criteria for the second category.

On the motion to strike in *Mathur*, Brown J accepted that the facts presented a distinction based on the enumerated ground of age without difficulty.[551] While she did not elaborate on her interpretation of the ground, she concluded that it would be possible for the applicants to prove that Ontario's conduct widens the gap between the disadvantaged group (youth and future generations) and the rest of society, rather than narrowing it (falling into category 3A above).[552] She stated that the distinction was self-evident: "[t]he adverse effects of climate change on younger generations – who presumably would have more years to live than current generations – may be considered self-evident."[553]

While the Superior Court in *Mathur* acknowledged that young people are disproportionately impacted by climate change because of their age, it rejected the claim for lack of causation and on the basis that section 15(1) does not require a court to remedy social inequalities (as discussed above). Respectfully, this last justification makes little sense in the context of a claim for present (and future) harms. The youth claimants are not asking courts to address past social inequalities but rather present harms that will worsen and compound over time. This is a different context from *Sharma*, where the court was dealing with over-representation of Indigenous women in the prison system due to inter-generational cycles of imprisonment from ongoing racism and historical disadvantage. The Ontario Court of Appeal found the application judge to have erred in her section 15 analysis due to her mischaracterization of the case as one seeking recognition of a positive obligation. Instead, speaking for the Court, Roberts J held that Vermette J should have considered whether Ontario chose a target that will create or contribute to a disproportionate harm on the basis of a protected ground.[554] However, neither the application nor the Court of Appeal judgments engaged with an analysis of the ground of age.

---

551 *Mathur v Ontario*, 2020 ONSC 6918 at para 179 [*Mathur 2020*].
552 *Ibid* at para 188.
553 *Ibid* at para 187.
554 *Mathur* CA, above note 7 at para 58.

In *La Rose*, the Federal Court of Appeal held that it was "beyond doubt that the burden of addressing the consequences will disproportionately affect Canadian youth."[555] However, Rennie J then characterized these harms as only prospective, future harms, stating that "[a]s distinct from their section 7 claims, the true nature of the appellants' equality argument is about how the legislation will affect them when they are older" and that there is no present harm to anchor the section 15(1) claim.[556] Respectfully, this conclusion is not based on the pleadings. In their original Statement of Claim, the *La Rose* appellants outlined clearly how climate change uniquely impacts children and youth generally due to their physiological development and then in great detail how each of the appellants is specifically impacted at present.[557] On the motion to strike decision, the Federal Court explained that "[t]he Statement of Claim describes each of the Plaintiffs' specific experiences with climate change," collectively describing how climate change "has negatively impacted their physical, mental and social health … threatened their homes, cultural heritage" and how as children and youth, they experience a particular vulnerability owing to their "state of development, increased exposure risk and overall susceptibility."[558] While the Statement of Claim also references risks of future impacts, it is erroneous to suggest the claim is fundamentally about how government conduct will affect them when they are older. While the intertemporal dimensions are part of the claim, the pleadings focus extensively on present harms.

The Federal Court of Appeal in *La Rose* disposed of the section 15(1) claim by characterizing it as one about future harms and intergenerational equity. While it acknowledged that "intergenerational considerations associated with climate change raise profound moral, social and economic questions" it held that "intergenerational equity is not within

---

555   *La Rose* FCA, above note 7 at para 76.

556   *Ibid at* para 86.

557   *La Rose* SoC, above note 59 at paras 78–89 and 94–221. The original statement of claim in *Misdzi Yikh* provided less detail about specific present harms but did identify a number of ways in which the plaintiffs are already experiencing harms from climate change (see *Misdzi Yikh* SoC above note 68 at para 74).

558   See *La Rose* FCTD above note 44 at para 2.

the scope of section 15(1), as the law currently stands."[559] Its justification for this conclusion was animated by concern that adjudicating claims about future inequalities would require the judiciary to participate in policy choices around resource allocation.[560] Influenced by the *Sharma*, Rennie J offered examples where broad policy decisions by governments today (e.g., incurring of debt or health care spending) affect different generations differently.[561] He concluded that the "true nature of the appellants' equality argument is about how the legislation will affect them when they are older,"[562] then (as noted above) held that allowing this would be "an unprecedented application of section 15, and not the kind of gradual, incremental change by which the law evolves."[563]

The conclusion that the case is about future harms confounds the first and third categories of distinction outlined above, with the effect that the Court focused solely on the third category and within that, exclusively on future harms despite the pleadings outlining present harms in all three categories. The Federal Court of Appeal conceded that the appellants in *La Rose* did claim a present harm in the form of psychological distress, acknowledging that this is "no doubt real, ongoing, and burdensome."[564] However, in yet another example of section 15(1) being a more difficult right to prove, the Court held that this distress is best analyzed under section 7 as a threat to security of the person.[565]

Justice Rennie ultimately concluded that the section 15 claim had no jurisprudential root and was "conceptually outside the scope of section 15, at least as it has been understood to date," noting that "not once in the jurisprudential journey has the question of whether intergenerational equity falls within the scope of section 15 arisen." Respectfully, the Court's refusal to engage with this possibility, considering the urgency of climate change, the existential nature of the threat it poses, and the global trend

---

559 *La Rose* FCA, above note 7 at para 82.
560 *Ibid* at para 83.
561 *Ibid* at para 86.
562 *Ibid*.
563 *Ibid* at para 84.
564 *Ibid* at para 125.
565 *Ibid* at para 125.

of holding governments to account for disproportionate harms on future generations, is disappointing. As Rennie J himself pointed out, the United Nations Committee on the Rights of the Child has urged states to take immediate action to address climate change given its threat to children's rights to life, survival, and development.[566]

Equality arguments in the climate cases outside Canada

While relatively fewer climate lawsuits have been grounded in equality rights as compared to the right to life or a healthy environment or related rights (such as the right to private and family life) in cases around the world, courts are increasingly engaging with equality arguments in the climate context. While the case was not grounded in equality rights, the Court in *Leghari* interpreted the plaintiff farmer's claim that the government's failure to meet its climate targets had impacts on his right to life through the lens of climate justice. The Court noted that climate change's impacts, which have led to food and water insecurity in Pakistan, create a "clarion call for the protection of fundamental rights of the citizens of Pakistan, in particular, the vulnerable and weak segments of the society who are unable to approach this Court."[567] In another decision from Pakistan, the Supreme Court held that climate considerations were properly included in a decision barring the construction of new cement plants in environmentally sensitive zones. The Court commented upon the intergenerational implications of climate change: "[t]his Court and the Courts around the globe have a role to play in reducing the effects of climate change for our generation and for the generations to come. Through our pen and jurisprudential fiat, we need to decolonize our future generations from the wrath of climate change, by upholding climate justice at all times."[568] In the *Future Generations v Ministry of the Environment and Others*,[569] the Colombian Supreme Court noted in its reasons that the

---

566 *General comment*, above note 537.
567 *Leghari v Pakistan*, above note 31 at para 6.
568 *DG Khan Cement Co v Government of Punjab* (2021), CP1290-L/2019 (SC Pakistan), online: climatecasechart.com [perma.cc/4LA9-SYML].
569 *Future Generations*, above note 35.

environmental rights of future generations create legal obligations on present generations to protect natural resources, and that the fundamental right of human dignity is substantially linked and dependent upon a healthy environment.

In the *Neubauer* case, the German Constitutional Court considered how the burden for reducing CO_2 emissions is distributed among different generations. Applying a proportionality analysis, it held that the government's current policies unfairly burdened future generations with a disproportionate share of CO_2 reductions and were thus unconstitutional.[570] It held that current decisions are unconstitutional under German law because they lock-in a path that leads to a disproportionate level of burden on future generations; decisions today must be made in a way that leads to a proportionate impact on future freedoms "while it is still possible to change course."[571] As a remedy, the Court in *Neubauer* remanded to the legislature the job of devising more specific emissions goals for 2031 to 2050, with a generous timeline for responding. This concept of one generation's decisions foreclosing fundamental choices for another generation is important for climate lawsuits, given the irreversible nature of climate change.

The Constitutional Court of South Korea recently followed the *Neubauer* approach in holding that the South Korean government's GHG reduction targets are set in a way that shifts an excessive burden to the future, contrary to that country's Constitution.[572] In an opinion provided to South Korea's Constitutional Court, the South Korean Human Rights Commission opined that it is discriminatory to shift the burden of GHG reduction unequally to future generations, and that it violates the constitutional principle of equality.[573]

In 2024, the European Court of Human Rights ruled in favour of senior Swiss women, finding that the Swiss government had violated their

---

570 *Neubauer*, above note 234 at para 192.

571 *Ibid.*

572 Constitutional Court of Korea, Press Release, "Case on the National Greenhouse Gas Reduction Targets Addressing the Climate Crisis" (29 August 2024), online: ccourt. go.kr [perma.cc/6G8F-YNX4].

573 National Human Rights Commission of the Republic of Korea, above note 547.

right to private and family life by not taking sufficient steps to address climate change.[574] While the Court acknowledged that senior women are disproportionately impacted by climate-related heat events, the decision did not turn on equality rights *per se*, but rather the European Convention's analogue to section 7.

Comparative and temporal aspects of climate harms a challenge for courts

It is the third category of age-related discrimination — **the disproportionate harms over time experience by youth as compared to others —** however that has caused the most confusion in the decisions. In *Enjeu*, the Quebec Court of Appeal recognized that young people "will undoubtedly feel the impact more," but "only because they will be affected for a longer period of time" (3A of Box 5 above).[575] This ignored the disproportionate impacts of climate change on children due to their physiological and psychological vulnerabilities (1) or their inability to vote (2).[576] The choice to frame the case as a class proceeding on behalf of claimants under thirty-five years of age necessarily focused the Court's attention on the intertemporal aspects of climate change in categories 3A and 3B of Box 5. The Court remarked that it was difficult to frame an equality claim in the absence of a specific measure or decision by the government, especially since climate change will impact everyone.[577] As such, the Court held it

---

574 See *Verein KlimaSeniorinnen Schweiz and Others v Switzerland* [GC], No 53600/20, [2024], ECLI:CE:ECHR:2024:0409JUD005360020. Note that the European Court of Human Rights issued two other rulings the same day, holding those cases inadmissible (see *Carême v France* [GC], No 7189/21, [2024], ECLI:CE:ECHR:2024:-0409DEC000718921; *Duarte Agostinho and others v Portugal and 32 others* [GC], No 39371/20, [2024], ECLI:CE:ECHR:2024:0409DEC003937120). The Inter-American Court of Human Rights has been asked for an advisory opinion on the climate emergency and human rights by Chile and Columbia (see Inter-American Court of Human Rights, "Observations on the Request for Advisory Opinion" (last visited 29 December 2024) online: corteidh.or.cr [perma.cc/7ZRN-XPQS]). Vanuatu, along with 130 other states, have requested an advisory opinion on climate justice from the International Court of Justice.

575 *Enjeu* CA above note 6 at para 43.

576 *Ibid.*

577 *Ibid.*

was justified in rejecting the group for the class proceeding as it would not include all potential victims.[578]

Enjeu is distinguishable from the other cases because of its nature as a proposed class action. Inherent in class proceedings is the need to identify all potential victims. The fact that climate change impacts everyone may have obscured the Court's attention to the point that younger people are disproportionately impacted in the ways outlined in category 3 of Box 5. Had the Court drawn a qualitative distinction between the kind of harms experienced by people thirty-five and under compared to those older based on these elements, it may have been more open to certifying the class.[579]

In *Mathur* SC, Vermette J held that the disproportionate and worsening burden of impacts on young people over time is not a distinction based on age but one that is temporal. She justified this finding on the basis that the impacts of climate change will be experienced by all age groups in the future, offering 2050 as a snapshot year wherein all alive at that time will experience the effects of climate change.[580] While she is correct that everyone will eventually be impacted, her finding ignores the comparative and disproportionate harm for young people. Her finding conflates generalized impacts with those that are disproportionately experienced by a protected group (i.e., children). It is the immutable characteristic of young age that makes them uniquely physiologically and psychologically vulnerable to the impacts of climate change. It is also a person's young age that subjects them to a longer duration of exposure to the irreversible, existential, and worsening features of the climate crisis (3A and 3B) and to the disproportionate burden of responsibility for the radical GHG reductions needed with the passage of time (3C).[581] As an American court recently found in *Held v Montana*, children born in 2020 will experience a

---

578 *Ibid* at para 44.

579 The Quebec Court of Appeal remarked that there was nothing in the facts alleged to justify the decision to limit the class to those under age thirty-five in the province (see *Enjeu* CA, above note 6 at para 44).

580 *Mathur* SC, above note 44 at para 180.

581 Mathur 2020, above note 551, Brown J ("[the] adverse effects of climate change on younger generations — who presumably would have more years to live than current generations — may be considered self-evident, especially if the Applicants are able

two to sevenfold increase in extreme events compared to people born in 1960.[582] This is not a temporal distinction but a disproportionate impact directly related to young age. Interestingly, the Ontario Court of Appeal did not comment upon the lower court's characterization of the distinction as temporal. However, it emphasized the evidence pointing to the target causing an increase in the risk of death and harms and the disproportionate impact on young people, suggesting it does not consider the distinction to be temporal but one based on age.[583]

Distinctions on a temporal basis have been held not to be ones that violate section 15, since they are not based on enumerated or analogous grounds.[584] Temporal distinctions are those based on a particular date or point in time, such as the date that someone dies or is injured. For instance, courts have held that a distinction based on differences in the dates of people's injuries under a workers compensation claim do not violate section 15 (whereas a distinction based on someone's disability, might).[585] Other examples of distinctions held to be temporal include those based on the date of remarriage[586] and those distinguishing between pre-1986 and post-1986 hepatitis C sufferers.[587] The key is to examine the substance of a case to see if the distinction is truly temporal or one based on differential treatment based on a prohibited ground. It is important that claims with temporal dimensions not obscure a genuine distinction

---

to present evidence of historical or sociological disadvantage the Applicants have experienced as a result of their age" at para 187).

582 *Held v Montana*, 2023 CDV 2020-307, at para 133 [*Held*] (decision upheld by the Montana Supreme Court: *Held v Montana*, 2024 MT 312).

583 See *Mathur* CA, above note 7 at paras 61–65.

584 *Canada (AG) v Hislop*, 2007 SCC 10 at para 37 [*Hislop*].

585 *Nova Scotia (Workers' Compensation Board) v Martin; Nova Scotia (Workers' Compensation Board) v Laseur*, 2003 SCC 54 at para 73. In *Vail*, the Prince Edward Island Court of Appeal dismissed a section 15 claim that was based on the date of the claimants' injuries (they were treated differently based on the statutory regime in place at the time of injury). This, the Court held, was a temporal distinction rather than one based on an analogous or enumerated ground. The claimants were not alleging differential treatment because of the nature of their disabilities, but rather the date of their injury (see *Vail & McIver v WCB (PEI)*, 2012 PECA 18 at para 25. See also *Vail v Prince Edward Island (Workers' Compensation Board)*, 2009 PECA 17.

586 *Bauman v Nova Scotia (AG)*, 2001 NSCA 51 at para 65.

587 *Hodge*, above note 419.

based on an enumerated or analogous ground. In *Hislop*, even though the case involved a temporal distinction (the date of one partner's death), the Supreme Court of Canada looked to the substance of the case and found it was effectively about differential treatment between same-sex versus opposite sex partners in the context of surviving spouse pension benefits.[588] The unanimous Court rejected the government's argument that the distinction was temporal, since the heart of the issue was that survivors were being treated differently based on whether they were same-sex partners.[589] Courts adjudicating climate cases must similarly take care not to let the temporal dimensions of climate change obscure the disproportionate harms experienced by young people.

Justice Vermette's characterization of the distinction as temporal may have stemmed from her misplaced reliance upon a decontextualized comparison between young and older persons. The Supreme Court of Canada has cautioned that relying too heavily on comparator groups may fail to capture substantive inequality.[590] Here, Vermette J's conclusion that in a given year (e.g., 2050), everyone will be impacted by climate change obscured the real question of whether certain groups are *disproportionately* impacted. Using a given "snapshot year" disregarded the essential existential and irreversible features of climate change. Even with the passage of time, young age is the key factor that determines *disproportionate* impact not only because of young people's physiological and psychological vulnerabilities but also due to the magnitude and duration of exposure (i.e., for most, if not all, of the lives of young people). The Court in *Enjeu* similarly missed the qualitatively different impacts on younger people when it concluded that all Canadians will be impacted by climate change. While this is true, it does not mean younger Canadians are not disproportionately impacted.

---

588 *Hislop*, above note 584 at para 37.

589 *Ibid* at para 37.

590 *Withler*, above note 411 at para 60. The Supreme Court of Canada explained that a reliance on comparator groups was found to obscure the real question of whether the claimant had equal access to education benefits (see *Moore v British Columbia (Education)*, 2012 SCC 61 at paras 30–31).

Justice Vermette's reliance on a "snapshot year" in *Mathur* also failed to account for the fact that inadequate climate targets burden young people and future generations with more radical future GHG reductions (3C of Box 5.). This is something the German Constitutional Court held justifies imposing stricter reductions now, pointing to the unfair burden delays in GHG reductions place on young people.[591] Similarly, the Constitutional Court of South Korea recently held that the government's GHG reduction targets are set in a way that shifts an excessive burden to the future, contrary to that country's Constitution.[592]

As explained above, Vernette J in *Mathur* justified characterizing the youth claim as temporal on the basis that the impacts of climate change will be experienced by all age groups in the future.[593] While true, this point sidesteps the fact that younger people will live a larger proportion of their lives on a compromised planet; that the government will have to divert a greater proportion of resources to deal with the growing spectrum of climate disasters; and that younger generations will bear a more significant and potentially radical burden of mitigation responsibility which is known to become costlier with every year of delay.[594] It also fails to address the point that younger bodies are more susceptible to the physical and mental impacts of climate change.

The Federal Court of Appeal's analysis of the section 15 claims in *La Rose* illustrates how the temporal dimensions of climate change are confounding the courts. As noted earlier, the Court recognized that the burden of addressing the consequences of climate change will fall disproportionately onto youth.[595] However, it then proceeded to qualify the claim as one about only future harms and anchored in the concept of intergenerational equity, something it then held was outside the scope of section 15(1).[596] The Court commented that the distinction between

---

591 *Neubauer*, above note 234 at para 192.

592 Constitutional Court of Korea, above note 596.

593 *Mathur* SC, above note 44 at para 180.

594 Canadian Climate Institute, "The Costs of Climate Change: A series of Five Reports" (last visited 29 December 2024), online: climateinstitute.ca [perma.cc/7BTB-6QKH].

595 *La Rose* FCA, above note 7 at para 76.

596 *Ibid* at para 82. Justice Rennie held that the claim had no jurisprudential root, since "not once in the jurisprudential journey has the question of whether international equity falls within the scope of section 15(1) arisen" (see *Ibid* at para 123).

"present but self-perpetuating harm" and "harm that lies in wait" (even if causing current psychological distress) may one day be relevant and able to sustain a cause of action under section 15(1), but that this was not the case today based on the state of the law on equality rights.[597] This obscured the existence of extensive present harms linked to young age identified in the pleadings, leading the Court to fail to analyze these in the context of the equality claim.

The Federal Court of Appeal acknowledged that the "international legal community is moving toward the recognition of youth climate rights and the promotion of intergenerational equity."[598] There are two different types of distinction referenced in this comment — the rights of youth and intergenerational equity. The Court concluded that intergenerational equity is not within the scope of section 15 as the law currently stands but did not adequately address the rights of youth for present harms.[599]

Future Generations

What if the claim is made on behalf of future generations? Many of the youth climate claims include reference to future generations. The question of whether (and if so how) future generations can be considered as part of the section 15(1) analysis has not been addressed squarely by Canadian courts to date. Only Brown J in the Mathur motion to strike decision referenced future generations directly when she concluded that the applicants had met the test for standing on behalf of future generations, at least at a preliminary level.[600] In the *La Rose/Misdzi Yikh* decision at the Federal Court of Appeal, Rennie J only tangentially referenced future generations when he discussed intergenerational equity and held it to be outside the scope of section 15(1). The Ontario Court of Appeal in *Mathur* similarly did not analyze the claim on behalf of future generations, simply referencing

---

597 *Ibid* at para 125.
598 *Ibid* at para 87.
599 *Ibid* at paras 82 and 87.
600 *Mathur* 2020, above note 550 at para 250.

the applicants' claims that Ontario's target causes harm to current and future generations.[601]

Is there a legitimate argument to be made that future generations are part of the ground of age or perhaps an analogous ground? The Supreme Court of Canada has remarked that it may take a flexible and generous approach to interpreting existing grounds. In *Fraser*, the Court decided the section 15 claim on the existing ground of sex rather than considering whether to recognize parental/family status as an analogous ground.[602] Justice Abella noted that a "robust and intersectional analysis of gender and parenting" can be carried out under the enumerated ground of sex.[603] A similarly robust and intersectional analysis of the concept of future generations could be undertaken under the enumerated ground of age. Courts are meant to take a purposive and contextual approach when deciding whether to recognize a new analogous ground.[604] Sometimes the fundamental nature of the characteristic is determinative, such as a trait that is inextricable from a person's identity or personhood. Other times, the decision centres on the fact that a trait is mutable or changeable at only an unacceptable personal cost.[605] Also central to the analysis is a claimant's lack of political power or vulnerability to having their interests overlooked.[606] Guided by these principles, the Supreme Court of Canada has recognized citizenship,[607] sexual orientation,[608] off-reserve members of a First Nation,[609] and, in some cases, marital status[610] as analogous

---

601 *Mathur* CA, above note 7 at para 29.

602 *Fraser*, above note 411 at para 114. The Court found that there was an insufficient evidentiary basis upon which to evaluate the analogous ground (see *ibid* at para 117).

603 *Ibid* at para 116.

604 *Ibid* at para 59.

605 *Ibid* at para 62.

606 *Ibid* at para 60.

607 *Andrews*, above note 413.

608 *Egan v Canada*, [1995] 2 SCR 513.

609 *Corbiere v Canada (Minister of Indian and Northern Affairs)*, [1999] 2 SCR 203 [*Corbiere*], McLachlin & Bastarache JJ ("[o]ff-reserve Aboriginal band members can change their status to on-reserve band members only at great cost, if at all" at para 14). Native language was raised as a possible analogous ground *in obiter* in *Gosselin*, above note 251 at para 12.

610 *Miron v Trudel*, [1995] 2 SCR 418 (SCC), L'Heureux-Dubé J ("[f]or a significant number of persons in so-called "non-traditional" relationships …the notions of 'choice' [to marry] may be illusory" at para 474).

grounds, while rejecting occupation or employment status,[611] province of residence,[612] persons charged with war crimes outside of Canada,[613] persons bringing claims against the Crown,[614] and marijuana use.[615] The key feature of the accepted analogous grounds has become their immutability; they are unchangeable or "changeable only at unacceptable cost to personal identity."[616]

One of the questions that will arise in the context of an argument that section 15(1) should apply to future generations is how to quantify or delimit the group. As a preliminary matter, youth litigants will need to differentiate future generations from the concept of the unborn (specifically fetuses) to distinguish a body of existing precedents related to abortion law and the rights of the unborn *vis-à-vis* the rights of pregnant women.[617] Beyond this basic starting point, there are a number of theoretical constructs that could assist the courts in delineating the concept of future generations. One would be to draw insight from Indigenous principles of sustainable design, where one must consider the impact of decisions on seven generations into the future.[618] Another would be to apply the principle of intergenerational equity in a way that helps courts understand what taking future generations into account means in practice.[619]

---

611 *Reference Re Workers' Compensation Act, 1983 (Nfld)*, [1989] 1 SCR 922; *Delisle v Canada (Deputy Attorney General)*, [1999] 2 SCR 989; *Health Services and Support — Facilities Subsector Bargaining Assn v British Columbia*, 2007 SCC 27 [*BC Health*].

612 *R v Turpin*, [1989] 1 SCR 1296; *Haig*, above note 469.

613 *R v Finta*, [1994] 1 SCR 701.

614 *Rudolph Wolff & Co v Canada*, [1990] 1 SCR 695.

615 *Malmo-Levine*, above note 193.

616 *Corbiere*, above note 608 at para 13.

617 For a discussion of this, see Chalifour, Earle & Macintyre, "Coming of Age," above note 468 at 62–66.

618 Pertaining to the role of the seventh-generation principle in sustainable design, see, e.g., Elizabeth Lewis et al, "Seventh Generation Principle" in Elizabeth Lewis, ed, *Sustainaspeak: A Guide to Sustainable Design Terms* (New York: Routledge, 2018) 232. For a discussion of the indigenous principles of "all my relations" and "seven generations," see Regna Darnell, "Reconciliation, Resurgence and Revitalization: Collaborative Research Protocols with Contemporary First Nations Communities" in John Borrows, Michael Asch & James Tully, eds, *Resurgence and Reconciliation: Indigenous-Settler Relations and Earth Teachings* (Toronto: University of Toronto Press, 2018) 229.

619 See Weiss, *In Fairness*, above note 525. The federal government's 2019–22 Sustainable Development Strategy makes reference to the principle of intergenerational equity.

Finally, the courts could look to international decisions for guidance. The Colombian *Future Generations* case, for instance, factored future generations into its interpretation of the rights of the Amazon.[620] The Maastricht Principles on the Human Rights of Future Generations,[621] created in 2023, would be an additional resource upon which to draw. Developed by leading international human rights experts from around the world, the principles clarify the state of international law and provide a progressive interpretation of existing human rights in a way that better reflects the interests of future generations. The principles can help inform courts as they grapple with the interpretation of human rights, including the right to equality, in the context of climate change.

In the context of climate rights cases, the rights of future generations might be conceived of as collective rights. Individual members of the "future generations" cohort would possess no individual, stand-alone *Charter* rights (such as the right to free speech); rather, the cohort, as a group, possesses only the narrow (yet critical) right to a stable climate. In other words, future generations are seeking a right that is narrow in both nature and scope and one that exists only by virtue of their membership in a generational cohort. This issue is novel and complex and, no doubt, merits further consideration.[622]

The Supreme Court of Canada has recognized collective *Charter* rights in other instances.[623] Canadian courts have also recognized and considered

---

The principle is said to inform the government's "commitment to conserve lands, water and wildlife and to address problems we face today — such as climate change — that threaten the well-being of future generations" (see Environment and Climate Change Canada, *Achieving a Sustainable Future: A Federal Sustainable Development Strategy for Canada 2019 to 2022* (Ottawa: ECCC, 2019) at 10, online: anada.ca/ [perma.cc/M4MN-DFUA].

620 *Future Generations*, above note 35.

621 Rights of Future Generations, "The Maastricht Principles on the Human Rights of Future Generations," online: perma.cc/88DE-6BF8.

622 Note that the German Constitutional Court in *Neubauer* held that the environmental associations had no standing to consider the rights of future generations — rather the claim was focused on the present personal and property rights of the youth litigants living in Germany.

623 For example, in *Loyola High School*, all seven judges acknowledged that freedom of religion has a communal — or collective — dimension to it, though it remains an individual right (see *Loyola High School v Quebec (AG)*, 2015 SCC 12). Chief Justice

future generations outside of *Charter* jurisprudence by looking at the collective rights enshrined in section 35(1) of the Constitution, which recognizes and affirms Aboriginal and treaty rights of Aboriginal people.[624] For example, in *Tsilhqot'in Nation v British Columbia*, the Supreme Court of Canada held that Aboriginal title is a "collective title held not only for the present generation but for all succeeding generations."[625] As a result, land subject to Aboriginal title cannot be used in a way that is "irreconcilable with the ability of succeeding generations to benefit from [it]." Additionally, the Federal Sustainable Development Strategy, developed following the implementation of the *Federal Sustainable Development Act*,[626] includes several references to future generations, recognizing it as a group worthy of consideration. Both the statute and the strategy define sustainable development as "development that meets the needs of the present without compromising the ability of future generations to meet their own needs." The strategy is also guided by the principle of stewardship, understood as the "management of resources in such a way that they can be passed on with integrity to future generations."[627]

A collective rights conception is tied to substantive equality in the context of climate change. The *Urgenda* decision picks up on this configuration, drawing upon objectives and principles set out in the *UNFCCC*, including the protection of the climate system for the benefit of both current and future generations.[628] The Court held that fairness was the animating value behind the intergenerational protection of the climate system, finding that this principle "means that [domestic climate policies]

McLachlin and Moldaver J have also written that freedom of religion has both "individual and collective aspects," which are "indissolubly intertwined." In their view, "the freedom of religion of individuals cannot flourish without freedom of religion for the organizations through which those individuals express their religious practices and through which they transmit their faith" (see *ibid* at para 94).

624 See, e.g., *R v Sparrow*, [1990] 1 SCR 1075 at 135; *R v Van der Peet*, [1996] 2 SCR 507 at paras 41 and 137; *Tsilhqot'in Nation v British Columbia*, 2014 SCC 44 at para 74 [*Tsilhqot'in*].

625 *Tsilhqot'in*, above note 623 at para 94.

626 *Federal Sustainable Development Act*, SC 2008, c 33.

627 See Environment and Climate Change Canada, above note 629 at 10 and 124.

628 *Urgenda* District Court, above note 2 at para 4.56. The appellate court took a similar approach with respect to the precautionary principle.

should not only start from what is most beneficial to the current generation at this moment, but also what this means for future generations, so that future generations are not exclusively and disproportionately burdened with the consequences of climate change."[629] Since Canadian courts have a history of drawing upon international law principles to guide their decision-making,[630] they could similarly utilize such *UNFCCC* principles to serve as an anchor for their adjudication.

Conclusion on Equality

While there have been glimmers of hope for the section 7 claim, the section 15(1) claims remain closed for all but two of the remaining cases — *Mathur* and *Dykstra*. Several factors are contributing to the difficulties, from the narrowed causation analysis applied by the Superior Court in *Mathur* (based on *Sharma*) to the confounding of the age-based claim considering the temporal dimensions of climate change (also in *Mathur*) and the emphasis on future harms and intergenerational equity in *La Rose*. Should the Supreme Court of Canada agree to hear the appeal in *Mathur*, it will have an opportunity to clarify many of these points and adapt equality analysis to the climate context. In the future, Canadian courts should take a more open-minded and purposive approach to s 15 analysis in the context of climate litigation.

---

629 *Ibid* at para 4.57. The judges further held the principle of fairness is reflected in the fact that the *UNFCCC* requires developed countries to make more aggressive GHG reductions compared to undeveloped countries, given that, historically, developed countries benefited the most from GHG emissions and were responsible for most of the emissions, and, as a consequence, they know enjoy greater economic prosperity that can be deployed to combat climate change.

630 Drawing upon the precautionary principle in the interpretation of a municipal bylaw, see, e.g., *114957 Canada Ltée (Spraytech, Société d'arrosage) v Hudson (Town)*, 2001 SCC 40. Collins and Boyd also explained the doctrine of non-regression and argued that it may "reasonably be viewed as a component of a *Charter* right to environment in Canada" (see Lynda M Collins & David Boyd, "Non-Regression and the Charter Right to a Healthy Environment" 2016 29 J Envtl L & Prac 285 at 300).

4.5 SECTION 1

The rights set out in the *Charter* are not absolute. According to section 1, the rights and freedoms guaranteed by the *Charter* are "subject only to such reasonable limits prescribed by law as can be demonstrably justified in a free and democratic society."[631] This section provides a balancing between the rights of individuals and the interests of society more broadly. Thus, even if a court finds there to be an infringement of section 7 or 15(1) in a climate case, that infringement may be permissible if it satisfies the test for section 1. However, recall that the courts have been strict in allowing state reliance on section 1 for sections 7 and 15.[632]

Whereas the claimants have the onus of proof to establish a rights infringement, the burden shifts to the defendant (usually the government) to justify the infringement under section 1 on a balance of probabilities.[633] There is a well-established test for determining if an infringement is reasonable and demonstrably justified as per section 1, originally articulated in *R v Oakes* and thus known as the "Oakes test."[634] The test asks whether (1) there is a pressing and substantial legislative goal and (2) there is proportionality between the objective and the means used to achieve it.[635] The second part of the test has three elements: first, the limit must be rationally connected to the objective; second, the limit must impair the right or freedom no more than reasonably necessary to accomplish the objective; and third, there must be proportionality between the deleterious

---

631   *Charter*, above note 5 s 1.

632   See, e.g., *Re British Columbia Motor Vehicle Act*, [1985] 2 SCR 486 at para 85 (where the SCC noted that section 1 could only arise in cases of exceptional circumstances such as war, natural disasters and epidemics). However, recent decisions of the Supreme Court of Canada suggest that infringements may be justified under section 1 in cases where the balancing of rights is against broader social interests that are also themselves protected by *Charter* rights. See, e.g., *R v Safarzadeh-Markhali*, 2016 SCC 14 at para 57, *Carter*, above note 10 at para 95. See also *Cambie Surgeries Corporation v British Columbia*, 2020 BCSC 1310 at para 2892.

633   *R v Oakes*, [1986] 1 SCR 103 at 136.

634   See *R v Oakes*, above note 635; *Canada v JTI-Macdonald Corp*, 2007 SCC 30 at paras 35–36 [*JTI-Macdonald*]; *Carter*, above note 10. See also Mathen & Mecklem, above note 12 at ch 17; Sharpe & Roach, above note 12 at ch 4; Hogg & Wright, above note 12 at ch 38.

635   *JTI-Macdonald*, above note 633 at para 36.

and salutary effects of the law.[636] Each element must be proven for section 1 to justify a rights infringement.

Section 1 has not been addressed in the climate *Charter* cases, since no case has yet reached the point where an infringement was found. However, the Ontario Court of Appeal in *Mathur* foreshadowed the possibility of a section 1 analysis in the re-hearing of the matter when it noted that it was being careful not to decide the case *nor limit the analysis to be undertaken, including the application of section 1, if pursued.*[637]

What would be the pressing and substantial objective in the climate *Charter* cases? The answer will depend upon how the claims were framed and analyzed.[638] The Ontario Court of Appeal in *Mathur* noted that the objective of the target enacted under the *Cap and Trade Cancellation Act* is to reduce GHG emissions to address and fight climate change.[639] Fighting climate change is clearly a pressing and substantial objective. In the absence of a legislated target, where a claim is framed as one for positive rights, the objective is more difficult to identify. The Federal Court in *Misdzi Yikh* observed this when it held that it would be impossible to conduct a section 1 analysis in the absence of a specific state law being challenged.[640] However, the Supreme Court of Canada's approach to the section 1 analysis in cases involving omissions is quite helpful for the climate context.

In *Vriend*, the Supreme Court of Canada was examining the omission from Alberta's human rights legislation of the ground of sexual orientation. In assessing the first part of the section 1 test, the Court held that the objective of the legislation as a whole, the impugned provisions, and

---

636 *Carter*, above note 10 at paras 102 and 122; *JTI-Macdonald*, above note 633 at para 45.

637 *Mathur* CA, above note 7 at para 8.

638 For critical commentary on this part of the test, see Graham Mayeda, "Between Principle and Pragmatism: The Decline of Principled Reasoning in the Jurisprudence of the McLachlin Court" in Sandra Rodgers & Sheila McIntyre, eds, *The Supreme Court of Canada and Social Justice: Commitment, Retrenchment or Retreat* (Toronto: LexisNexis, 2010) 41 at 54–66.

639 *Mathur* CA, above note 7 at para 50.

640 *Misdzi Yikh*, above note 44 at paras 55, 60, 61 & 62. The Federal Court of Appeal did not mention section 1 when it overturned the motion to strike in *La Rose/Misdzi Yikh* (see *La Rose* FCA, above note 7).

the omission itself should all be considered.[641] While there may not be an objective for the omission, it should be considered as part of the objective to further the overall goal of the legislation or relevant provision. In *M v H*, the Supreme Court of Canada noted that legislation often strikes a balance between several goals, some of which may be in tension.[642] In *Vriend*, the Court held that the legislative omission was on its face the very antithesis of the principles embodied in the legislation as a whole. As such, the legislation included no discernible objective for the omission that could be considered so pressing and substantial to justify overriding constitutionally protected rights.

In the climate cases, the overall objective of any climate plan or target will likely be to reduce GHG emissions to combat climate change. A defendant government may then argue that this is clearly a pressing and substantial objective. However, the analysis would need to consider not only the overall goal of the measure but the objective of what is alleged to constitute an infringement — such as an inadequate target. Is there a pressing and substantial objective to setting a target lower than what is justified scientifically?

A defendant government might argue that it is choosing a lower target to reduce economic costs or preserve jobs in affected sectors or that it cannot afford to reduce emissions at the level determined scientifically to be sufficient since this might have a negative impact on the fossil fuel sector and the economy. Governments might argue that that there is a pressing and substantial objective relating to addressing climate change in a way that safeguards jobs and the economy or allows a fair transition for vulnerable sectors, such as the oil and gas industry. Claimants might respond that there is abundant evidence to show that the economic costs of insufficient action far outweigh the costs of acting now.[643] The claimants could also argue that, although energy development is important to Canadians, fossil fuels are not the only means of producing energy. In fact, the growth in renewable energy is so great that countries that fail to

---

641 *Vriend*, above note 279 at para 109. See also *M v H*, [1999] 2 SCR 3 at paras 100–1.
642 *M v H*, above note 640 at para 100.
643 Canadian Climate Institute, above note 593.

transition to clean forms of energy in a timely manner risk being left in a cloud of (fossil-fuel) dust. Claimants could also argue that setting a significantly deficient target is antithetical to the purpose of the legislation.

The first step of the proportionality analysis considers whether there is a rational connection between the pressing and substantial objective identified at the first stage of the test and the impugned government conduct.[644] If the government relied on protecting jobs or reducing energy costs as the pressing and substantial objectives at the first stage of the section 1 analysis, it would have to demonstrate that its climate target is in fact connected to those purposes — in other words, that the government crafted its climate target to safeguard jobs and/or keep energy prices down. The Supreme Court of Canada has consistently characterized the rational connection test as "not particularly onerous."[645] The government need only demonstrate that it is "reasonable to suppose" that its rights-infringing behaviour "may further the [pressing and substantial] goal, not that it will do so."[646] Therefore, since an inadequate climate target could be characterized as having the goal of protecting jobs (e.g., in GHG intensive sectors) or keeping energy prices low, it might be possible to show that the goal is rationally connected to their choice of target.

The second step of the proportionality analysis considers whether the impugned government conduct is minimally impairing. Here, the question is whether the government conduct, which has already been found to infringe the claimants' rights, limits the right or freedom "no more than reasonably necessary to accomplish the objective."[647] The government would need to show that there are no other less impairing ways of achieving the objective.[648] At this point in the analysis, a court will consider both the size of the cohort affected — in this case, millions of Canadian youth, Indigenous Peoples, and future generations — and the

---

644   *Oakes*, above note 632 at para 70.

645   *Little Sisters*, above note 457 at para 228; *Trociuk v British Columbia (AG)*, 2003 SCC 34 at para 34; *BC Health*, above note 612 at para 148; *Mounted Police Association of Ontario v Canada (AG)*, 2015 SCC 1 at para 143 [*Mounted Police*].

646   *Hutterian Brethren*, above note 475 at para 48.

647   *AM v Benes* (1998), DLR (4th) 658 at para 64 (ONSC); *Mounted Police*, above note 647 at para 149, citing *RJR-MacDonald Inc v Canada*, [1995] 3 SCR 199 at para 160.

648   *Hutterian Brethren*, above note 474 at paras 54–55.

scope and significance of the burdens imposed. A court will have already accepted the legitimacy of the claimants' assertions that their right to life, security of the person and/or equality have been infringed. The government would then be required to show that the infringement is no more than reasonably necessary and that there are no other less impairing ways of achieving the objective.

Depending on the specifics of the claim, the claimants in climate cases may point to evidence of options for achieving GHG emissions reductions in a way that mitigates any increases in energy prices and which safeguards jobs (or even creates them). At this stage of the analysis, a court is also entitled to consider other countries' legislative responses to emissions reductions. The abundant evidence that governments in other jurisdictions with similar objectives of safeguarding jobs and their economies have chosen far more ambitious targets would weigh against the Canadian government's argument that their climate targets are minimally impairing. For instance, dozens of states have substantially reduced their GHG emissions between 1990 and 2024 without significant adverse economic or employment impacts.[649]

The third and final part of the proportionality test is a balancing exercise. Here, a court considers how the harmful effects of the impugned government conduct measure up against the salutary effects that flow from the government's pressing and substantial objective. The focus of the inquiry is the severity of the impacts of the government's conduct on individuals or groups as compared to the benefits derived therefrom.[650] Youth claimants will invite the court to consider the gravity, irreversibility, and extent of harms facing them to demonstrate that they will be subject to consequences that far outweigh the potential short-term cost savings associated with the government's climate conduct. It is likely here that the government will have the hardest time justifying an insufficient climate

---

649 For example, consider the following GHG reductions between 1990 and 2024 in these countries: Czechia (-40 percent), Denmark (-36 percent), Estonia (-69 percent), Finland (-33 percent), Germany (-39 percent), Hungary (-42 percent), Latvia (-59 percent), Lithuania (-58 percent), Romania (-63 percent), Slovakia (-44 percent), and the United Kingdom (-47 percent).

650 *Oakes*, above note 632 at paras 70–71.

target. Jurisdictions around the world are faced with the same imperative to reduce emissions, and Canada (and its sub-national governments) are relatively privileged and wealthy jurisdictions that should be able to do their part in the global effort to rapidly reduce emissions.

To summarize, the section 1 analysis will of course depend upon the particulars of a given case, including how the claim is framed. The analysis is more challenging in the case of claims based on inaction or the insufficiency of a target, since the analysis is conceived around the examination of an impugned law or overt conduct. However, the courts have shown a willingness to conduct a purposive analysis with some flexibility to adapting the section 1 test to different contexts. The context of climate change is important, in that many of the competing public objectives will be insufficient to justify the existential infringements of climate harms. Also, a lot of the heavy lifting on section 7 claims will take place in the principles of fundamental justice analysis. While there has yet to be a section 1 analysis in the climate cases, it is likely just a matter of time.

4.6 REMEDIES

The issue of remedies in *Charter* cases is a broad topic that I only touch on briefly here.[651] However, it is important because the nature of climate change is such that remedies could have a systemic impact. The respondent governments in the Canadian climate cases have raised concerns that the remedies requested would bring the judiciary into areas for which it does not have institutional capacity or legitimacy to address.[652] Interestingly, it is the systemic nature and magnitude of climate change that renders the cases difficult not only in terms of adjudicating the rights violations but also determining the appropriate remedy.

There is a lot of variety in terms of how remedies have been addressed in climate cases around the world, from courts sustaining broad remedies and retaining jurisdiction over the execution of those remedies to courts

---

651 See, generally, Mathen & Mecklem, above note 13 at ch 25; Sharpe & Roach, above
     note 13 at ch 18; Hogg & Wright, above note 13 at ch 37.
652 See, e.g., *Mathur* CA, above note 8 at para 67.

holding claims non-justiciable due in part to the breadth of remedies sought.[653] Requests for remedies range from strong, immediate requests (such as the request to invalidate three EU directives and an injunction in the *Carvalho and Others v Parliament and Council* case[654]) to an advisory opinion that seeks no immediate or enforceable remedies (such as that issued by the International Tribunal on the Law of the Sea in 2024, holding that states have an obligation to reduce marine pollution from carbon dioxide).[655]

Requests may also take the form of interim relief and applicants may ask a court to retain jurisdiction. Requests for interim relief such as injunctions often garner public attention and may be relatively lower cost as compared to full hearings.[656] There are interesting examples of interim relief being granted in climate-related cases. An American federal court in Texas granted preliminary relief based on human rights requiring prison officials to address the extreme heat that threatened the health of inmates. The threat of irreparable harm justified the injunction. In 2019, the Supreme Court of India granted an injunction to stop an airport from being built given concerns it would jeopardize the country's Paris commitments. The injunction was lifted when the government committed to a carbon-neutral airport.[657]

The retention of jurisdiction may be important in the context of climate change in light of the rapidly evolving science and evidence. When

---

653 As Kent Roach notes, "[j]udges do not decide whether a claim is justiciable or whether rights have been violated without worrying about remedies" (see Kent Roach, "Judicial Remedies for Climate Change" (2021) 17:1 JL & Equality 105 at 105.

654 The claim was rejected on the basis of standing, as the plaintiffs from multiple countries did not suffer special harm distinct from the public at large (see *Carvalho and Others v Parliament and Council*, [2021] C-565/19 P, ECLI:EU:C:2021:252).

655 International Tribunal for the Law of the Sea, *"Request for an Advisory Opinion Submitted by the Commission of Small Island States on Climate Change and International Law: Advisory Opinion"* (21 May 2024) online: itlos.org [perma.cc/L6SV-HSUC].

656 Roach, above note 652 at 120, citing Daniel Bodansky, Jutta Brunnée & Lavanya Rajamani, *International Climate Change Law* (Oxford: Oxford University Press, 2017) at 300.

657 Roach, above note 652 at 120, citing *Cole v Collier*, Case No 4:14-CV-1698 (SD Tex 2017) (preliminary injunction); Roach, above note 652 at 120, citing *Aroskar v India*, [2019] INSC 386 (AsianLII) at para 147 (SC India); Roach, above note 652 at 120, citing *Aroskar v Union of India*, [2020] INSC 42 (AsianLII) at paras 46–49 (SC India).

granting injunctions, courts often retain jurisdiction so they may change relief in response to changing circumstances. This may be relevant in the context of climate change, where targets may need to be adjusted considering the science.[658] In *Leghari*, the court's retention of jurisdiction allowed it to set deadlines for the preparation of action plans by a variety of government ministries. When officials noted some ministries were not cooperating, the court ordered a commission that required the ministries to work together. In the *Future Generations* case, the Colombian Supreme Court set deadlines for various ministries, municipalities, and authorities to develop action plans with the active participation of plaintiffs, affected communities, scientific organizations and ENGOs, and interested population.[659]

If there is any pattern to discern, it is that cases which have found the climate claims to be justiciable were characterized by remedial modesty and an emphasis on the desired outcome of remedies, rather than dictating the means of achieving the outcome.[660] In other words, courts are more apt to declare a rights violation and order a government to enact a constitutionally-compliant target than they are to dictate the precise level of target or how the target should be met. This is unsurprising considering the need to respect the separation of powers and the respective roles of the judiciary, legislature, and executive.

A helpful way to think about remedies in climate cases is to consider the judiciary's role as being to ascertain the appropriate standard the government must meet to respect *Charter* rights and evaluate whether the government is meeting that standard. With this frame of reference, an appropriate remedy would then be a simple declaration of this violation accompanied by an order requiring the government to establish a particular science-driven target but appropriately leaving the choice of how to meet that standard to the executive. This is the approach taken in *Urgenda* and followed in many cases since.

---

658 Roach, above note 655 at 109 and 112.
659 *Ibid* at 115.
660 *Ibid* at 109.

It is possible to be even more deferential, as the Belgian Court in the *Klimaatzaak* case was initially by declaring that the government had violated the rights of the claimants but declining to order specific reduction targets (effectively requiring the government to set a target in line with the latest credible science). The lower court had initially left it to the legislative and executive branches to determine the appropriate targets, holding only that the current target violated constitutional rights.[661] The Supreme Court of Canada took a similar approach in *Khadr 2010*, for example, declaring that Khadr's section 7 rights had been violated but leaving it to the government to decide how best to respond.[662] Interestingly, on appeal in the *Klimaatzaak* case, the Brussels Court of Appeal followed the stricter approach in *Urgenda*, opting to quantify the target and ordered the government to reduce its GHG emissions by at least 55 percent by 2030.[663]

In the Canadian context, the question of remedies has been most relevant in terms of its potential role in influencing the outcome on justiciability or a motion to strike. While the breadth of remedies claimed initially created a barrier for justiciability, the courts have clarified that this is no longer the case.[664] The most recent climate *Charter* decisions suggest that remedies alone will not be a barrier for claimants. In *La Rose/Misdzi Yikh*, Rennie J for the Federal Court of Appeal held that remedies are not necessarily determinative of justiciability, especially at the outset of litigation.[665] A request for remedies that "push the boundaries of a court's competence" should not, alone, cause a matter to be non-justiciable in part because courts can exercise "principled discretion" in tailoring remedies if a breach is found.[666] In *Mathur*, the Ontario Court of Appeal similarly held that the

---

661 *Klimaatzaak*, above note 24.

662 *Khadr*, above note 223 at para 39.

663 *Klimaatzaak*, above note 24.

664 For example, in *Misdzi Yikh* on the motion to strike, McVeigh J found the remedies sought (which included not only several declarations but also an order requiring the government to prepare an accounting of cumulative GHG emissions in a format allowing the emissions to be set in the context of a global carbon budget) to be inappropriate and render the claim non-justiciable (see *Misdzi Yikh*, above note 44 at paras 71–73.

665 *La Rose* FCA, above note 7 at para 48.

666 *Ibid* at para 51.

remedies requested by the applicants were not impossible nor too vague or imprecise. The Court started by noting that simple declaratory relief which flags a violation without necessarily telling Ontario what to do is a valid option.[667] The Court further noted that it would be appropriate to order Ontario to establish a science-based target consistent with Ontario's share of reductions, without having to specify the exact target. As the Court noted, "there are clear international standards based on accepted scientific consensus that can inform what a constitutionally compliant Target and Plan should look like."[668]

In closing, the comments of a dissenting judge in an American case are worth reflecting upon. In her dissent in the *Juliana* case (which failed in part due to the remedy issue and the particular standing test in the US Constitution), Justice Stanton remarked that the relief sought in that case was appropriate and offered an analogy to the desegregation orders and statewide prison injunctions which the United States Supreme Court had sanctioned in the past — remedies that were not designed to solve the entire problem but which went some distance to move the needle in the right direction.[669] She powerfully noted that "considering the plaintiffs seek no less than to forestall the Nation's demise, even a partial and temporary reprieve would constitute meaningful redress."[670]

4.7 CONCLUSION

The recent decisions by the Federal Court of Appeal in *La Rose/Misdzi Yikh* and the Ontario Court of Appeal in *Mathur* signal a shift in Canadian climate *Charter* litigation. While these cases have yet to reach final resolution and may be bound for the Supreme Court of Canada, the question of

---

667 *Mathur* CA, above note 7 at para 69.

668 *Ibid* at para 70.

669 *Juliana*, above note 2 at 1176.

670 *Ibid*. For a comment about the *Juliana* decision's analysis of remedies, see "Federal Courts — Justiciability — Ninth Circuit Holds that Developing and Supervising Plan to Mitigate Anthropogenic Climate Change Would Exceed Remedial Powers of Article III Court. — Juliana v. United States, 947 F.3d 1159 (9th Cir. 2020)" (2021) 134:5 Harv L Rev 1929.

whether government actions and inactions in relation to climate change violate the *Charter* remains alive and continues to evolve.

Justiciability, once a major obstacle in climate cases, is diminishing in significance, particularly for section 7 claims. Although broad claims without legislative anchors may still face challenges, the substantive issues raised by such claims are increasingly addressed during the analysis of sections 7 or 15. Also, there is far more climate legislation now to provide a legislative anchor. It appears that the range of cases that will be characterized as positive rights claims is narrow, which leaves open the possibility that courts will either apply existing principles of fundamental justice or recognize new ones. The fate of section 15 remains uncertain, but the Ontario Court of Appeal's critique of the lower court errors in *Mathur* — including its acknowledgement of unaddressed issues like the *Charter* rights of Indigenous Peoples — offers a glimmer of hope for equality-based climate arguments.

There are encouraging signs on the horizon. The suite of policies enacted in recent years is setting the stage for meaningful reductions in GHG emissions for the first time since Canada signed the UNFCCC. There has been significant growth in renewable energy, including a number of Indigenous-led projects that exemplify climate leadership.[671] In parallel, ongoing *Charter*-based climate lawsuits by youth concerned for their future exert pressure on governments to sustain and deepen climate action. These cases have profound legal and environmental implications, shaping both constitutional law and Canada's role in global climate governance. As courts around the world increasingly hold governments accountable for inadequate mitigation and adaptation efforts, Canada stands at a critical juncture: will it remain an outlier or is a transformative judicial decision imminent? The analysis in this chapter suggests the latter.

Additional Resources

Jamie Benidickson, *Environmental Law*, 5th ed (Toronto: Irwin Law 2019).
    Michael Berger & Maria Antonia Tigre, *Global Climate Litigation Report: 2023*

---

671 See, e.g., Indigenous Clean Energy, online : Indigenous Clean Energy [perma.cc/
    FJF9-BXVT].

Status Review (2023, Sabin Centre for Climate Change Law), online: https: //scholarship.law.columbia.edu/sabin_climate_change/202/.

Lisa Benjamin & Sara L Seck, "Mapping Human Rights-Based Climate Litigation in Canada" (2022) 13:1 J Hum Rts & Env't 178.

Maxim Bönnemann & Maria Antonia Tigre, The Transformation of European Climate Litigation.

David R Boyd, *The Right to a Healthy Environment: Revitalizing Canada's Constitution* (Vancouver: UBC Press, 2012).

David R Boyd, "The Constitutional Right to a Healthy Environment" *Environment* (July-August 2012).

Camille Cameron & Riley Weyman, "Recent Youth-Led and Rights-Based Climate Change Litigation in Canada: Reconciling Justiciability, Charter Claims and Procedural Choices" (2022) 34:1 J Envtl L 195.

Camille Cameron, Riley Weyman & Claire Nicholson, "Legal Hurdles and Pathways: The Evolution (Progress?) of Climate Change Adjudication in Canada" (2024) 47:2 Dalhousie LJ 1 at 3.

Jamie Cameron, "Positive Obligations Under Sections 15 and 7 of the Charter: A Comment on Gosselin v. Quebec" (2003) 20 SCLR (2d) 65.

Mark Carter, "Fundamental Justice in Section 7 of the Charter: A Human Rights Interpretation" (2003) 52 UNB LJ 243.

Nathali J Chalifour, "Environmental Justice and the Charter: Do Environmental Injustices Infringe Sections 7 and 15 of the Charter?" (2015) 28:1 JELP 89.

Nathalie J Chalifour & Jessica Earle, "Feeling the Heat: Climate Litigation under the Canadian Charter's Right to Life, Liberty and Security of the Person" (2018) 42 Vermont L Rev 688.

Nathalie J Chalifour & Dayna N Scott, "Environmental Justice" in Alastair Lucas et al, eds, *Environmental Law and Policy*, 4th ed (Emond Montgomery, 2019)

Nathalie J Chalifour, Jessica Earle & Laura Macintyre, "Coming of Age in a Warming World: The Charter's Section 15(1) Equality Guarantee and Youth-Led Climate Litigation" (2021) 17 JL & Equal 1.

Catherine Choquette, Dustin Klaudt & Laura Shay Lynes, "Climate Change Litigation in Canada" in Comparative Climate Change Litigation: Beyond the Usual Suspects (2021) pp 153–19.

Lynda M Collins, "An Ecologically Literate Reading of the Canadian Charter of Rights and Freedoms" (2009) 26 Windsor Rev Legal Soc Issues 7.

Lynda M Collins, "Safeguarding the Longue Durée: Environmental Rights in the Canadian Constitution" (2015) 71 SCLR (2d) 519.

Lynda M Collins, *The Ecological Constitution: Reframing Environmental Law* (Abingdon: Routledge, 2021).

Lawrence David, "A Principled Approach to the Positive/Negative Rights Debate in Canadian Constitutional Adjudication" (2014) 23:1 Const F 41.

Erin Dobbelsteyn, "Fundamental (In)justice and the Existential Threat of Climate Change" (2025) CJWL (accepted).

Meinhard Doelle & Chris Tollefson, *Environmental Law: Cases and Materials* (Thomson Reuters, 2019).

Elizabeth Donger, "Children and Youth in Strategic Climate Litigation: Advancing Rights through Legal Argument and Legal Mobilization" (2022) 11:2 Transnational Envtl L 263.

Meinhard Doelle & Sara Seck, *Research Handbook on Climate Change Law and Loss & Damage* (Eward Elgar, 2021).

David Estrin, "Limiting Dangerous Climate Change: The Critical Role of Citizen Suits and Domestic Courts – Despite the Paris Agreement" (May 2016, CIGI Papers).

Colin Feasby, David Devlieger, & Mathhew Huys "Climate Change and the Right to a Healthy Environment in the Canadian Constitution" (2020) 58:2 Alta L Rev 213.

Suzy Flader, "Fundamental Rights for All: Toward Equality as a Principle of Fundamental Justice under Section 7 of the Charter" (2020) 25 Appeal: Rev Current L & L Reform 43.

Kerri A Froc, "Constitutional Coalescence: Substantive Equality as a Principle of Fundamental Justice" (2012) 42:3 OLR 411.

Mari Galloway, "The Unwritten Constitutional Principles and Environmental Justice: A New Way Forward" (2021) 52:2 OLR 199.

Patricia Galvao Ferreira, "'Common but Differentiated Responsibilities' in the National Courts: Lessons from Urgenda v Netherlands" (2016) 5:2 Trans'l Env Law.

Alexandra Guillot, "Environmental Justice in Pollution Hotsports and Sections 7&15 of the Charter: The Case of the Aamjiwnaang Community in 'Chemical Valley'" (2023) 53:2 Envtl L 273.

Julia Hernandez & Anne Levesque, "Movement Lawyering and the Caring Society Litigation"(2023) 15:2 Journal of Human Rights Practice, 395–413.

Peter Hogg & Wade Wright, *Constitutional Law of Canada*, 5th ed (Toronto: Carswell, 2019).

Katelyn Horne, Maria Antonia Tigre & Michael B. Gerrard, Status Report on Principles of International and Human Rights Law Relevant to Climate Change.

Martha Jackman, "Constitutional Castaways: Poverty and the MacLachlin Court" in Sandra Rodgers & Sheila McIntyre, eds, *The Supreme Court of Canada and Social Justice* (LexisNexis Canada, 2010) pp 297–328.

Martha Jackman, "Wizened Stump or Living Tree: Section 7 Principles of Fundamental Justice" in Howard Kislowicz, Kerri Froc & Richard Moon, eds, *The Surprising Constitution* (Vancouver: UBC Press, 2024) 260.

Orla Kelleher, "Incorporating Climate Justice into Legal Reasoning: Shifting toward a Risk-Based Approach to Causation in Climate Litigation" (2022) 13:1 J Hum Rts & Env't 290.

Dustin W Klaudt, "Can Canada's 'Living Tree' Constitution and Lessons from Foreign Climate Litigation Seed Climate Justice and Remedy Climate Change?" (2018) 31:3 JELP 185.

Jennifer Koshan & Jonnette Watson Hamilton, "Tugging at the Strands: Adverse Effects Discrimination and the Supreme Court Decision in Fraser" (9 November 2020), online: https://ablawg.ca/2020/11/09/tugging-at-the-strands-adverse-e"ects-discrimination-and-the-supreme-courtdecision-in-fraser / [perma.cc/CKK2-8JHR].

Louis Kotzé, *Global Environmental Constitutionalism in the Anthropocene* (Hart Publishing, 2016).

Emmett Macfarlane, "Dialogue, Remedies, and Positive Rights: Carter v Canada as a Microcosm for Past and Future Issues Under the Charter of Rights and Freedoms" (2018) 49:1 OLR 107.

Jason MacLean, "You Say You Want an Environmental Rights Revolution: Try Changing Canadians' Minds Instead (of the Charter) (2018) 49:1 OLR 183.

Carissima Mathen & Patrick Mecklem, eds, *Canadian Constituitonal Law*, 6th ed (Toronto: Emond Publishing, 2022).

Benoit Mayer, "The Contribution of Urgenda to Mitigation of Climate Change (2023) 35:2 J Env L 167.

Martin Olszynski et al, "A Landmark Decision in Canadian Charter-based Climate Litigation: Mathur v Ontario, 2024 ONCA 762" (31 Oct 2024), online: http://ablawg.ca/wp-content/uploads/2024/10/Blog_MOJKNBJWH _Mathur.pdf.

Larissa Parker, "Climate Litigation and Emerging Environmental Dimensions of Human Rights: An Opportunity in Canada" (2021), 5 PKI Global Justice Journal 42.

Larissa Parker et al, "When the Kids Put Climate Change on Trial: Youth-Focused-Rights-Based Climate Litigation around the World" (2022) 13:1 J Hum Rts & Env't 64.

Jacqueline Peel & Hari M Osofsky, "A Rights Turn in Climate Change Litigation?" (2018) 7:1 Transnational Envtl L 37.

Dianne Pothier, "Tackling Disability Discrimination at Work: Toward a Systemic Approach" (2010) 4:1 McGill JL & Health 17.

Kent Roach, "Judicial Remedies for Climate Change" (2021) 17: 1 Journal of Law and Equality, 105.

Dayna Scott & Garance Malival, "Intergenerational Environmental Justice and the Climate Crisis: Thinking With and Beyond the Charter" (2021) 17 J L& Equality 165.

Joana Setzer & Catherine Higham, *Global Trends in Climate Change Litigation: 2024 Snapshot* (LSE Policy Report).

Robert J Sharpe & Kent Roach, *Charter of Rights and Freedoms*, 7th ed (Toronto: Irwin Law, 2021).

Jocelyn Stacey, *The Constitution of the Environmental Emergency* (Hart Publishing, 2018).

Maria A Tigre & Joana Setzer, "Human Rights and Climate Change for Climate Litigation in Brazil and Beyond: An Analysis of the Climate Fund Decision" 54(4) Geo. J. Int'l L. 593 (2023).

Maria Antonia Tigre & Margaret Barry, Climate Change in the Courts: A 2023 Retrospective.

Maria Antonia Tigre & Natalia Urzola, "The 2017 Inter-American Court's Advisory Opinion: Changing the Paradigm for International Environmental Law in the Anthropocene" 12:1 Journal of Human Rights and the Environment 24–50.

Stephan Wood, "Mathur v Ontario: Grounds for Optimism about the Recognition of a Constitutional Right to a Stable Climate System in Canada? (2024) 69:1 McGill LJ 3.

David W L Wu, "Embedding Environmental Rights in Section 7 of the Canadian Charter: Resolving the Tension Between the Need for Precaution and the Need for Harm" (2014) 33:2 Toronto 191–225.

Margot Young, "Unequal to the Task: 'Kapp'ing the Substantive Potential of Section 15" (2010) 50 Sup Ct L Rev (2d) 183.

Margot Young, "'A Code Red for Humanity': Judicial Relevance in a Time of Climate Emergency" (2021) 17 JL&Equality 151.

CONCLUSION[*]

Climate instability is irreversibly transforming the conditions that sustain life on Earth. For decades, warnings based on a growing body of science predicted a looming crisis: raging wildfires, deadly heat waves, devastating floods, and rising seas, to name a few. Today, these once-distant forecasts are grim realities as we witness humanity's march toward an uncertain and frightening future. Yet, hope remains. There is still time to avert the worst outcomes through decisive, rapid decarbonization. In this pivotal moment, the judiciary has a vital role to play. The legal interpretations rendered in the next few years will be influential in determining the planet's future.

Climate change presents a range of novel challenges for courts. It is a uniquely existential crisis, where the cumulative daily actions of industrialized society threaten the very systems that have allowed humanity to thrive. As Wagner CJ aptly noted in the Supreme Court of Canada's *GGPPA References* decision, climate change is unique in that it has no boundaries; its effects do not have a direct connection to the source of GHG emissions; and, no single province, territory, or country can address

* The author wishes to thank Professor Lynda Collins and Professor Heather McLeod-Kilmurray for their insightful comments on this conclusion.

climate change on its own.[1] The Supreme Court of Canada's decision in that case demonstrated a judicial capacity to adapt constitutional principles to the complexities and urgency of climate change. However, only two years later, the Court issued a cautious and narrow interpretation of Parliament's ability to account for transboundary GHG emissions in impact assessment decisions.[2] While no *Charter* climate rights case has yet arrived at the Supreme Court of Canada, this is almost certain to change in the next few years. Perhaps the Chief Justice's opening comment in the *IAA Reference* hints at the potential success of the youth climate cases when he states that "[t]he Canadian judiciary, in tandem with the other branches of government, has an important role to play in protecting the 'right to a safe environment.'"[3]

A CLIMATE-READY CONSTITUTION

This book explored the fundamental question of whether the Canadian Constitution, as currently interpreted and applied, is ready to confront the disruptions of climate change. Can it adapt to support all orders of government in responding to the climate emergency? Can it safeguard the rights of all Canadians, especially youth and future generations who are poised to inherit a legacy of irresponsible levels of emissions that breached the safe and just operating space for humanity in the blink of a geological eye? Can the Constitution evolve in a way that decolonizes and embraces a new era of prosperous co-leadership with Indigenous Peoples?

I believe it can, but doing so will require judges to be courageous in adapting legal interpretations to the stark realities of climate change. It will require more than cautious, incremental shifts in applying principles and doctrine ill-suited to the unique circumstances of the climate threat. It will require a spirit of trust, honesty, and openness to create and adapt to new constitutional realities that will safeguard core values such as federalism and fundamental rights with equanimity and an eye to survival.

---

1   *References re Greenhouse Gas Pollution Pricing Act*, 2021 SCC 11 [*GGPPA References*].

2   *Reference re Impact Assessment Act*, 2023 SCC 23 [*IAA Reference*]

3   *Ibid* at para 1, citing *Ontario v Canadian Pacific Ltd*, [1995] 2 SCR 1031 at para 55.

We are part way there. Canadian courts have already laid important groundwork, having consistently, unanimously and unequivocally accepted the science of climate change,[4] its gravity,[5] urgency and irreversibility,[6] and its disproportionate impacts on Indigenous Peoples, youth, and vulnerable regions.[7] They have acknowledged the necessity of reducing GHG emissions[8] and the collective action required to address this global problem.[9] They have consistently rejected the *de minimis* defence to deflect responsibility for emissions in light of other contributors.[10]

However, differences remain on critical questions such as the justiciability of *Charter* claims, particularly section 15 claims and those framed around the failure to act;[11] the extent of positive obligations on governments (and the concomitant characterization of claims as positive or negative);[12] the explicit recognition of the right to a safe environment – including a stable climate, the relevance of the public trust doctrine[13] and international law principles such as intergenerational equity.[14] Canadian

---

4   See, e.g., *GGPPA References*, above note 1 at para 2; *Environnement Jeunesse c Canada (PG)*, 2021 QCCA 1871 [*Enjeu* CA] at para 6.

5   *GGPPA References*, above note 1 at para 2; *Mathur v His Majesty the King in Right of Ontario*, 2023 ONSC 2316 [*Mathur* SC] at para 120.

6   *Enjeu* CA, above note 4 at para 22; *Mathur v Ontario*, 2024 ONCA 762 [*Mathur* ONCA] at para 11; see also Staton J's dissent in *Juliana* where she notes that what sets the harms of climate change apart "from all others is not just its magnitude, but its irreversibility" given how emission today "lock in" catastrophic damage. See *Juliana v United States*, 947 F (3d) 1159 at 1175 (9th Cir 2020) [*Juliana*] at p 34.

7   *Mathur* SC, above note 5 at para 178; *Mathur* ONCA, above note 6 at para 13; *La Rose v Canada*, 2023 FCA 241 [*La Rose* FCA] at para 76.

8   *GGPPA References*, above note 1 at para 2; *Enjeu* CA, above note 4 at para 2.

9   See *Mathur* SC above note 5 at para 149; *GGPPA References*, above note 1 at para 188; *La Rose v Canada*, 2020 FC 1008 [La Rose FCTD] at para 74.

10   See *GGPPA References*, above note 1 at para 189; *Mathur SC*, above note 5 at paras 147–48.

11   See, e.g., *Enjeu CA*, above note 4 (dismissing the case on the grounds of justiciability); *La Rose* FCA, above note 7 at para 21 (dismissing the section 15 claim).

12   Compare *Mathur* ONCA, above note 6 and *La Rose* FCA, above note 7.

13   See *La Rose* FCA, above note 7 at para 21. See also *Mathur* ONCA, above note 6 at para 6.

14   While the Federal Court of Appeal held that intergenerational equity is not within the scope of section 15 as the law currently stands, it also recognized that the international community is moving towards the recognition of intergenerational equity. *La Rose* FCA, above note 7 at paras 82 and 87. In overturning the lower court decision,

courts have also differed on the scope of remedies that may be available in the event of a successful claim[15] and the interpretation of age-related discrimination in the climate context.[16]

Additionally, Canadian courts have not addressed, or have insufficiently considered, a number of important questions. What is the role of unwritten constitutional principles in the interpretation of the division of powers and the *Charter*?[17] To what extent can and should rights impose positive obligations on governments to reduce GHG emissions?[18] To what extent are the rights of future generations encompassed within age-related

---

the Ontario Court of Appeal in *Mathur* noted that the "application of international law, including international environmental law" had not been adequately addressed. *Mathur* ONCA, above note 6 at para 6.

15 While the lower court decisions in *La Rose*, *Misdzi Yikh*, and *Mathur* each raised concern about the breadth of remedies claimed, the Federal and Ontario Courts of Appeal clarified that remedies are not necessarily determinative, especially at the outset of litigation. See, e.g., *La Rose* FCA, above note 7 at para 48; *Mathur* ONCA, above note 6 at para 74.

16 Compare, e.g., *La Rose* FCA, above note 7, which struck the section 15 claim, and *Mathur* SC, above note 5, where the lower court struck the section 15 claim as not related to age-discrimination (*inter alia*), and *Mathur* ONCA, above note 6 at paras 56–57, where the Ontario Court of Appeal remitted the question for re-hearing.

17 The Ontario Court of Appeal in *Mathur* noted that the "recognition and impact of certain unwritten constitutional principles, including societal preservation and ecological sustainability" had not been adequately addressed by the application judge. *Mathur* ONCA, above note 6 at para 6.

18 Globally, the jurisprudence has clearly established that human rights can create positive obligations on states to take steps to reduce GHG emissions and safeguard citizens from the risks of climate change, even if those risks have not yet manifested. See, e.g., Rechtbank Den Haag [The Hague District Court], The Hague, 24 June 2015, *Urgenda Foundation v The State of the Netherlands (Ministry of Infrastructure and the Environment)*, ECLI:NL:RBDHA:2015:7196, C/09/456689 / HA ZA 13-1396 (Netherlands), online (unofficial English translation): uitspraken.rechtspraak.nl [perma.cc/LU7H-XU7X]. See also Gerechtshof Den Haag [The Hague Court of Appeal], The Hague, 9 October 2018, *The State of the Netherlands v Urgenda Foundation*, ECLI:NL:GHDHA:2018:2610, 200.178.245/01 (Netherlands), online (unofficial English translation): uitspraken.rechtspraak.nl [perma.cc/T3WL-EDRG]; Hoge Raad [Supreme Court of the Netherlands], The Hague, 20 December 2019, *The State of the Netherlands v Urgenda Foundation*, ECLI:NL:HR:2019:2007, 19/00135 (Netherlands), online (unofficial English translation): uitspraken.rechtspraak.nl [perma.cc/3D52-RRU5] [*Urgenda* Supreme Court].

discrimination?[19] The list of under-explored questions continues, including the *Charter* rights of Indigenous Peoples and the role of section 35 rights,[20] the section 1 analysis,[21] and the possible recognition of new principles of fundamental justice. What is the role of Indigenous science, knowledge, and lived experiences in either division of powers or *Charter* cases? How might temporary measures under the emergency branch of POGG be interpreted, and does POGG need to be further adapted to incorporate a proportionality assessment that attenuates the sharp edges of national concern and the emergency branch?

THE WAY FORWARD

Scholarship and emerging jurisprudence offer a number of pathways forward. They emphasize the need to prioritize substance over form, remaining firmly committed to the fundamental principles of justice and equality, and embracing a living constitutional framework that can weather the storms of climate change. This vision requires reimagining the Canadian federation as one rooted in genuine partnership with Indigenous Peoples and fostering cohesion and co-creation among all orders of government.

The climate crisis calls for unity and resolve, transcending political divisions that often define federalism. Disasters such as wildfires, heatwaves and floods remind us of our shared humanity, inspiring generosity, and solidarity in the face of devastation. These moments remind us of what truly matters: our collective resilience and commitment to a promising future. Recent meaningful progress on reducing GHG emissions in Canada – after years of broken promises, delay and denial – is encouraging.[22] Similarly, powerful decisions from around the world, including recently the International Tribunal on the Law of the Sea make it clear

---

19   See, e.g., *La Rose* FCA, above note 7 at para 86 (raising concerns about how safeguarding the rights of future generations raises concerns about the separation of powers).

20   *Mathur* ONCA, above note 6 at para 6.

21   *Ibid.*

22   Federal government projections suggest Canada will reach 36% below 2005 levels by 2030 if all modelled measures are fully implemented. Government of Canada, *2023 Progress Report on the 2030 Emissions Reduction Plan*.

that states have legal responsibilities to act decisively and urgently address climate change.

My hope is that Canadians — especially those in the judiciary and positions of power and influence — rise above fear, vitriol, and short-term interests to embrace a shared vision of sustainability and equity. This is not merely about survival but about creating a Canada where all living beings and communities — young and old — can thrive. There is still a small window of time left when we can turn the ship around to avoid the most catastrophic harms. The decisions made in the next several years — including in the courtroom — will define our legacy as stewards of our constitutional democracy and our planet. May those decisions be made with wisdom, courage, and conviction and land on the right side of history alongside the judgments of a growing number of courts around the world. Although time is running out, hope endures.

TABLE OF CASES

TABLE OF STATUTES

TABLE OF INTERNATIONAL AGREEMENTS

INDEX

ABOUT THE AUTHOR

Nathalie Chalifour is a Full Professor with the Centre for Environmental Law and Global Sustainability at the University of Ottawa. Professor Chalifour is a leading expert on the rapidly evolving legal framework for climate change. She has published extensively about the constitutional authority to enact climate laws and youth-led climate litigation under the *Canadian Charter of Rights and Freedoms*. Professor Chalifour has represented interveners before courts in a number of climate-related cases. Her work has been cited by courts, including the Supreme Court of Canada, and she is a frequent commentator in the media. She is a member of the Royal Society of Canada's College of New Scholars and has won a number of awards, including the *Excellence in Teaching, Excellent in Research* and *Public Engagement Award for Media Relations* for the Law Faculty. Professor Chalifour obtained her PhD in Law at Stanford University and has a Masters in Law obtained as a Stanford Fellow and Fulbright Scholar.